Multivariate Statist
for Wildlife and
Ecology Research

Springer
*New York
Berlin
Heidelberg
Barcelona
Hong Kong
London
Milan
Paris
Singapore
Tokyo*

Kevin McGarigal   Sam Cushman
Susan Stafford

Multivariate Statistics for Wildlife and Ecology Research

With 57 Figures

Springer

Kevin McGarigal
Department of Natural Resources
  Conservation
University of Massachusetts
Amherst, MA 01003-5810
USA
mcgarigalk@forwild.umass.edu

Sam Cushman
Organismic and Evolutionary Biology
University of Massachusetts
Amherst, MA 01003-5810
USA
cushman@nsm.umass.edu

Susan Stafford
Department of Forest Sciences
Colorado State University
Fort Collins, CO 80523-1470
USA

Cover illustration: Trivariate probability surfaces showing the distribution of three rodent species on three environmental variables (see Figure 1.4). Photos of iguana, sulphur-crested cockatoo, and mongoose by David Alexander.

Library of Congress Cataloging-in-Publication Data
McGarigal, Kevin.
  Multivariate statistics for wildlife and ecology research / Kevin
McGarigal, Sam Cushman, Susan Stafford.
     p.   cm.
  Includes bibliographical references.
  ISBN 0-387-98891-2 (alk. paper). — ISBN 0-387-98642-1 (pbk.: alk.
paper)
  1. Animal ecology—Research—Statistical methods. 2. Multivariate
analysis.   I. Cushman, Sam.   II. Stafford, Susan G., 1952–   .
III. Title.
QH541.2.M39   2000
577'.07'27—dc21   99-16036

Printed on acid-free paper.

© 2000 Springer-Verlag New York, Inc.
All rights reserved. This work may not be translated or copied in whole or in part without the written permission of the publisher (Springer-Verlag New York, Inc., 175 Fifth Avenue, New York, NY 10010, USA), except for brief excerpts in connection with reviews or scholarly analysis. Use in connection with any form of information storage and retrieval, electronic adaptation, computer software, or by similar or dissimilar methodology now known or hereafter developed is forbidden. The use of general descriptive names, trade names, trademarks, etc., in this publication, even if the former are not especially identified, is not to be taken as a sign that such names, as understood by the Trade Marks and Merchandise Marks Act, may accordingly be used freely by anyone.

Production managed by Francine McNeill; manufacturing supervised by Jeffrey Taub.
Photocomposed copy prepared from the authors' WordPerfect and Microsoft Word files using FrameMaker.
Printed and bound by Maple-Vail Book Manufacturing Group, York, PA.
Printed in the United States of America.

9 8 7 6 5 4 3 2 1

ISBN 0-387-98891-2 Springer-Verlag New York Berlin Heidelberg   SPIN 10732829   (hardcover)
ISBN 0-387-98642-1 Springer-Verlag New York Berlin Heidelberg   SPIN 10695695   (softcover)

Preface

This book is intended to serve as an introduction to the use and interpretation of the most common multivariate statistical techniques used in wildlife ecological research. Specifically, this book is designed with three major functions in mind: (1) to provide a practical guide to the use and interpretation of selected multivariate techniques; (2) to serve as a textbook for a graduate-level course on the subject for students in wildlife ecology programs; and (3) to provide the background necessary for further study of multivariate statistics.

It is important to acknowledge upfront that it was not our intention to provide any "new" information in terms of multivariate statistics. Our primary purpose was to synthesize and summarize the current body of literature on multivariate statistics and to present it in a simplified form that could be understood by most wildlife researchers and graduate students. Consequently, we have drawn heavily upon an excellent array of more comprehensive books on the subject (e.g., Morrison 1967; Harris 1975; Gauch 1982; Dillon and Goldstein 1984; Digby and Kempton 1987; Hair, Anderson, and Tatham 1987; see Chapter 1 for citations).

So why do we need another textbook on the subject? First, most of the comprehensive books are so laden with mathematical and theoretical detail that most wildlife researchers, particularly graduate students, find these books difficult, if not impossible, to read and understand. It was our intention to draw from these sources only the particularly relevant material pertaining to the application of these procedures in wildlife ecological research. Second, few of these books focus on ecological applications, and those that do, generally focus on applications in community ecology or are limited in scope (e.g., focus on a single family of techniques). None of the treatments that we are aware of focus on wildlife habitat relationships.

In developing this book, our intent was to present multivariate statistical techniques in a manner that could be understood by wildlife researchers with an *intermediate* knowledge of statistics. Because this book is targeted for wildlife researchers, the discussions are focused on applications in wildlife research, but the comments are equally valid for most other ecological applications. In addition, this book emphasizes the practical aspects of each technique—that is, what each technique is designed to accomplish and how it is used and interpreted—rather than the detailed mathematics and underlying theory. Nevertheless, some background in statistics is necessary in order to understand all the material presented here. Because this book is intended to serve as an introduction to selected multivariate techniques, we strongly recommend that other texts be consulted (see Chapter 1 bibliography) to obtain additional insight on the performance of these techniques under varying conditions.

It is beyond the scope of this (or any) book to provide an exhaustive and in-depth review of all multivariate statistical techniques. Therefore, we include only those techniques commonly used in wildlife ecology research. Even so, it is not possible to present all the techniques used by wildlife researchers. For example, there are several prominent ordination techniques used in ecological studies, but we focus on principal components analysis because it is the most well-understood and most widely used ordination method in wildlife ecology, and provide a brief overview of other common ordination techniques for comparison. This does not mean that principal components analysis is the "best" choice of methods in all, or even most, cases, but rather, it is one of the more commonly used methods. Furthermore, we include only the true or classical multivariate techniques in this book; methods such as multiple regression, which involve multiple variables, but not multiple dependent variables, are not included.

This book is divided into three parts. In the first part (Chapter 1), we introduce the field of multivariate data analysis common in wildlife ecological applications and clarify some of the terminology in order to avoid confusion in subsequent chapters. Each of the multivariate techniques covered in the text is succinctly described here.

In the second part (Chapters 2 to 5), we focus on each of four techniques (or families of techniques), emphasizing the general concepts and methods involved in the proper application of each technique. Here we focus on *how to use each technique and how to interpret the results*. Each chapter is divided into sections of practical importance, including a conceptual and geometric overview of the technique, types of suitable data, assumptions and diagnostics for testing the assumptions, sample size requirements, and important output resulting from the procedure. An illustrative wildlife example using the SAS system for personal computers is incorporated into each chapter. The example is intended to represent the kind of real-world data sets that most wildlife ecologists collect. In other words, the assumptions are not always met, and the results are not always straightforward and easy to interpret. Each chapter in the second part is intended to stand alone without too much dependence on the material in other chapters. For a researcher interested in a specific technique, this has the advantage that each chapter can be read and understood without reading the entire book. Unfortunately, this also results in some unavoidable redundancy, especially in the case of common statistical assumptions associated with the techniques.

In the third and final part of the book (Chapter 6), we summarize and compare the various multivariate techniques. The focus here is on *when to use each technique*. We compare the various techniques with respect to the types of research questions and data sets that are appropriate for each technique. This part is intended to establish a conceptual relationship among multivariate techniques and thereby serve as a guide for choosing the appropriate technique(s) in any particular application.

Each chapter ends with a bibliography of selected publications. References on each technique (Chapters 2 to 5) are divided into two groups: (1) those related to the statistical procedure, and (2) those on applications of the technique from the wildlife literature. In this format, the bibliography can serve those who desire additional technical information, as well as those who wish to review how others have applied the techniques to answer wildlife research questions.

Finally, although there remains much within the field of multivariate data analysis that is not included here, we hope this book is effective in portraying the types of wildlife research questions that can be addressed using multivariate techniques. Moreover, while statistics is often a fearful subject to many wildlife researchers—multivariate statistics especially so—we hope that our treatment provides enough of a conceptual framework so that books more technical than this one can be approached without apprehension.

Acknowledgments. We are grateful to the many quantitative wildlife ecology students who provided myriad comments and useful suggestions during the development of this book. We thank Robert G. Anthony for providing the initial impetus for developing this book, and several anonymous reviewers for providing useful suggestions.

Kevin McGarigal
Sam Cushman
Susan Stafford

Contents

Preface			v
1	**Introduction and Overview**		**1**
	1.1	Objectives	1
	1.2	Multivariate Statistics: An Ecological Perspective	2
	1.3	Multivariate Description and Inference	9
	1.4	Multivariate Confusion!	10
	1.5	Types of Multivariate Techniques	14
		1.5.1 Ordination	14
		1.5.2 Cluster Analysis	15
		1.5.3 Discriminant Analysis	15
		1.5.4 Canonical Correlation Analysis	16
	Bibliography		16
2	**Ordination: Principal Components Analysis**		**19**
	2.1	Objectives	19
	2.2	Conceptual Overview	20
		2.2.1 Ordination	20
		2.2.2 Principal Components Analysis (PCA)	23
	2.3	Geometric Overview	24
	2.4	The Data Set	25
	2.5	Assumptions	27
		2.5.1 Multivariate Normality	28
		2.5.2 Independent Random Sample and the Effects of Outliers	31
		2.5.3 Linearity	33

	2.6	Sample Size Requirements	34
		2.6.1 General Rules	34
		2.6.2 Specific Rules	35
	2.7	Deriving the Principal Components	35
		2.7.1 The Use of Correlation and Covariance Matrices	35
		2.7.2 Eigenvalues and Associated Statistics	37
		2.7.3 Eigenvectors and Scoring Coefficients	39
	2.8	Assessing the Importance of the Principal Components	41
		2.8.1 Latent Root Criterion	41
		2.8.2 Scree Plot Criterion	41
		2.8.3 Broken Stick Criterion	43
		2.8.4 Relative Percent Variance Criterion	43
		2.8.5 Significance Tests	45
	2.9	Interpreting the Principal Components	50
		2.9.1 Principal Component Structure	50
		2.9.2 Significance of Principal Component Loadings	51
		2.9.3 Interpreting the Principal Component Structure	53
		2.9.4 Communality	56
		2.9.5 Principal Component Scores and Associated Plots	57
	2.10	Rotating the Principal Components	58
	2.11	Limitations of Principal Components Analysis	61
	2.12	R-Factor Versus Q-Factor Ordination	61
	2.13	Other Ordination Techniques	63
		2.13.1 Polar Ordination	63
		2.13.2 Factor Analysis	64
		2.13.3 Nonmetric Multidimensional Scaling	66
		2.13.4 Reciprocal Averaging	67
		2.13.5 Detrended Correspondence Analysis	68
		2.13.6 Canonical Correspondence Analysis	69
	Bibliography		73
	Appendix 2.1		78
3	**Cluster Analysis**		**81**
	3.1	Objectives	81
	3.2	Conceptual Overview	82
	3.3	The Definition of Cluster	85
	3.4	The Data Set	86
	3.5	Clustering Techniques	89
	3.6	Nonhierarchical Clustering	91
		3.6.1 Polythetic Agglomerative Nonhierarchical Clustering	92
		3.6.2 Polythetic Divisive Nonhierarchical Clustering	93
	3.7	Hierarchical Clustering	94
		3.7.1 Polythetic Agglomerative Hierarchical Clustering	95

		3.7.2	Polythetic Divisive Hierarchical Clustering	120
	3.8		Evaluating the Stability of the Cluster Solution	121
	3.9		Complementary Use of Ordination and Cluster Analysis	122
	3.10		Limitations of Cluster Analysis	123
	Bibliography			124
	Appendix 3.1			127
4	**Discriminant Analysis**			**129**
	4.1		Objectives	129
	4.2		Conceptual Overview	130
		4.2.1	Overview of Canonical Analysis of Discriminance	133
		4.2.2	Overview of Classification	134
		4.2.3	Analogy with Multiple Regression Analysis and Multivariate Analysis of Variance	136
	4.3		Geometric Overview	137
	4.4		The Data Set	138
	4.5		Assumptions	141
		4.5.1	Equality of Variance-Covariance Matrices	141
		4.5.2	Multivariate Normality	144
		4.5.3	Singularities and Multicollinearity	146
		4.5.4	Independent Random Sample and the Effects of Outliers	151
		4.5.5	Prior Probabilities Are Identifiable	152
		4.5.6	Linearity	153
	4.6		Sample Size Requirements	153
		4.6.1	General Rules	153
		4.6.2	Specific Rules	154
	4.7		Deriving the Canonical Functions	155
		4.7.1	Stepwise Selection of Variables	155
		4.7.2	Eigenvalues and Associated Statistics	158
		4.7.3	Eigenvectors and Canonical Coefficients	159
	4.8		Assessing the Importance of the Canonical Functions	161
		4.8.1	Relative Percent Variance Criterion	161
		4.8.2	Canonical Correlation Criterion	162
		4.8.3	Classification Accuracy	163
		4.8.4	Significance Tests	167
		4.8.5	Canonical Scores and Associated Plots	169
	4.9		Interpreting the Canonical Functions	169
		4.9.1	Standardized Canonical Coefficients	171
		4.9.2	Total Structure Coefficients	171
		4.9.3	Covariance-Controlled Partial F-Ratios	173
		4.9.4	Significance Tests Based on Resampling Procedures	175
		4.9.5	Potency Index	175

	4.10	Validating the Canonical Functions	176
		4.10.1 Split-Sample Validation	177
		4.10.2 Validation Using Resampling Procedures	178
	4.11	Limitations of Discriminant Analysis	179
	Bibliography	180
	Appendix 4.1	185

5 Canonical Correlation Analysis — 189

5.1	Objectives	189
5.2	Conceptual Overview	190
5.3	Geometric Overview	195
5.4	The Data Set	196
5.5	Assumptions	198
	5.5.1 Multivariate Normality	198
	5.5.2 Singularities and Multicollinearity	199
	5.5.3 Independent Random Sample and the Effects of Outliers	202
	5.5.4 Linearity	203
5.6	Sample Size Requirements	204
	5.6.1 General Rules	204
	5.6.2 Specific Rules	204
5.7	Deriving the Canonical Variates	205
	5.7.1 The Use of Covariance and Correlation Matrices	205
	5.7.2 Eigenvalues and Associated Statistics	206
	5.7.3 Eigenvectors and Canonical Coefficients	208
5.8	Assessing the Importance of the Canonical Variates	209
	5.8.1 Canonical Correlation Criterion	209
	5.8.2 Canonical Redundancy Criterion	212
	5.8.3 Significance Tests	216
	5.8.4 Canonical Scores and Associated Plots	218
5.9	Interpreting the Canonical Variates	220
	5.9.1 Standardized Canonical Coefficients	220
	5.9.2 Structure Coefficients	221
	5.9.3 Canonical Cross-Loadings	223
	5.9.4 Significance Tests Based on Resampling Procedures	225
5.10	Validating the Canonical Variates	225
	5.10.1 Split-Sample Validation	226
	5.10.2 Validation Using Resampling Procedures	227
5.11	Limitations of Canonical Correlation Analysis	227
Bibliography		228
Appendix 5.1		230

6	**Summary and Comparison**		**233**
	6.1	Objectives	233
	6.2	Relationship Among Techniques	234
		6.2.1 Purpose and Source of Variation Emphasized	234
		6.2.2 Statistical Procedure	236
		6.2.3 Type of Statistical Technique and Variable Set Characteristics	237
		6.2.4 Data Structure	238
		6.2.5 Sampling Design	239
	6.3	Complementary Use of Techniques	241

Appendix: Acronyms Used in This Book   **249**

Glossary   **251**

Index   **279**

1
Introduction and Overview

1.1 Objectives

By the end of this chapter, you should be able to do the following:

- Recognize the types of research questions that are best handled with a multivariate analysis.
- List seven advantages of multivariate statistical techniques over univariate statistical techniques.
- Explain how multivariate statistics can be used for both descriptive and inferential purposes.
- Recognize the difference between categorical and continuous variables and list examples of the following types within each class:
  — Categorical
  — Dichotomous versus polytomous
  — Nominal versus ordinal
  — Continuous
  — Ratio versus interval
- Contrast, with examples, the difference between independent and dependent variables.
- Describe the general differences among four commonly used multivariate analysis techniques:
  — Principal Components Analysis (PCA)
  — Cluster Analysis (CA)
  — Discriminant Analysis (DA)
  — Canonical Correlation Analysis (CANCOR)

1.2 Multivariate Statistics: An Ecological Perspective

Nature is complicated. In any real world ecological system, the patterns we see are typically driven by a number of interacting ecological processes which vary in space and time. The distribution, abundance, or behavior of an organism, for example, are affected simultaneously by many biotic and abiotic factors. Moreover, there is often synergism and feedback between biotic and abiotic processes and the patterns they create. This multiplicity and interaction of causal factors makes it exceptionally difficult to analyze ecological systems. As a result, researchers have long since abandoned sole reliance on the classic univariate design (Table 1.1).

TABLE 1.1 Advantages of multivariate statistical techniques for ecological data.

Reflect more accurately the true multidimensional, multivariate nature of natural ecological systems.

Provide a way to handle large data sets with large numbers of variables by summarizing the redundancy.

Provide rules for combining variables in an "optimal" way.

Provide a solution to a kind of multiple comparison problem by controlling the experimentwise error rate.

Provide for post-hoc comparisons which explore the statistical significance of various possible explanations of the overall statistical significance of the relationship between the independent and dependent variables.

Provide a means of detecting and quantifying truly multivariate patterns that arise out of the correlational structure of the variable set.

Provide a means of exploring complex data sets for patterns and relationships from which hypotheses can be generated and subsequently tested experimentally.

Univariate methods are extremely powerful in situations where the response of a single variable is of sole interest and other factors can be controlled. However, in ecological research, it is more often the case that the question at hand can be answered only by considering a number of variables simultaneously. As the name implies, *multivariate statistics* refers to an assortment of descriptive and inferential techniques that have been developed to handle just these situations—where sets of variables are involved either as predictors or as measures of performance. The emphasis here is on *sets* of variables rather than on single, individual variables.

Multivariate approaches are required whenever more than one characteristic is needed to meaningfully describe each sampling entity and when the relationships among these characteristics require that they be analyzed simultaneously. For example, it has become abundantly clear that a given experimental manipulation will affect many different, but partially correlated, aspects of an organism's behavior. Similarly, a variety of pieces of information about an organism may be of value in explaining the organism's performance under varying conditions, and therefore it is necessary to consider how to combine all of these pieces of information into a single "best" description of performance.

To illustrate the types of ecological questions that lend themselves to multivariate approaches, let us consider an idealized, simple ecological system. This system consists of a simple forest community. In this community, suppose there are 10 environmental factors that vary spatially across the study area. In addition, suppose there are 10 herbaceous, 10 shrub, and 10 tree species. The distribution of these plant species depends in part on a combination of environmental gradients (e.g., soil texture and fertility, microclimate, light availability) and the interaction of several processes (e.g., disturbance, herbivory, competition). On top of this, suppose there are 20 arthropod species, 10 bird species, and 5 mammal species. The distribution and abundance of these species are influenced by the interaction of environmental gradients, vegetation, and several interspecific factors (e.g., predation, competition). This community is clearly a very simplified abstraction. The real world is much more complex than this. But even in this simple case, analyzing the nature of ecological patterns and determining the processes that drive them can be very challenging. Let us consider a few specific examples.

Suppose that three of the five mammal species in the community are ground-dwelling rodents, and that researchers set out to determine how these apparently similar species can coexist in the same area. They record 10 niche variables from the suite of environmental and vegetation variables for 30 individuals of each of the three species. Collectively, these variables are used to define each species' niche.

These data can be tabulated into a two-way data matrix with 90 rows and 10 columns, where each row of the data matrix represents an individual and each column represents one of the niche variables recorded for each individual (Fig. 1.1). In this case, the individuals (rows) are grouped into three species. Each individual has a value, or score, on each variable. This data matrix also can be conceptualized as a multidimensional data space (or niche space, in this case), where each niche variable represents an axis and each individual has a unique location in this space defined by its value on each axis. Collectively, the sample entities (individuals) form a *data cloud* in this 10-dimensional niche space. The shape, clumping, and dispersion of this cloud describe the distribution and overlap of these three species in ecological space. Multivariate techniques can optimally summarize, order, or partition this data cloud to elucidate the structure of each species' niche and degree of separation from the other species.

Taking each variable singly, the scientists can describe the range and overlap of the three rodents on the single niche dimension represented by that variable (Fig. 1.2). When only one niche variable is considered, the individuals are distributed along that single axis and any patterns in the distribution of species are immediately apparent and easily tested (e.g., using an F-test). Analyzing each niche variable separately in this manner can be quite effective when there are only a few variables to consider, and when the variables are uncorrelated (i.e., independent). In this case, for example, species A and C clearly partition resources along a gradient defined by variable 1, but not variable 2; whereas species B overlaps broadly with species A and C on all variables except variable 2, where it has a relatively low overlap with species A.

		Niche Variables				
Individual	Species	Variable 1	Variable 2	Variable 3	...	Variable 10
1	A	12	8	22	...	.
2	A	10	11	25		.
.	.	.	.	.		.
.	.	.	.	.		.
.	.	.	.	.		.
30	A	16	6	21		.
31	B	15	22	12		.
32	B	14	21	9		.
.	.	.	.	.		.
.	.	.	.	.		.
.	.	.	.	.		.
60	B	17	19	13		.
61	C	26	15	6		.
62	C	22	13	3		.
.	.	.	.	.		.
.	.	.	.	.		.
.	.	.	.	.		.
90	C	27	17	8		.

FIGURE 1.1 A hypothetical two-way data matrix containing niche variable scores for 30 individuals of each of three rodent species across ten environmental variables. In this case, the score on each environmental variable equals the average value of that particular variable across all sites at which the individual in question was observed. For example, if variable 1 was snag basal area, then the score of individual 1 of species A would be the average snag basal area across all sites in which this individual was recorded.

When two niche variables are considered simultaneously, each individual animal has a score on each axis and the data are distributed in a two-dimensional niche space. Instead of a univariate frequency distribution, the niche data in this case are presented as a bivariate density *surface* (Fig. 1.3). The simultaneous analysis of the two variables reveals patterns that are not immediately apparent from inspection of each variable individually. In this case, for example, all three species are fairly well separated on the variable 1,2-plane. Species A is represented by the peak in the middle left of the diagram and overlaps minimally with the other two species. Species B and C are separate over most of their distributions on these two variables, but overlap somewhat at high values of variables 1 and 2. In contrast, the species are much less clearly separated on the variable 2,3- and variable 1,3-planes. For example, on the variable 1,3-plane, species A and species C are widely separated, with species A most abundant at low values of variable 1 and high values of variable 3, and species C most abundant at high values of variable 1 and low values of variable 3. Species B, however, is very broadly distributed on variable 1 and overlaps broadly both species A and C on variable 3.

When three niche variables are taken together, the data form a cloud in a three-dimensional niche space. In this case, the distribution of each species is plotted as a trivariate isoprobability surface (Fig. 1.4). This data cloud describes the niche structure and overlap of the three rodent species on three dimensions of their niche space.

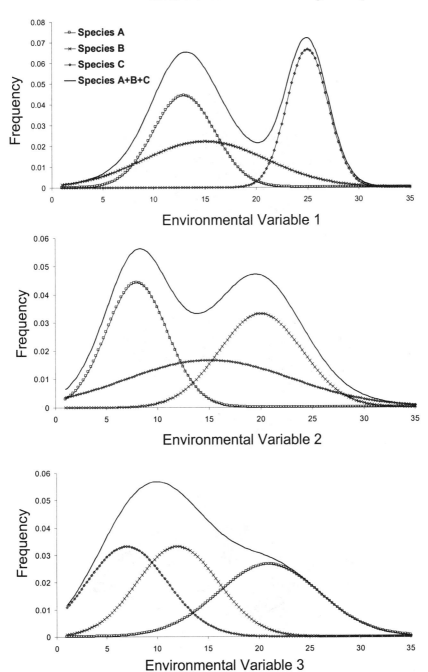

FIGURE 1.2 Univariate frequency distributions showing the individual and cumulative distributions of the three rodent species for the first three environmental variables depicted in Figure 1.1. The curves show the distribution of each species on each environmental variable and overlap between species.

6   1. Introduction and Overview

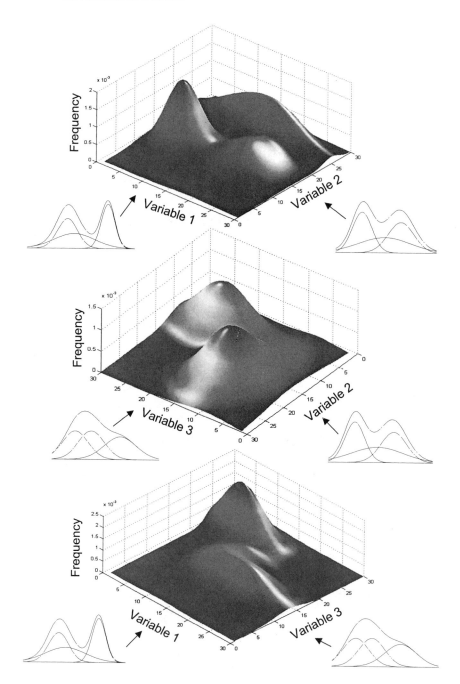

FIGURE 1.3 Bivariate density surfaces showing the distribution of the three rodent species across each of the three bivariate pairs of the first three environmental variables depicted in Figure 1.1. The height of the surface is proportional to the density of observations at that bivariate combination of conditions.

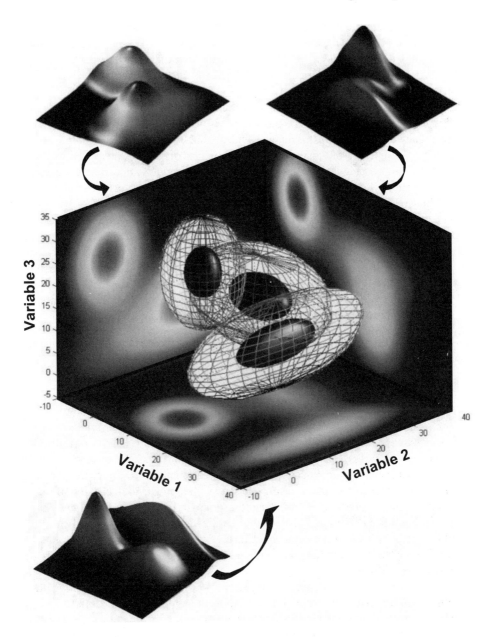

FIGURE 1.4 Trivariate probability surfaces showing the distribution of the three rodent species on the first three environmental variables depicted in Figure 1.1. The distribution of each species is plotted using two concentric surfaces. The inner surface, shown in solid black, is the 75% density surface. Seventy-five percent of observations of each species are located within the respective spheres. The outer mesh surface is the 95% density surface, containing 95% of the observations of each species. The bivariate probability surfaces are projected onto the three bivariate planes defining the three-dimensional space.

The three-dimensional objects represent the surface of the data cloud containing a specified proportion of the total records of each species. In the present case, the solid black surface corresponds to the 75 percent probability surface, and the mesh surface the 95 percent probability surface. Thus, 75 percent of the observations of each species falls within the portion of the data space depicted by the solid black objects. In this case, the niche structure and separation among species is readily apparent. With three or fewer niche variables, it is possible to visualize the niche structure and overlap of these three species in this manner, and, although cumbersome, a combination of univariate and bivariate analyses can suffice to describe these patterns. However, when the number of niche variables exceeds three, it becomes impossible to visualize the patterns in the data space and so multivariate approaches are essential.

Many other ecological questions can be conceptualized in the same way as the niche structure example above. For example, suppose instead that researchers want to describe the microhabitat relationships of one of the bird species in this community. Suppose they employ a method that allows them to precisely locate individuals of this species within the study area. Suppose they locate each of 30 individuals 10 times, and at each location they record values of 10 habitat variables. These variables might include a combination of environmental factors (e.g., slope, aspect, elevation) and vegetation variables (e.g., shrub density, canopy cover, canopy height). To avoid pseudoreplication, the researchers take the average value of each individual across its 10 sample locations. To compare this sample of utilized habitat with a sample of habitat available at random, the researchers also select 30 locations at random within the study area and measure each of the habitat variables.

As before, these data can be tabulated into a two-way data matrix, in this case with 60 rows and 10 columns, where each row represents a vector of habitat measurements for an individual or random location, and each column represents one of the habitat variables. In this case, the rows are partitioned into two groups (utilized or random locations). Each individual and random point has a value, or score, on each variable, and therefore occupies a unique location in this 10-dimensional data space (or habitat space, in this case). The shape, clumping, and dispersion of this cloud in relation to the individual and random groups describe the structure and variability of utilized habitat, and differences of utilized habitats from randomly available habitat. Multivariate techniques can be used to optimally describe these multidimensional patterns and help determine whether the species utilizes habitat in a nonrandom fashion.

One last example. Suppose that researchers wish to determine whether the arthropod community is structured in relation to environmental factors; that is, whether the arthropod community changes in composition or structure along important environmental gradients. Suppose they sample 30 locations distributed randomly or systematically throughout the study area, and at each location they measure the abundance of each of the 10 arthropod species and a suite of 10 environmental variables.

As in the previous examples, these data can be tabulated into a two-way data matrix, in this case with 30 rows and 20 columns, where each row represents a site (or sampling location) and each column represents one of the species or environmental variables. In contrast to the previous examples, in this case the columns (not rows) are partitioned into two groups of variables (species' abundances and environmental factors). Now, it makes more sense to consider these as two separate data spaces. Each site has a value on each variable and simultaneously occupies a unique location in both 10-dimensional spaces. The shape, clumping, and dispersion of each data cloud individually and in relation to each other describe the ecological patterns of this study system. Multivariate techniques exist to describe the multidimensional patterns in each data space and to determine if the patterns are jointly distributed.

We could pose many other scenarios where multivariate techniques are required to answer the research question. These few examples suffice to illustrate the multivariate data structure common to many ecological research questions. In all these cases, a number of variables are measured to describe each sampling entity. In each case, the patterns and coordination across the variables convey the ecological message, and multivariate methods are the tools through which this message is transcribed and interpreted.

1.3 Multivariate Description and Inference

Multivariate statistical techniques accomplish two general functions, corresponding roughly to the distinction between *descriptive* and *inferential* statistics. On the descriptive side, they provide rules for combining the variables in an optimal way. Unfortunately, what is meant by "optimal" may vary from one technique to the next, but this will be made explicit in the following chapters where we discuss the techniques in greater detail.

On the inferential side, multivariate techniques provide explicit control over the experimentwise error rate. Many situations in which multivariate techniques are applied could be analyzed through a series of univariate significance tests (e.g., t- or F-tests), with one such univariate test for each possible combination of one of the independent variables with one of the dependent variables. However, since each of the univariate tests is designed to produce a significant result with a specified probability (significance level) when the null hypothesis is incorrect, the probability of having at least one of the tests produce a significant result due to chance variation alone (Type I error) increases rapidly as the number of tests increases. It is thus desirable to explicitly control the experimentwise error rate; multivariate statistical techniques provide this control.

The descriptive and inferential functions of multivariate statistics are by no means independent. Indeed, one whole class of multivariate techniques bases its tests of significance on the sampling distribution of the *composite variable*, which results when the original variables are combined according to the criterion of optimality employed by that particular technique. Consequently, when we achieve statistical significance (inferential function), we also automatically know which combination of the original set of variables (descriptive function) provided the strongest evidence against the overall null hypothesis.

An important difference between descriptive and inferential uses of multivariate techniques is that the former requires no assumptions whatsoever about the distributions from which the observations are sampled, while most of the commonly used multivariate significance tests, like their univariate counterparts, are based on the usual parametric assumptions (e.g., homogeneity of variance-covariance matrices and multivariate normality). Little is known about the robustness of the multivariate techniques to failures of the assumptions, but the evidence suggests that they are generally quite robust.

Finally, it is important to note that multivariate techniques are most frequently employed as descriptive and exploratory tools; that is, as a means of exploring complex data sets involving many variables for patterns and relationships from which hypotheses can be generated and subsequently tested using controlled experiments. This is particularly true in ecological research, where experiments are often designed to test one or more specific hypotheses using univariate techniques (e.g., analysis of variance) and multivariate techniques are used in a post hoc manner to explore other secondary relationships, often in an attempt to suggest possible causal mechanisms, and always as an aid in developing hypotheses for subsequent experimentation. Consequently, rarely are multivariate techniques the sole analytical tool used in ecological studies. More often, they are used in a complementary manner with the more standard univariate statistical techniques. This does not mean, though, that multivariate techniques should never be the sole basis for study design and analysis. Indeed, planning and designing a study within a particular multivariate analytical framework is the best way to ensure proper and effective use of the multivariate procedure(s), and is often the best or only way to answer a particular research question.

1.4 Multivariate Confusion!

Unfortunately, distinguishing among multivariate techniques can be somewhat confusing. This is in part due to the fact that many of the techniques differ in only subtle ways. The same research question may be handled by several different techniques and the same data set may be analyzed with different techniques to answer different research questions. Mathematically, many of the techniques are very similar. The differences among them will become clear as you learn the details of each technique in the following chapters. In Chapter 6, we will revisit this problem of differentiating among techniques.

Unfortunately, the confusion among techniques is also partly due to the inconsistent use of terminology promulgated by the common statistical analysis packages (e.g., SAS, SPSS, BMDP, SYSTAT). Table 1.2 lists a sample of alternative terms used to refer to the techniques discussed in this book; we will adopt the terminology most commonly used in the wildlife ecology literature. For example, depending on the context in which it is used, the term *classification* may refer to several different methods or combination of methods, including cluster analysis, classification (i.e., predictive discriminant analysis), both classification and discriminant analysis (i.e., predictive and descriptive discriminant analysis), or all three methods as a group. In this book, we use classification exclusively to refer to the subset of discriminant analysis dealing

TABLE 1.2 Alternative terminology for multivariate techniques. The labels used in this book are given in the left-hand column.

Technique	Alternative Names
Ordination	
Principal components analysis	Factor analysis
Polar ordination	Bray and Curtis ordination
Factor analysis	None
Nonmetric multidimensional scaling	None
Reciprocal averaging	Correspondence analysis
Detrended correspondence analysis	None
Canonical correspondence analysis	None
Cluster analysis	Botryology
	Classification
	Clumping
	Grouping
	Morphometrics
	Nosography
	Nosology
	Numerical taxonomy
	Partitioning
	Q-analysis
	Segmentation analysis
	Systematics
	Taximetrics
	Taxonorics
	Typology
	Unsupervised pattern recognition
Discriminant analysis	
Canonical analysis of discriminance	Discriminant analysis
	Discriminant function analysis
	Multiple discriminant analysis
	Descriptive discriminant analysis
	Canonical variates analysis
	Fisher's linear discriminant function analysis
Classification	Discriminant analysis
	Discriminant function analysis
	Multiple discriminant analysis
	Predictive discriminant analysis
	Fisher's linear discriminant function analysis
Canonical correlation analysis	Canonical analysis

with predictive discrimination. Similarly, the terms *factor analysis* and *principal components analysis* are sometimes used interchangeably, when in fact they are distinctly different techniques (albeit techniques that accomplish a similar end).

Equally confusing is the terminology used to define or classify variables. In this book, variable labels are used repeatedly to describe techniques and, more importantly, to compare and contrast various techniques. Consequently, it is very important that you fully understand the meaning of the various labels. Variables are labeled

12   1. Introduction and Overview

in two ways: (1) by the type of data and measurement scale they represent, and (2) by their relationship to other variables.

With respect to the type of data and measurement scale represented (Table 1.3), variables are termed *categorical*, *discrete*, or *nonmetric* if the values are assigned for convenience only and are not useful for quantitative expressions. In this book, we use the term "categorical" exclusively. If a categorical variable has only two possible values (i.e., it is binary), then it is referred to as a *dichotomous* variable. Sex (male or female) and species occurrence (present or absent) are examples of dichotomous variables. If more than two states exist (i.e., multistate), then the variable is referred to as *polytomous*. If the order of the categories in a polytomous variable is arbitrary, then it is a *nominal scale* variable: the numbers or symbols assigned to categories have no quantitative meaning beyond indicating the presence or absence of the attribute or characteristic under investigation. Residency status (e.g., year-round, breeding only, winter only, seasonal migrant), animal behavior (e.g., feeding, roosting, brooding), and habitat type (e.g., meadow, forest, pond) are examples of nominally scaled, polytomous variables. If there is a natural ordering of the categories, then it is referred to as an *ordinal scale* variable: the numbers assigned to categories represent an inherent ordering or ranking of categories and reflect the relative amount of the attribute possessed. Age class (e.g., juvenile, sub-adult, adult), seral stage (e.g., grass-forb, sapling, pole, sawtimber), and canopy cover class (e.g., less than 20 percent, 20–40 percent, 40–60 percent, 60–80 percent, greater than 80 percent canopy cover) are examples of ordinally scaled, polytomous variables. Categorical variables are also sometimes referred to as *qualitative* variables if the categories are determined by qualitative criteria (e.g., habitat types) and they have no numerical significance (i.e., nominal scale). Categorical variables are also sometimes referred to as *manipulated* variables, although this term is misleading and its usage should be avoided.

Variables are termed *continuous*, *quantitative*, or *metric* if the values they assume are useful for quantitative expressions and occur at any point in a continuum of possible values; that is, at any point along the real number line. In this book, we use the term "continuous" exclusively. If there is an arbitrary zero point, then the variable is referred to as an *interval scale* variable (e.g., temperature, time). If there is an absolute zero point (i.e., true origin), then the variable is referred to as a *ratio scale* variable. Most ecological measurements are ratio scale variables. Continuous variables are also sometimes referred to as *measurement* variables, although this term is misleading and its usage should also be avoided.

Count data (also referred to as *meristic data*) possess characteristics of both categorical and continuous variables. Count variables contain values in discrete, indivisible units of the same size and are typically integer-valued (e.g., number of individuals of each species). However, count variables often assume the properties of a continuous variable; hence, statistical methods designed for continuous variables (e.g., regression) often can be used quite successfully to analyze count data. Count variables also can be converted into categorical variables by grouping all values greater than a certain value, or all values within prespecified intervals into catego-

1.4 Multivariate Confusion!

TABLE 1.3 Variable labels based on type of data and measurement scale.

Variable label/measurement scale	Definition
Categorical variable	Discrete values are assigned for convenience only and are not useful for quantitative expressions.
Dichotomous	Only two possible values.
Polytomous	More than two states exist.
Nominal scale	Order of categories is meaningless; numbers utilized in nominal scale are arbitrary.
Ordinal scale	Order of categories is meaningful and indicates relative differences; numbers utilized in ordinal scale, however, are nonquantitative.
Continuous variable	Values are useful for quantitative expressions and can assume values at any point in a continuum of possible values.
Ratio scale	Absolute zero point (i.e., true origin).
Interval scale	Arbitrary zero point (i.e., no true origin).
Count variable	Values assigned in discrete, indivisible units of the same size (e.g., integer-valued).

ries. Indeed, *any* variable can be categorized, but categorical data cannot be converted into continuous data.

Variables are also named with respect to their relationships to other variables (Table 1.4). Most statistical techniques analyze the relationship between one set of variables, referred to as *independent*, *predictor*, *regressor*, *explanatory*, or simply X variables, and a second set of variables, referred to as *dependent*, *response*, *outcome*, or simply Y variables. Independent variables are presumed to represent factors that cause any change in a dependent variable; they are often regarded as fixed, either as in experimentation or because the context of the data suggests they play a causal role in the situation under study. Dependent variables are presumed to represent factors that respond to a change in an independent variable; they are free to vary in response

TABLE 1.4 Variable labels based on their relationship to other variables.

Variable label	Definition
Independent variable (X)	Variable presumed to be a cause of any change in a dependent variable; often regarded as fixed, either as in experimentation or because the context of the data suggests they play a causal role in the situation under study.
Dependent variable (Y)	Variable presumed to be responding to a change in an independent variable; free to vary in response to controlled conditions.

to controlled conditions. The choice of terms is partly a matter of personal preference and partly dependent on the technique being discussed. For example, in regression, the variable labels *predictor/regressor* and *response/outcome* accurately reflect the variables' respective functions; however, the choice of "predictor" over "regressor" and "response" over "outcome" is somewhat arbitrary. Unfortunately, no single pair of labels is without misleading connotation for all techniques. Consequently, in practice, all of these labels are commonly employed. For purposes of consistency, in this book we use the terms "independent" (X) and "dependent" (Y) to describe variables—even though this is somewhat awkward in places. In some techniques (e.g., ordination, cluster analysis), there is no distinction between dependent and independent variables. In these cases, we simply refer to them as "variables" without any other designator.

1.5 Types of Multivariate Techniques

Multivariate analysis includes a broad array of techniques. Those commonly used in wildlife ecological research include: (1) ordination techniques, such as principal components analysis and canonical correspondence analysis, (2) cluster analysis, (3) discriminant analysis, and (4) canonical correlation analysis. Several other procedures are less well known and have only been tentatively applied by researchers. In the following sections, we will introduce each of these techniques, briefly defining the technique and its major purpose. We will investigate the detailed relationships among these techniques more fully in Chapter 6.

1.5.1 Ordination

Ordination comprises a family of techniques, including principal components analysis (PCA), whose main purpose is to organize sampling entities (e.g., sites, individuals, species) along a meaningful continuum or gradient. Ordination techniques are employed to quantify the interrelationships among a large number of interdependent variables and to explain those variables in terms of a smaller set of underlying dimensions (e.g., components). The approach involves condensing the information contained in the original variables into a smaller set of dimensions (e.g., components) such that there is minimal loss of information. Principal components analysis accomplishes this by creating linear combinations of the original variables (i.e., principal components), which are oriented in directions that describe maximum variation among individual sampling entities. In other words, the procedure maximizes the variance of the principal components. In doing this, the procedure organizes entities along continuous gradients defined by the principal components and seeks to describe the sources of greatest variation among entities. Entities are generally assumed to represent a single random sample (labeled N, meaning "of size N") of a known or unknown number of populations (in the statistical sense of the word). The data set must consist of a single set of two or more continuous, categorical, and/or count variables, and no distinction exists between independent and dependent variables.

1.5.2   Cluster Analysis

Cluster analysis (CA) is a family of analytic procedures whose main purpose is to develop meaningful aggregations, or groups, of entities based on a large number of interdependent variables. Specifically, the objective is to classify a sample of entities into a smaller number of usually mutually exclusive groups based on the multivariate similarities among entities. The procedure creates groups using one of many different clustering strategies that, in general, maximizes within-group similarity (i.e., minimizes within-group distances in multidimensional space) and minimizes between-group similarity (i.e., maximizes between-group distances in multidimensional space) based on the variables. Entities are generally assumed to represent a single random sample (N) of an unknown number of populations (in the statistical sense of the word). The data set must consist of a single set of two or more continuous, categorical, and/or count variables, and there is no distinction between independent and dependent variables.

1.5.3   Discriminant Analysis

Discriminant analysis (DA) comprises a couple of closely related analytic procedures whose main purpose is to *describe* the differences among two or more well-defined groups and *predict* the likelihood that an entity of unknown origin will belong to a particular group based on a suite of discriminating characteristics. Groups are defined on the basis of one categorical grouping variable. *Canonical analysis of discriminance* (CAD), the descriptive part of DA, is an analytic procedure that creates linear combinations (i.e., canonical functions) of the original independent variables that maximize differences among prespecified groups; that is, the procedure maximizes the F-ratio (or t-statistic for two groups) on the canonical function(s). The procedure organizes entities along continuous gradients defined by the canonical functions and emphasizes differences among groups of sampling entities. *Classification*, the predictive part of DA, is an analytic procedure that classifies entities into groups using a classification criterion that, in general, maximizes correct classification of entities into prespecified groups. Consequently, the procedure emphasizes differences among groups of sampling entities. Samples for CAD and classification are generally assumed to represent either a single random sample (N) of a mixture of two or more distinct populations (groups) (i.e., multiple regression analysis analogy), or several independent, random samples ($N_1, N_2, ..., N_k$) of two or more distinct populations (i.e., multivariate analysis of variance analogy). The data set must consist of one categorical grouping variable with at least two levels, and at least two continuous, categorical, and/or count (usually continuous) discriminating variables. The distinction of the grouping variable as the dependent or independent variable is partly dependent on sampling design and the particular research situation.

1.5.4   Canonical Correlation Analysis

Canonical correlation analysis (CANCOR) is an analytic technique used to describe the relationship between two (or more) sets of variables; CANCOR can be viewed as a logical extension of multiple regression analysis. The objective is to correlate simultaneously two (or more) sets of usually continuous variables in which one set is often logically treated as the independent set and the opposing set is treated as the dependent set. Thus, whereas multiple regression analysis involves a single dependent variable, CANCOR involves multiple dependent variables. The overall objective is to develop linear combinations (i.e., canonical variates) of variables in each set (both dependent and independent) such that the correlation between the composite variates is maximized. The procedure involves finding the sets of weights for the dependent and independent variables that result in the highest possible correlation between the composite variates. The procedure organizes sampling entities along pairs of ecological gradients defined by the canonical variates. Samples are generally assumed to represent a single random sample (N) of a population (in the statistical sense of the word). The data set must consist of two (or more) sets of at least two continuous, categorical, and/or count (usually continuous) variables. Technically, the distinction between independent and dependent variables is arbitrary, since both sets of variables can serve either function. In practice, however, one set is often logically treated as the independent set and the other as the dependent set.

Bibliography

Afifi, A.A., and Clark, V. 1984. *Computer-Aided Multivariate Analysis*. Belmont: Lifetime Learning.

Anderson, T.W. 1958. *An Introduction to Multivariate Statistical Analysis*. New York: Wiley and Sons.

Capen, D.E., ed. 1981. *The Use of Multivariate Statistics in Studies of Wildlife Habitat*. U.S. Forest Service Gen. Tech. Report RM–87.

Cooley, W.W., and Lohnes, P.R. 1971. *Multivariate Data Analysis*. New York: Wiley and Sons.

Digby, P.G.N., and Kempton, R.A. 1987. *Multivariate Analysis of Ecological Communities*. New York: Chapman and Hall.

Dillon, W.R., and Goldstein, M. 1984. *Multivariate Analysis: Methods and Applications*. New York: Wiley and Sons.

Gauch, H.G. 1982. *Multivariate Analysis in Community Ecology*. Cambridge: Cambridge University Press.

Gnanadesikan, R. 1977. *Methods for Statistical Data Analysis of Multivariate Observations*. New York: Wiley and Sons.

Green, R.H. 1979. *Sampling Design and Statistical Methods for Environmental Biologists*. New York: Wiley and Sons.

Hair, J.F., Jr., Anderson, R.E., and Tatham, R.L. 1987. *Multivariate Data Analysis*, 2nd ed. New York: Macmillan.

Harris, R.J. 1975. *A Primer of Multivariate Statistics*. New York: Academic Press.

James, F.C., and McCulloch, C.E. 1990. Multivariate analysis in ecology and systematics: panacea or pandora's box. *Annual Review of Ecology and Systematics* 21:129–166.

Johnson, D.H. 1981. The use and misuse of statistics in wildlife habitat studies. In *The Use of Multivariate Statistics in Studies on Wildlife Habitat*, ed. D.E. Capen, pp 11–19. U.S. Forest Service Gen. Tech. Report RM–87.

Johnson, D.H. 1981. How to measure—a statistical perspective. In *The Use of Multivariate Statistics in Studies on Wildlife Habitat*, ed. D.E. Capen, pp 53–58. U.S. Forest Service Gen. Tech. Report RM–87.

Kachigan, S.K. 1982. *Multivariate Statistical Analysis: A Conceptual Introduction*. New York: Radius Press.

Kendall, Sir M. 1980. *Multivariate Analysis*. New York: MacMillan.

Marriott, F.H.C. 1974. *The Interpretations of Multiple Observations*. London: Academic Press.

McDonald, L.L. 1981. A discussion of robust procedures in multivariate analysis. In *The Use of Multivariate Statistics in Studies on Wildlife Habitat*, ed. D.E. Capen, pp 242–244. U.S. Forest Service Gen. Tech. Report RM–87.

Morrison, D.F. 1967. *Multivariate Statistical Methods,* 2nd ed. New York: McGraw-Hill.

Nie, N.H., Hull, C.H., Jenkins, J.G., Steinbrenner, K., and Bent, D.H. 1975. *Statistical Package for the Social Sciences*. New York: McGraw-Hill.

Noy-Meir, I., and Whittaker, R.H. 1977. Continuous multivariate methods in community analysis: some problems and developments. *Vegetation* 33:79–98.

Orloci, L. 1975. *Multivariate Analysis in Vegetation Research*. The Hague: W. Junk.

Pimentel, R.A. 1979. *Morphometrics. The Multivariate Analysis of Biological Data*. Dubuque: Kendall/Hunt.

Press, S.J. 1972. *Applied Multivariate Analysis*. New York: Holt, Rinehart and Winston.

Rexstad, E.A., Miller, D.D., Flather, C.H., Anderson, E.M., Hupp, J.W., and Anderson, D.R. 1988. Questionable multivariate statistical inference in wildlife habitat and community studies. *Journal of Wildlife Management* 52:794–798.

Seal, H. 1966. *Multivariate Statistical Analysis for Biologists*. London: Methuen.

Shugart, H.H. Jr. 1981. An overview of multivariate methods and their application to studies of wildlife habitat. In *The Use of Multivariate Statistics in Studies on Wildlife Habitat*, ed. D.E. Capen, pp 4–10. U.S. Forest Service Gen. Tech. Report RM–87.

Tatsuoka, M.M. 1971. *Multivariate Analysis: Techniques for Educational and Psychological Research*. New York: Wiley and Sons.

2
Ordination: Principal Components Analysis

2.1 Objectives

By the end of this chapter, you should be able to do the following:

- Define *ordination* and give examples of how it can be used to reduce the dimensionality of a research problem.
- List eight important characteristics of ordination techniques.
- Draw a diagram of the first and second principal components for a hypothetical situation involving three variables to demonstrate a geometric understanding *of principal components analysis* (PCA).
- Explain the difference between using the covariance and correlation matrix in an eigenanalysis.
- Determine the appropriate number of principal components using five criteria:
  — Latent root criterion
  — Scree test criterion
  — Broken stick criterion
  — Percent trace criterion
  — Significance tests based on resampling procedures
- Given a matrix of principal component loadings, recognize the important variables and interpret the principal components.
- Explain from both a geometric and practical standpoint, the rotation of principal components and the potential benefits.
- List five limitations of PCA.
- Differentiate between R-factor and Q-factor ordination.
- Contrast the following ordination techniques with PCA:
  — Polar ordination (PO)
  — Factor analysis (FA)

- Nonmetric multidimensional scaling (NMMDS)
- Reciprocal averaging (RA)
- Detrended correspondence analysis (DCA)
- Canonical correspondence analysis (CCA)

2.2 Conceptual Overview

Ecologists are generally interested in understanding patterns in the distribution and abundance of organisms and the ecological or environmental factors that control those patterns. Often this involves a two-step process: (1) identifying and describing patterns in the distribution of organisms; (2) identifying ecological or environmental factors that help explain those patterns. Alternatively, this goal can be accomplished by first detecting patterns in the environment, and then determining the response of organisms to those patterns. In either case, the process involves identifying and describing important patterns in the data. One common approach involves sampling organisms and the environment at numerous sites distributed in some random or systematic fashion throughout the study area. For example, we might sample bird species and vegetation characteristics across a wide range of forest environments (e.g., plant communities in varying stages of successional development). The resulting data set would contain information on each species' presence or abundance and a suite of vegetation characteristics for each site. If there are consistent patterns in the distribution of species in relation to vegetation, then we should be able to organize sites along a meaningful continuum or gradient such that the relative positioning of sites along the continuum offers insight into relationships among sites.

2.2.1 Ordination

Ordination is a family of mathematical techniques whose purpose is to simplify the interpretation of complex data sets by organizing sampling entities along gradients defined by combinations of interrelated variables (Table 2.1). Essentially, ordination seeks to uncover a more fundamental set of factors that account for the major patterns across all of the original variables. The hope is that a few major gradients will explain much of the variability in the total data set. In such cases, the patterns in the data can be interpreted with respect to this small number of major factors or gradients, without a substantial loss of information.

The fundamental principle behind using ordination in ecology is that much of the variability in a multivariate ecological data set often is concentrated on relatively few dimensions, and that these major gradients are usually highly related to certain ecological or environmental factors. In other words, a few ecological factors usually drive a number of patterns and processes simultaneously. As a result, there is usually a great deal of redundancy (i.e., identical information shared by two or more variables) in the distribution of ecological data. Ordination techniques use this redundancy to extract and describe the major independent gradients in multivariate data sets.

2.2 Conceptual Overview

TABLE 2.1 Important characteristics of ordination techniques.

Organizes sampling entities (e.g., species, sites, observations) along continuous ecological gradients.

Assesses relationships within a single set of variables; no attempt is made to define the relationship between a set of independent variables and one or more dependent variables.

Extracts dominant, underlying gradients of variation (e.g., principal components) among sampling units from a set of multivariate observations; emphasizes variation among samples rather than similarity.

Reduces the dimensionality of a multivariate data set by condensing a large number of original variables into a smaller set of new composite variables (e.g., principal components) with a minimum loss of information.

Summarizes data redundancy by placing similar entities in proximity in ordination space and producing a parsimonious understanding of the data in terms of a few dominant gradients of variation.

Defines new composite variables (e.g., principal components) as weighted linear combinations of the original variables.

Eliminates noise from a multivariate data set by recovering patterns in the first few composite dimensions (e.g., principal components) and deferring noise to subsequent axes.

Consider the previous example and the following hypothetical data set involving the abundances of five bird species at six sites (rank order of sites for each species is shown in parentheses):

Sites	Species A	Species B	Species C	Species D	Species E
1	0 (1)	5 (1)	1 (1)	10 (4)	10 (4)
2	2 (3)	8 (3)	4 (3)	12 (6)	20 (6)
3	8 (6)	20 (6)	10 (6)	1 (2)	3 (2)
4	4 (5)	11 (5)	8 (5)	11 (5)	14 (5)
5	1 (2)	6 (2)	2 (2)	2 (3)	6 (3)
6	3 (4)	10 (4)	6 (4)	0 (1)	0 (1)

Although they differ in absolute abundances, species A, B, and C have an identical rank order of abundance among sites. Consequently, their distribution patterns are redundant and can be explained by a single ordering (or gradient) of sites. The same can be said of species D and E. Hence, the distribution patterns of these five species can be explained succinctly by two ordination gradients. If these two gradients can be related to certain ecological or environmental factors, then we can succinctly describe the major patterns in the data. Of course, real ecological data sets usually contain a more complex structure. For example, rarely would a single ordering of sites perfectly describe the distribution patterns of several species. Usually, finding the order of sites that "best" describes the overall structure of a complex multivariate data set is a nontrivial problem.

Ordination techniques offer different approaches for solving this problem. Specifically, the ordination family includes a number of different techniques (e.g., principal components analysis, polar ordination, reciprocal averaging, detrended correspondence analysis, canonical correspondence analysis, nonmetric multidimensional scaling, and factor analysis), which vary in their mathematical and geometric approaches, assumptions, properties, and intuitive appeal. No single method is robust enough to handle all data sets. The properties of each method should be evaluated with respect to each individual data set, so that the method or combination of methods selected is best suited for the data and questions under study. It is therefore necessary to understand the uses and limitations of the various techniques before deciding on a particular strategy.

Most ordination techniques are referred to as *unconstrained ordination* methods. An ordination is conducted on a single set of interdependent variables with the purpose of extracting the major patterns among those variables, irrespective of any relationship to variables outside that set. In other words, the ordination of sampling entities is not constrained by any relationship to variables outside the set. In the previous example, we discussed the ordination of sites on the basis of the species abundance variables. In this case, the ordination was not constrained by any relationships between species distribution and vegetation characteristics. Alternatively, we could have ordinated sites on the basis of vegetation characteristics alone without consideration of any relationship to species distribution patterns.

Unconstrained ordination methods, therefore, do not inform as to the possible mechanisms creating the observed patterns; they simply describe the major gradients inherent to the data set. This, of course, can be very informative, but tells us little about what those gradients mean and nothing about what causes them. For us to interpret their ecological meaning, the gradients must be associated with other ecological factors. Unconstrained ordination methods leave this to a separate, secondary analysis.

Ecologists often want to directly relate the patterns in one set of variables to patterns in a second set. For example, they might want to associate the major patterns in the composition of the plant community directly to environmental conditions. Or they might want to associate the abundance of a several animal species (i.e., community structure) directly with the structure of the plant community. Or they might wish to determine the influence of landscape structure on the composition of the animal community. These and other similar questions could be approached in a two-step process using unconstrained ordination methods. First, the researcher would use unconstrained ordination methods such as *principal components analysis* (PCA) or *detrended correspondence analysis* (DCA) to extract the dominant patterns in the data. Then they would attempt to associate these patterns with a second set of explanatory (independent) variables. Unfortunately, the association of ordination axes with patterns in a second set of variables often produces ambiguous results. Often it is more advantageous to directly ordinate the first set of variables on axes that are combinations of the second set of variables; that is, to constrain the ordination of the first set of variables by their relationships to a second set. This is called *constrained ordination*. In this manner, the gradients in the first variable set can be described directly in terms of the second (explanatory) set.

Unfortunately, it is not within the scope of this book to present a detailed description of every ordination method—constrained or unconstrained. Rather, we focus on PCA because it is the most well-understood and most widely used ordination method in wildlife ecology, and we provide a brief overview of other common ordination techniques for comparison. Furthermore, PCA provides the conceptual background for ordination that is helpful to understanding the other techniques.

2.2.2 Principal Components Analysis (PCA)

Unlike many multivariate techniques, PCA assesses relationships within a single set of interdependent variables, regardless of any relationships they may have to variables outside the set. As such, PCA is an unconstrained ordination technique; it does not attempt to define the relationship between a set of independent variables and one or more dependent variables, leaving this to subsequent analyses.

The main purpose of PCA is to condense the information contained in a large number of original variables into a smaller set of new composite dimensions, with a minimum loss of information. It reduces the P original dimensions of the data set, where each dimension is defined by one variable, into fewer new dimensions, where each new dimension is defined by a linear combination of the original P variables. These linear combinations are called *principal components*. Inherent in this approach is the assumption that the data structure is intrinsically low-dimensional and that this relatively simple fundamental structure can be extracted from the more complex multidimensional ecological data set.

Principal components analysis summarizes data redundancy by placing similar entities in proximity in ordination space and producing an economical description of the data in terms of a few dominant gradients of variation. If the original data set contains a lot of redundancy, then PCA will extract most of the variation in the fewest possible dimensions. Principal components analysis creates new composite variates (or components) out of the original variables that maximize the variation among sampling entities along their axes. That is, principal components are weighted linear combinations of the original variables that represent gradients of maximum variation within the data set. The ecological "meaning" of each component is reflected in the importance of each variable defining the component (i.e., higher importance is placed on those variables with larger weights). Thus, PCA provides a meaningful interpretation of each principal component based on the variables that are most important in defining the dimension. The relationships among sampling entities are evaluated by the entities' relative positions on the newly defined gradients. If the gradients have meaningful ecological interpretations, then the ecological relationships among entities can be defined based on the entities' relative positions on the ordination axes.

Principal components analysis offers a means of handling linear dependencies among variables that are troublesome for most multivariate techniques. It takes advantage of these dependencies by eliminating redundancy and generating new, fully uncorrelated variables. These independent variables then can be used in other multivariate techniques, such as multiple regression and multivariate analysis of variance, that are less capable of handling linear dependencies.

Finally, PCA can be an effective means of eliminating noise in a data set. Simulations of community data sets demonstrate that PCA selectively recovers patterns in early ordination axes, while selectively deferring noise to later axes (Gauch 1973).

2.3 Geometric Overview

The multivariate data set used in PCA can be depicted as a multidimensional cloud of sample points (Fig. 2.1). Each dimension, or axis, of the data set is defined by one of the P original variables. Each sample point has a position on each axis and therefore occupies a unique location in this P-dimensional space. Collectively, all the sample points form a cloud of points in P-dimensional space. If the cloud of points is concentrated in certain directions (i.e., not spherical), as is usually the case, then we can project new axes (principal components) that define the multidimensional structure of the data in fewer than P dimensions.

Geometrically, principal components can be viewed as projections through the cloud of sample points at orientations that maximize variation along each axis. The first principal component axis is drawn through the centroid of the cloud's longest

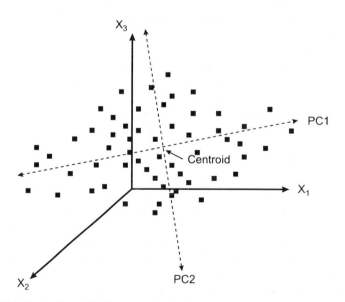

FIGURE 2.1 Derivation of the first two principal component axes (PC1 and PC2) from three original axes (X_1, X_2, and X_3) by translation of the origin to the centroid, followed by rotation about the centroid to maximize variances along the principal component axes.

direction. Hence, the first component explains the maximum amount of variation possible in a single dimension. The second principal component axis is constrained by orthogonality (i.e., independence) and maximization of the remaining variance. Geometrically, this means that the second component axis is perpendicular to the first in the direction of the cloud's greatest width. In practical terms, this means that the second axis explains the maximum remaining variation not explained by the first axis, and is thus statistically independent of the first. Subsequent axes follow the same constraints. Thus, in theory, there is no redundancy in the principal components; they are completely independent and complementary (i.e., orthogonal).

Geometrically, each principal component axis can be viewed in relation to the original P axes (corresponding to the original P variables). Some of the original P axes may be oriented in the same general direction as the new principal component axes, while others may not. The degree of similarity in orientation between the original axes and principal component axes indicates which of the original P variables are most important in the new principal components.

For PCA to be successful, a large portion of the variability within the data must be concentrated on a relatively small number of dimensions. If each variable were independent of the other variables, then the data cloud would be spread in an uncoordinated fashion across all P dimensions of the data space. None of the many possible projections of the data cloud would be better than any other, and none would give a better summarization than does projection on the original variables. When the variables are uncoordinated, no summarization will be successful and all variables must be retained to describe the patterns in the data. However, when the variables are highly coordinated in their responses—that is, when they change consistently with respect to one another—the data will be concentrated in a relatively low-dimension subspace within the P-dimensional data cloud. When the sampling entities are concentrated along two or three dimensions, then projection onto these dimensions will provide a much better summary of the total data structure than projection on any other dimensions, and PCA will be successful.

2.4 The Data Set

Unconstrained ordination techniques address problems involving a single set of interdependent variables. There is no distinction between independent and dependent variables; although, in a sense, each of the original variables can be considered as a dependent variable that is a function of some underlying latent (i.e., not easily observed) and hypothetical set of factors (components). Conversely, one can look at each component as a dependent variable that is a function of the originally observed independent variables. This is in marked contrast to constrained ordination techniques, such as canonical correspondence analysis, where the ordination of one data set (dependent variables) is constrained by its relationships to a second set (independent variables).

Preferably, variables should either be all continuous or all categorical. There is no consensus on the validity of mixed data sets, but most experts advise against using mixed data sets. Two other aspects of the data set should be noted. First, every sampling entity must be measured on the same set of variables. Second, as a general rule, there must be more sampling entities (rows) than number of variables (columns).

Ordination algorithms usually require a two-way data matrix as input. The two-way data can take any form. Some of the more common ecological two-way data matrices include:

- Sites-by-environmental parameters
- Species-by-niche parameters
- Species-by-behavioral characteristics
- Sites-by-species

Applications of ordination in the wildlife literature are many and varied (see Sec. 2.14.2). A substantial proportion of these concern the assessment of species-habitat relationships. The data typically consist of a set of observations for which there are associated vectors of environmental variables (e.g., habitat characteristics), where each observation may represent a species, individual, or site. For example, a sites-by-environmental parameters matrix consisting of N sites and P environmental variables would look as follows:

$$\begin{array}{c|ccccc} & X_1 & X_2 & X_3 & \cdots & X_P \\ 1 & x_{11} & x_{12} & x_{13} & \cdots & x_{1P} \\ 2 & x_{21} & x_{22} & x_{23} & \cdots & x_{2P} \\ 3 & x_{31} & x_{32} & x_{33} & \cdots & x_{3P} \\ \vdots & \vdots & \vdots & \vdots & \ddots & \vdots \\ N & x_{N1} & x_{N2} & x_{N3} & \cdots & x_{NP} \end{array}$$

In this book, we discuss PCA on a samples-by-variables matrix, where "samples" represent any type of sampling unit (e.g., sites, species, specimens) for which there is an associated vector of measurements, and "variables" represent any set of parameters recorded for each entity (e.g., habitat features, morphological characteristics, behavioral characteristics, species abundances). Unless otherwise noted, we refer to the samples (rows) as "entities" or "sampling entities" to reflect the generality of PCA to any type of sampling unit. However, our comments are applicable to any data of suitable form.

Empirical Example. The empirical example we present here was taken from a study on bird-habitat relationship in mature, unmanaged forest stands in the central Oregon Coast Range (McGarigal and McComb, unpublished data). In this chapter, we use PCA to investigate intrariparian gradients in vegetation composition along second- and third-order streams. Specifically, in this mountainous landscape subject to high rainfall, riparian vegetation can vary dramatically from one location to another in response to physical environmental gradients and disturbance and successional processes. These complex riparian environments provide habitat for myriad organisms.

In order to understand how streamside vegetation affects the distribution and abundance of animal populations and communities, it would be helpful to know something

OBS	ID	FORB	GRASS	FERN	SALAL	GRAPE	SALMON	CURRANT	HUCKLE	THIMBLE	DEVIL	VINE	ELDER	HAZEL	PLUM	OCEAN	ALDER	MAPLE	FIR	HEMLOCK	CEDAR	HARDWD	CONIFER
1	1S0	20	1	5	0	0	55	4	2	0	0	16	25	0	0	0	55	0	5	0	0	55	5
2	1S1	35	1	5	0	0	90	0	2	0	0	20	0	10	0	0	60	0	0	0	0	60	0
3	1S2	20	10	25	0	0	25	3	6	0	0	45	0	0	3	1	21	0	43	0	0	21	43
4	1S3	10	1	40	1	0	25	4	3	0	0	41	0	0	5	0	7	0	72	15	15	7	99
.	.	.	.	.	.	.	.	.	.	.	.	.	.	.	.	.	.	.	.	.	.	.	.
48	6S7	30	3	30	0	0	40	2	2	0	0	45	5	5	1	0	21	0	52	3	0	21	55

FIGURE 2.2 The example data set listing observation points (OBS labeled 1 to 48) and associated values for each of 22 floristic variables.

about how the vegetation varies from one location to another, and if this variation is inherently continuous or discrete (i.e., whether riparian vegetation can be classified into discrete types or not). We might hypothesize that each of the many plant species growing in these streamside environments responds to a few dominant environmental gradients (e.g., soil moisture, light, and disturbance regime) and that, consequently, we might be able to effectively describe the variation in streamside vegetation in a simplified way by identifying these dominant gradients. Principal component analysis provides us with a means to identify and quantitatively describe the existence of dominant gradients in streamside vegetation, and to qualitatively determine whether that variation is inherently continuous or discrete. Specifically, in this example, we use PCA to reduce the dimensionality of the data set by summarizing 22 floristic variables into a few principal components, and to describe major intrariparian gradients in floristic composition; that is, to describe the floristic characteristics that vary most dramatically among locations within these streamside areas.

The data set consists of 23 variables, including 22 floristic variables and a single variable (ID) that identifies the sampling point (Fig. 2.2). The floristic variables are all continuous and represent percent cover of 20 different plant species; HARDWD and CONIFER represent total hardwood and conifer tree cover, respectively. All variables were measured at 48 sampling points (OBS) distributed evenly among 6 different streams at 100-m intervals along 800-m long transects. Note that these 48 observations do not represent completely independent samples, as they were distributed among 6 streams (i.e., blocks). However, for the purposes of this chapter, we will treat these samples as independent. See Appendix 2.1 for an annotated SAS script for all of the analyses presented in this chapter.

2.5 Assumptions

For PCA to be strictly applicable, a data set must meet several assumptions of the PCA model (Morrison 1967; Harris 1975; Gauch 1982). Of course, ecological data sets rarely ever meet the requirements precisely. For applications involving confirmatory or inferential testing of hypotheses, the assumptions must be met rather well;

for merely descriptive purposes, larger departures from ideal data structure are tolerable (Gauch 1982). Even for descriptive purposes, however, it must be remembered that PCA has an underlying mathematical model and, consequently, may be applicable to one data set but not another (Gauch 1982).

In wildlife research, PCA is generally used as a method for describing and exploring covariance patterns in complex multivariate data sets from which testable hypotheses can be generated. Therefore, rigorous concern over meeting the assumptions is generally not warranted. However, if the PCA results are to be used in statistical inference, the assumptions should be met rather closely. The primary assumptions are listed below. Of these, failure to meet the linearity assumption is most likely to lead to erroneous interpretation of the results. Understanding the degree to which a given data set adheres to the assumptions will allow you to determine how cautiously to interpret the results.

2.5.1 Multivariate Normality

Principal components analysis assumes that the underlying structure of the data is *multivariate normal*. Geometrically, a multivariate normal distribution exists when the data cloud (in P-dimensional space) is hyperellipsoidal with normally varying density around the centroid (Beals 1973). Such a distribution exists when each variable has a normal distribution about fixed values on all others (Blalock 1979). If the data structure is multivariate normal, then linear axes (i.e., principal components) will adequately display the data. Otherwise, if the data follow some other distribution, then linear axes may fail to adequately describe the data structure.

In addition, second and subsequent ("later") principal component axes maintain strict independence (i.e., orthogonality) only when the underlying structure of the data is multivariate normal. Because most data sets are not completely multivariate normal, there is usually some redundancy in the principal components. In practical terms, this means that later principal components (i.e., those associated with smaller eigenvalues) will often define similar gradients to those defined by earlier components, but they will be considerably less pronounced. Unfortunately, there is no objective way to fully evaluate the assumption of multivariate normality. The best reassurance is a sufficiently large sample; although even an infinitely large sample will not normalize an inherently nonnormal distribution.

Diagnostics

(a) Univariate Normality of Original Variables

Assess skewness, kurtosis, and normality and construct stem-and-leaf, box, and normal probability plots of each original variable. Normal probability plots are plots of sample points against corresponding percentage points of a standard normal variable. If the data are from a normal distribution, the plotted values will lie on a straight line. However, whereas multivariate normality implies univariate normality, univariate normality does not imply multivariate normality. Hence, these diagnostics are of limited use only. In practice, these diagnostics are often used to determine whether the variables should be transformed prior to the PCA. In addition, it is usually assumed that univariate normality is a good step toward multivariate normality.

(b) Normality of Principal Component Scores

Assess skewness, kurtosis, and normality and construct stem-and-leaf, box, and normal probability plots for each of the principal component variables. These plots are generally more appropriate for evaluating multivariate normality than the univariate diagnostics, but even normally distributed principal component scores do not guarantee a multivariate normal distribution.

Empirical Example. In the example, we assessed each of the 22 floristic variable for skewness, kurtosis, and normality. As is the case in most ecological studies, most of the variables do not exhibit normal distributions. In Figure 2.3, for exam-

```
Variable = ALDER                                    Variable = ALDER

        Moments                                     Stem Leaf              #       Boxplot
                                                     9 59                  2
N                 48       Sum Wgts      48          9
Mean         40.4375       Sum         1941          8 66                  2
Std Dev      27.1621       Variance  737.7832        8 01                  2
Skewness      0.4596       Kurtosis   -0.8182        7 55                  2
USS           113165       CSS      34675.8100       7 01                  2
CV           67.1707       Std Mean    3.9205        6 6                   1
T:Mean = 0   10.3143       Prob >|T|   0.0001        6 0                   1
Sgn Rank         564       Prob >|S|   0.0001        5 55                  2       +-----+
Num ^= 0          47                                 5 1111                4       |     |
W:Normal      0.9370       Prob < W    0.0177        4 6666                4       |     |
                                                     4                             |  +  |
        Quantiles (Def = 5)                          3 556                 3       *-----*
                                                     3 0111                4       |     |
100% Max       99          99%        99             2 556                 3       |     |
 75% Q3      57.5          95%        86             2 0111                4       |     |
 50% Med       35          90%        81             1 566                 3       +-----+
 25% Q1        18          10%         7             1 123                 3
  0% Min        0           5%         4             0 777                 3
                            1%         0             0 034                 3
                                                    ----+----+----+----+
Range          99                                   Multiply Stem.Leaf by 10**+1
Q3-Q1        39.5
Mode           46

        Extremes                                            Normal Probability Plot

Lowest   Obs     Highest    Obs                  97.5+                           *  ++
   0    (32)       81      (15)                      |                              ++
   3    (36)       86      (18)                      |                          * *++
   4    (35)       86      (23)                      |                         ** ++
   7    (37)       95      (17)                      |                        ** ++
   7    (29)       99      (47)                      |                        * +
                                                     |                       *++
                                                     |                       *+
                                                     |                     ***
                                                     |                     ***
                                                     |                    ++
                                                     |                   ++**
                                                     |                  +**
                                                     |                 +**
                                                     |                +**
                                                     |               +*
                                                     |              ***
                                                     |             ***
                                                  2.5+   *   * * +
                                                     +----+----+----+----+----+----+----+
                                                         -2       -1       0       +1      +2
```

FIGURE 2.3 Univariate summary statistics used to assess normality for the variable ALDER.

```
Variable = PC1                                          Variable = PC1

              Moments                                   Stem Leaf                    #       Boxplot
                                                         3 01                        2          |
N                48      Sum Wgts      48                2                                      |
Mean              0      Sum            0                2                                      |
Std Dev           1      Variance       1                1                                      |
Skewness     1.1896      Kurtosis  1.6331                1 01124                      5         |
USS              47      CSS           47                0 677888889                  9      +-----+
CV                .      Std Mean  0.1443                0 113                        3      |  +  |
T:Mean = 0        0      Prob >|T| 1.0000               -0 4322111                    7      *-----*
Sgn Rank        -46      Prob >|S| 0.6420               -0 99999988776665555         17      +-----+
Num ^= 0         48                                     -1 41100                      5         |
W:Normal     0.8826      Prob < W  0.0001                 ----+----+----+----+

              Quantiles (Def = 5)                                       Normal Probability Plot
                                                       3.25+                                     *   *
100% Max    3.0598      99%    3.0598                      |                                   +++++
 75% Q3     0.7941      95%    1.3536                      |                                  +++++
 50% Med   -0.2457      90%    1.1333                      |                                 ****+*
 25% Q1    -0.8049      10%   -0.9948                  1.75+                                ******
  0% Min   -1.4285       5%   -1.0628                      |                             ******
                         1%   -1.4285                      |                         ++++**+
Range       4.4883                                     0.25+                     +++*****
Q3-Q1       1.5989                                         |                ************
Mode       -1.4285                                         |          ***********
                                                      -1.25+   *     *  *+*+++
              Extremes                                     +----+----+----+----+----+----+----+----+
                                                              -2        -1         0        +1       +2
Lowest     Obs       Highest   Obs
-1.4285   (18)       1.1333   (35)
-1.0907   (41)       1.1627    (3)
-1.0628   (15)       1.3538   (40)
-1.0005    (2)       3.0305    (4)
-0.9948   (26)       3.0598   (32)
```

FIGURE 2.4 Univariate summary statistics used to assess normality of the first principal component.

ple, the kurtosis statistic indicates that the variable ALDER has a flatter than normal distribution. Also, the Shapiro–Wilk statistic is significant ($P = 0.018$), indicating a distribution significantly different from normal. The plots all reflect these conclusions. Based on these findings, we suspect that the multivariate distribution is nonnormal, but we are unable to conclude this based on these results alone. Nevertheless, given these results, it is advisable to attempt data transformations in an attempt to improve the distribution of each variable. Therefore, we transformed each variable using several standard transformations, including the arcsine, square root, logit, and rank transformations—all standard transformations for percentages. The transformations did not normalize the variables, although slight improvements in the distributions of most variables were realized and a few were dramatically improved. Despite these improvements, the results of the PCA did not change in any substantive way. Therefore, we chose to retain the untransformed data for purposes of the final analysis.

We assessed the distribution of the scores for the first three principal components (Fig. 2.4). The first principal component has a skewed (skewness) and more peaked (kurtosis) than normal distribution; the Shapiro–Wilk statistic is significant ($P < .001$), indicating a nonnormal distribution. Again, the plots reflect these conclusions. Overall, these diagnostics indicate that the multivariate normality assumption is probably not met well. However, rigorous concern over multivariate normality is not warranted here, since we are using PCA solely as a means to explore and describe the gradients of variation in the data set, and not to generate statistical inferences concerning the underlying population.

2.5.2 Independent Random Sample and the Effects of Outliers

Principal components analysis assumes that random observation vectors (i.e., the variable measurements) have been drawn independently from a P-dimensional multivariate normal population; that is, observations represent an independent, random sample of the multidimensional space (Beals 1973; Gauch, Whittaker and Wentworth 1977). Since there is no test to determine whether a true random sample has been taken, the best reassurance is an intelligent sampling plan. Furthermore, while there are procedures for evaluating the statistical independence of samples, in practice, the best way to ensure independence is to consider it in the design of the study.

Most statistical procedures assume independence among samples, and PCA is not an exception. Failure to collect independent samples is common in wildlife research studies—often for good reasons. However, the potential effect of failure to meet this assumption on the statistical properties of the procedure being used is often overlooked. Fortunately, we can use PCA for exploratory or descriptive purposes whether this assumption is strictly met or not.

In field work, an ecologist may try to collect a random sample of a community series over a given area, but this does not ensure a random sample of the hyperellipsoid space. If there is a predominance in the field of certain conditions (e.g., vegetation types), then, by random sampling, these will be oversampled so far as PCA is concerned, and the calculated centroid will deviate from the true center of the hyperellipsoid. In addition, these point groups (i.e., clusters of similar entities) may distort the PCA by exerting undue influence on the direction of the principal component axes. Consequently, the longer underlying gradients may go undetected, or at least be substantially distorted. Under such conditions, stratified random sampling may be a more effective strategy.

A random sample of a community series is likely to include outliers (i.e., points that lie distinctly apart from all others) resulting either from unusual environmental or historical conditions, or from aberrations of the organism responses. These exert undue pull on the direction of the component axes and therefore strongly affect the ecological efficacy of the ordination. True outliers should be eliminated prior to the final component analysis. Thus, some form of screening for outliers is advisable (Grubbs 1950; Bliss, Cochran and Tukey 1956; Anscombe 1960; Kruskal 1960). However, it is important to distinguish between extreme observations and true ecological outliers. Extreme observations deviate considerably from the group mean, but may still represent meaningful ecological conditions. Thus, caution should be exercised in the deletion of suspect points, since careless elimination of outliers may result in the loss of meaningful information.

Diagnostics for Detecting Outliers

(a) Standard Deviations

Perhaps the simplest way to screen for outliers is to standardize the data by subtracting the mean and dividing by the standard deviation for each variable (i.e., standardize to zero mean and unit variance) and then inspect the data for entities with any value more than, for example, 2.5 standard deviations from the mean on any variable.

(b) Univariate Plots of Original Variables

Construct univariate stem-and-leaf, box, and normal probability plots for each variable and check for suspected outliers (i.e., points that fall distinctly apart from the rest).

(c) Plots of Principal Component Scores

Construct stem-and-leaf, box, and normal probability plots of the principal component scores for each principal component and check for suspected outliers. In addition, construct scatter plots for each pair of principal components and check for suspect points. In most cases (i.e., if the correlation matrix is used in the PCA or the data are standardized, as in the example below), the principal components are standardized variables; that is, they have a mean of zero and variance of one, and the values are in standard deviation units. Consequently, the principal component value (score) of each sampling entity represents the number of standard deviations from the principal component mean (or overall centroid).

Empirical Example. In the example, we inspected univariate stem-and-leaf, box, and normal probability plots of each variable for potential outliers. In Figure 2.3, for example, there are no apparent outliers with respect to ALDER. A few points deviated noticeably from the norm with respect to other floristic variables, but after careful assessment, we did not consider any to be true outliers. Rather, we considered these points to be real and meaningful variations in streamside vegetation composition.

We also looked for potential outliers in the stem-and-leaf, box, and normal probability plots of the first three principal components (Fig. 2.4) and in the scatter plots of each pairwise combination of the first three components (Fig. 2.5). There are two suspect points with respect to the first principal component, located 3.0 and 3.1 standard deviations from the mean (Fig. 2.4). These two points are also revealed in the plot of PC1 against PC2 (Fig. 2.5). As discussed previously, we did not consider these sample points true outliers. Therefore, they were retained in all further analyses.

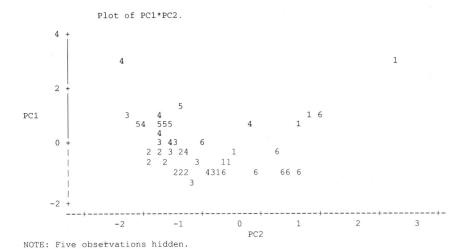

FIGURE 2.5 Scatter plot of principal component 1 versus principal component 2 used to assess the linearity assumption. Site ID (stream number) is used as the point label.

2.5.3 Linearity

Principal components analysis assumes that variables change linearly along underlying gradients (Austin and Noy-Meir 1971; Beals 1973; Austin 1976; Johnson 1981; Gauch 1982) and that linear relationships exist among variables such that the variables can be combined in a linear fashion to create principal components (Johnson 1981). This assumption is particularly troublesome when attempting to ordinate species in environmental space. It is known, for example, that the success (e.g., frequency distribution) of a species tends to exhibit a Gaussian (or normal, bell-shaped) curve along an environmental gradient. In this case, there will not be a linear or even unidirectional (i.e., monotonic) relationship between the species and the environment, except over short ranges of the environmental gradient or where the maximum for a species is near the edge of the environmental range sampled. Therefore, a species dimension often has a complex relationship to an environmental dimension, making the relationship difficult to portray in a linear, monotonic fashion. Pronounced nonlinearities in the data set usually result in what are described as the *arch effect* and the *compression effect* (see Gauch 1982 for a discussion of these phenomena). When this occurs, the relative positioning of sampling entities on the ordination axes becomes distorted, and the second and subsequent axes become difficult, and sometimes impossible, to interpret. Alternative ordination methods, such as detrended correspondence analysis (see Sec. 2.13.5) or detrended principal components analysis (Ludwig and Reynolds 1988), handle the nonlinearity problem in a more effective, though somewhat more complicated, manner.

Diagnostics

(a) Scatter Plots of Principal Component Scores

Construct scatter plots for each pair of principal components and examine them for nonlinear patterns. Arched or curvilinear configuration of sample points often indicates pronounced nonlinearities.

Empirical Example. In the example, we inspected scatter plots of the first three principal components. Figure 2.5, for example, portrays a slight U-shaped distribution of sample points in the plot of PC1 versus PC2. Note, however, that much of the curvilinear appearance is due to the two potential outliers discussed previously. Nevertheless, it is quite possible that nonlinearity is affecting the efficacy of the PCA; it would be instructive to compare the results of PCA with those from procedures less sensitive to nonlinearity (e.g., detrended correspondence analysis).

2.6 Sample Size Requirements

There is no absolute rule for determining the appropriate sample size in PCA; it largely depends on the complexity of the environment being investigated. Obviously, the more homogeneous the environment, the fewer samples needed to uncover any patterns that exist. Beyond this, there are some general rules to follow as well as some specific rules proposed by various authors.

2.6.1 General Rules

As a minimum, there must be at least as many sample entities as variables. Beyond this, in PCA enough entities should be sampled to adequately describe each distinctive community; as few as three replicate samples in each community type may suffice for descriptive purposes, or as many as fifty replicate samples in each community type may be necessary to satisfy statistical considerations (Gauch 1982). Furthermore, enough entities should be sampled to ensure that the variance structure of the population is estimated accurately and precisely by the sample data set. This can be done in several ways, for example, by sampling sequentially until the mean and variance of the parameter estimates (e.g., the eigenvalues and eigenvectors; see below) stabilize. Once the data have been gathered, the stability of the results can be assessed using a resampling procedure (see Sec. 2.8.5), and the results can be used to determine the sample size needed to estimate the parameters with a specified level of precision in future studies.

2.6.2 Specific Rules

The following specific rules have been suggested by different authors (where P = number of variables):

Rule A	$N = 20 + 3P$	(Johnson 1981)
Rule B	$N = 4P$	(Hair, Anderson, and Tatham 1987)
Rule C	$N = 10P$	(unknown author)

These represent somewhat conservative rules, and in many instances the researcher is forced to analyze a set of variables with only a 2:1 ratio of samples to variables. In general, at a minimum a 3:1 ratio should be maintained. When dealing with smaller sample sizes and lower ratios, the importance of evaluating the stability of the parameter estimates (e.g., by using a resampling procedure, Sec. 2.8.5) increases and the results should be interpreted cautiously.

Empirical Example. In the example, the data set contains 48 observations and 22 variables. Based on the specific rules [Rule A: $N = 20 + (3 \cdot 22) = 86$; Rule B: $N = 4 \cdot 22 = 88$; Rule C: $N = 10 \cdot 22 = 220$], the example data set contains far too few observations. However, since the data set contains better than a 2:1 ratio of samples to variables, it can be analyzed, but it must be interpreted cautiously. Furthermore, because in this case we are using PCA solely for descriptive purposes, we can forego rigorous requirements on sample size.

2.7 Deriving the Principal Components

The first major step in PCA is to derive the principal components. This derivation step represents the nuts and bolts of the analysis; it includes several small steps that must be understood before we can consider how to interpret the principal components. A number of decisions must be made about how to analyze the data. These include: (1) whether to use the correlation matrix or covariance matrix to summarize the variance structure of the original data matrix; (2) how to compute the eigenvalues from the correlation or covariance matrix; and (3) how to compute the eigenvectors from the correlation or covariance matrix.

2.7.1 The Use of Correlation and Covariance Matrices

Principal components are not derived directly from the original data matrix; instead, they are derived from a secondary data matrix computed from the original data. Either the sample covariance matrix (also referred to as the variance-covariance matrix), the sample correlation matrix, or any suitable dissimilarity matrix derived from the original data can be used in the analysis. In all cases, the secondary data matrix is a $P \times P$ symmetrical matrix (where P equals the number of original variables), in which the diagonal elements represent a measure of internal association (i.e., the relationship between a variable and itself) and the off-diagonals represent the relationship between each pairwise combination of variables. Thus, in the correlation matrix all the diagonal elements have a value of one (i.e., because a variable's correlation with itself is equal to one) and the off-diagonals represent Pearson product-moment correlation coefficients between each pair of variables. In the covariance matrix, the diagonal elements correspond to the variances of the variables and the off-diagonals represent covariances. Note that the correlation matrix is equivalent to the covariance matrix computed from standardized data (variables with zero mean and unit variance). Hence, to avoid confusion we will refer to the covariance matrix computed from the raw, unstandardized data as the *raw covariance matrix*.

Principal components analysis seeks to summarize the structure of the secondary data matrix. Thus, it is important to understand the differences between the correlation matrix and raw covariance matrix. Recall from our geometric overview that PCA seeks to identify axes (i.e., gradients of variation) that effectively summarize the overall structure (i.e., shape) of the data cloud, where the dimensions of the data cloud are defined by the original variables. A variable with a large variance will have observations distributed widely in space relative to a variable with a smaller variance, and will therefore contribute more to the shape of the data cloud. As a result, this variable will have greater weight (i.e., pull) in defining the new axes (i.e., principal components) if the raw variance-covariance structure is analyzed. A component analysis of the raw covariance matrix, therefore, gives more weight to variables with larger variances (i.e., it assigns weights to variables proportionate to their variances). Computationally, this occurs because the diagonal elements of the raw covariance matrix are equal to the variances of the corresponding variables. A component analysis of the correlation matrix, on the other hand, gives equal weight to all variables. Computationally, this occurs because the diagonal elements of the correlation matrix are all equal to one. Note that the principal component solutions obtained from the correlation and raw covariance matrices will be different.

In most cases, the correlation matrix is preferred and is always more appropriate if the scale or unit of measurement differs among variables (Noy-Meir et al. 1975) because we usually have little a priori basis for deciding if one variable is more ecologically important than another. And the fact that one variable may exhibit a greater spread in absolute values than another (i.e., greater absolute variance) does not necessarily mean that it expresses a more important or meaningful ecological gradient. However, there may be circumstances when the raw covariance matrix is more desirable, particularly when the variables all share a common measurement scale. If, for example, the data set consists of sites-by-species abundances, it is often the case that the more abundant species will have greater absolute variances. One obvious consequence of using the raw covariance matrix is that the variables (species, in this case) with larger variances will tend to dominate the first component. Statistical algorithms generally use the correlation matrix in the eigenanalysis unless otherwise specified.

The correlation matrix is useful in another way in PCA; it provides the product-moment correlation coefficients for all pairwise combinations of the original variables. Correlations indicate how parsimoniously the PCA will be able to summarize the data. Higher correlations mean greater redundancy, and greater redundancy results in more variation extracted in fewer components. However, a large number of high correlations may indicate that several variables will weight heavily in each principal component, thus complicating interpretation.

Empirical Example. In the example, all floristic variables have the same measurement scale. Thus, it would be acceptable to use the covariance matrix in the component analysis. However, because we sought to give equal emphasis to all species regardless of their absolute variances, we elected to use the correlation matrix. In this manner, relatively uncommon plant species, which tend to have smaller absolute variances, receive equal weight as the more common species. Consider, for example, two of the more widely distributed species: huckleberry (HUCKLE) and Douglas-fir (FIR). The former is much less abundant than the latter in terms of percent cover, yet both species exhibit similar variability in their distribution patterns. The absolute variance in HUCKLE is much less than the variance in FIR. Use of the covariance matrix, therefore, would place a disproportionate emphasis on FIR to derive the principal components. Yet there is no reason to assume that Douglas-fir distribution is any more indicative of underlying environmental gradients than the less abundant, but equally well-distributed huckleberry. Overall, the pairwise correlations between variables (Fig. 2.6) were generally low ($r < 0.5$), indicating that there is not a lot of redundancy in the data set. Hence, we would not expect PCA to be extremely effective in reducing the dimensionality of the data set—a conclusion later borne out.

Correlations

	FORB	GRASS	FERN	SALAL	GRAPE	SALMON
FORB	1.00000	0.39742	-0.27762	-0.22513	-0.09335	0.40029
GRASS	0.39742	1.00000	-0.06556	-0.12450	-0.12450	0.13504
FERN	-0.27762	-0.06556	1.00000	0.49756	0.17372	-0.36834
SALAL	-0.22513	-0.12450	0.49756	1.00000	-0.02128	-0.19710
GRAPE	-0.09335	-0.12450	0.17372	-0.02128	1.00000	-0.31622

FIGURE 2.6 Correlation matrix containing pairwise correlations between the original variables. Only a portion of the correlation matrix is shown.

2.7.2 Eigenvalues and Associated Statistics

Computationally, PCA is essentially an eigenanalysis problem, although the details of the relevant matrix algebra are not presented here. For an $N \times P$ data set, where N equals the number of sampling entities (rows of the data matrix) and P equals the number of variables (columns), there exist P eigenvalues, also referred to as *latent roots* or *characteristic roots*. The eigenvalues are determined by solving the *characteristic equation*

$$|R - \lambda I| = 0$$

where R is the correlation or covariance matrix, λ is the vector of eigenvalue solutions, and I is the identity matrix. Note that either the correlation or raw covariance matrix (or any dissimilarity matrix) can be used in the characteristic equation. Eigenvalues represent the variances of the corresponding principal components; they measure the extent of variation among sampling entities along the dimension specified by the principal component. Each eigenvalue is associated with one principal component. Consequently, for an $N \times P$ data set there exist P principal components.

All of the P eigenvalues will be positive or zero, and the larger the eigenvalue is, the greater the sample variation on that principal component. Thus, the component with the largest eigenvalue (always the first component) does the best job of capturing the sample variance-covariance structure, while the component with the smallest eigenvalue does the worst. Eigenvalues near zero indicate that the corresponding component has minimal explanatory power.

The first eigenvalue is always the largest. Therefore, the first principal component defines the dimension or gradient with the single highest variance (i.e., the maximum variance among entities). The second eigenvalue measures the variance along the second principal component; it represents the largest variance in a dimension orthogonal to (i.e., independent of) the first dimension. Thus, the second component provides the greatest explanation of sample variance after the first has done its best. The third eigenvalue corresponds to the component with the next greatest variance, given that it is orthogonal to the first and second components; it provides the greatest additional explanation of sample variance after the first and second have done their best. And so on for P eigenvalues and components.

The actual numbers representing the eigenvalues will differ depending on whether the correlation or raw covariance matrix is used. The sum of the eigenvalues will always equal the *trace* (the sum of the diagonal elements) of the secondary matrix used. For example, if you use the correlation matrix, then the sum of the eigenvalues will equal the number of variables (because the correlation matrix is a $P \times P$ matrix with ones along the diagonal). If you use the raw covariance matrix, then the sum of the eigenvalues will equal the sum of the variances for all P variables. In the latter case, the absolute values of the eigenvalues do not have any comparative value from one data set to another because they reflect the scale and variance of the original variables.

Eigenvalues of the Correlation Matrix: Total = 22   Average = 1

	PC1	PC2	PC3	PC4	PC5	PC6
Eigenvalue	5.3095	3.4170	2.1397	1.5348	1.4392	1.3300
Difference	1.8925	1.2772	0.6048	0.0956	0.1091	0.2420
Proportion	0.2413	0.1553	0.0973	0.0698	0.0654	0.0605
Cumulative	0.2413	0.3967	0.4939	0.5637	0.6291	0.6896

FIGURE 2.7 Eigenvalues and associated statistics for the first six principal components.

Empirical Example. In the example, we computed the 22 eigenvalues of the correlation matrix; the first 6 are displayed in Figure 2.7. The eigenvalue associated with the first principal component is equal to 5.3. This value represents the variance associated with PC1, but the absolute magnitude has no *direct interpretive value*.

2.7.3 Eigenvectors and Scoring Coefficients

The other product of eigenanalysis is an eigenvector associated with each eigenvalue. The eigenvectors are determined by solving the following system of equations:

$$[R - \lambda_i I]v_i = 0$$

where R is the correlation or covariance matrix, λ_i is the eigenvalue corresponding to the ith principal component, I is the identity matrix, and v_i is the eigenvector associated with the ith eigenvalue. Again, note that either the correlation or raw covariance matrix (or any dissimilarity matrix) can be used in the equation above. The number of coefficients in an eigenvector equals the number of variables in the data matrix (P). These are the coefficients of the variables in the linear equations that define the principal components and are referred to as *principal component weights*.

If the correlation matrix is used in the eigenanalysis, as is usually the case, then the principal component weights are directly proportional to the correlations between the corresponding variables and the principal components (i.e., the principal component loadings; see below). In this case, the principal component weights represent the relative importance of the original variables in the principal components, and thus provide one way of generating an ecological interpretation of the components. However, principal component weights are not the actual correlations between the original variables and principal components, even though they have the same utility if the correlation matrix is used. Because of this, interpretation of the principal components is almost always based on the actual correlations and not the variable weights as given by the eigenvectors.

If the raw covariance matrix is used in the eigenanalysis, then the principal component weights are not proportional to the correlations between the corresponding variables and the principal components, and thus cannot be used to infer any ecological meaning to the principal components. In this case, comparison of the relative magnitude of the combining weights as given by the elements of each eigenvector is inappropriate because these are weights to be applied to the variables in raw score scales and are therefore affected by the particular unit (i.e., measurement scale) and variance of each variable.

A simple adjustment to the eigenvectors results in coefficients that give the principal components more desirable properties. When we adjust the eigenvector coefficients, we refer to them as *standardized* principal component weights. These coefficients represent the weights that would be applied to the variables in standardized form to generate standardized principal component scores, and thus we often refer to them as *standardized scoring coefficients*. The adjustment means that the principal component scores over all the sampling entities will have a mean of zero and a standard deviation of one. Each axis is stretched or shrunk such that the score for an entity represents the number of standard deviations it lies from the grand centroid. In this manner, the score for an entity immediately describes how far it is from the average score on that component. Although the standardized scoring coefficients are valuable for producing useful component scores, they generally serve little purpose in interpreting the components, since this is usually accomplished by means of interpreting the principal component loadings (see Section 2.9.1).

Empirical Example. In the example, we computed the eigenvectors corresponding to the 22 principal components; we retained only the first 3 for interpretation (Fig. 2.8). Since we used the correlation matrix in the eigenanalysis, the principal component weights are directly proportional to the correlations between the original variables and principal components. Hence, we could use the principal component weights (i.e., eigenvector coefficients) to interpret the ecological meaning of each principal component, but we will reserve this task to the principal component loadings described below.

Eigenvectors

	PC1	PC2	PC3
FORB	-0.16944	-0.25605	0.26181
GRASS	-0.11608	0.00735	0.22125
FERN	0.22392	0.32496	0.01292
SALAL	0.19387	0.25826	0.30333
GRAPE	0.19574	-0.10587	-0.27739
SALMON	-0.27122	-0.08300	0.22570
CURRANT	0.20022	-0.20554	-0.20348
HUCKLE	0.27967	-0.02785	-0.01490
THIMBLE	0.02904	-0.10617	0.09580
DEVIL	0.16227	-0.01713	-0.38963
VINE	0.01253	0.42662	-0.06504
ELDER	-0.06388	0.11763	-0.12626
HAZEL	-0.06833	0.26611	-0.33231
PLUM	0.20842	0.30346	0.36479
OCEAN	-0.01508	0.07693	0.14998
ALDER	-0.34090	0.11286	0.00613
MAPLE	-0.13153	0.21036	-0.28273
FIR	0.24961	0.33417	0.10368
HEMLOCK	0.16179	-0.26127	0.21247
CEDAR	0.25627	-0.24751	-0.14405
HARDWD	-0.35390	0.16145	-0.06974
CONIFER	0.38374	-0.02219	0.11040

FIGURE 2.8 Eigenvectors for the first three principal components.

2.8 Assessing the Importance of the Principal Components

An important decision in PCA is determining how many principal components to retain and interpret. Recall that the larger the eigenvalue, the greater the explanatory power of that component. Components with small eigenvalues are more likely to describe error variance or represent influences which affect only one or very few of the variables in the system. Several heuristic and statistical approaches (often referred to as *stopping rules*) have been developed to help judge the importance of each principal component and determine the number of significant components to retain and interpret. In this section, we review some of the more common approaches.

2.8.1 Latent Root Criterion

The most common heuristic approach is the *latent root criterion*, also referred to as the Kaiser–Guttman criterion (Guttman 1954; Cliff 1988). The latent root criterion is simple to apply: When the correlation matrix is used in the eigenanalysis, components with eigenvalues less than one are dropped from further scrutiny. These components represent less variance than is accounted for by a single original variable and therefore should not be interpreted. Recall that when using the correlation matrix, all variables have unit variance. The rationale for the latent root criterion is that any individual component should account for at least the variance of a single variable if it is to be retained for interpretation.

The latent root criterion is probably most reliable when the number of variables is between 20 and 50. In instances where the number of variables is less than 20, there is a tendency for this method to extract a conservative number of components, and where there are more than 50 variables, it is common for too many components to be extracted (Hair, Anderson, and Tatham 1987). In practice, most researchers use this criterion to determine the maximum number of components to retain. Unfortunately, this method has been criticized because a PCA of randomly generated, uncorrelated data will produce eigenvalues greater than one.

2.8.2 Scree Plot Criterion

A *scree plot* is derived by plotting the eigenvalues (latent roots) against the component number in the component's order of extraction; the shape of the resulting curve is used to evaluate the appropriate number of components to retain (e.g., see Fig. 2.9; Cattell 1966). Recall that eigenvalues decrease sequentially from the first through the last component. Typically, the scree plot slopes steeply down initially and then asymptotically approaches zero. The point at which the curve first begins to straighten out indicates the maximum number of components to extract. As a general rule, the scree test criterion results in at least one, and sometimes two or three,

42   2. Ordination: Principal Components Analysis

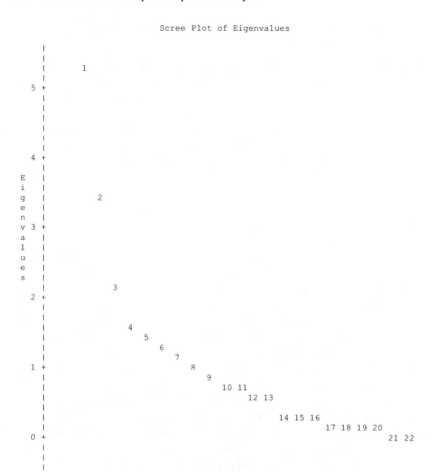

FIGURE 2.9 Scree plot for the eigenvalues associated with the 22 principal components.

more components to place under consideration as significant than does the latent root criterion (Hair, Anderson, and Tatham 1987).

The scree plot method has been criticized because occasionally there is no obvious break point, or there are multiple apparent break points (Horn 1965; Jackson 1993). Horn proposed a modification to the scree plot to address these drawbacks. First, an ordinary scree plot is constructed for the ordination. Then, a number of matrices of the same rank but containing uncorrelated variables are generated. A scree plot is then constructed from the mean values of the eigenvalues of these simulated random matrices. The point at which these two scree plots intersect indicates the maximum limit of interpretable components. This modification provides an objective means for deciding where to "break" the scree plot.

2.8.3 Broken Stick Criterion

Frontier (1976) proposed a *broken stick* method that is based on eigenvalues from random data. The broken stick model assumes that if the total variation (i.e., sum of the eigenvalues) is randomly distributed among components (i.e., there is no real structure), then the scree plot will exhibit a broken stick distribution. Observed eigenvalues that exceed the eigenvalues expected under the broken stick distribution are considered meaningful and are retained for interpretation. Broken stick methods of estimating the number of components to retain have produced the best results of any heuristic approach in tests on simulated data with known distributions (Jackson 1993). In fact, in a study comparing a number of methods, broken stick methods performed as well or better than any statistical technique, but with the advantage of easy calculation. The eigenvalues under the broken stick model are readily calculated as

$$\lambda_i^* = \sum_{k=i}^{P} \frac{1}{k}$$

where k is the number of the ith principal component, and λ_i^* is the size of the eigenvalue for the ith principal component under the broken stick model (Jackson 1993).

2.8.4 Relative Percent Variance Criterion

As we have already discussed, the actual numbers representing the eigenvalues do not have a great deal of interpretive value; this is because they reflect the number of variables in the data set (when the correlation matrix is used) or the scale and variance of the original variables (when the raw covariance matrix is used). However, we can compare the relative magnitudes of the eigenvalues to see how much of the total sample variation in the data set is accounted for by each principal component. To do this, we simply convert the eigenvalues into relative percentages. *Relative percent variance* (Φ) (also referred to as *percent trace*) is defined as follows:

$$\Phi_i = \frac{\lambda_i}{\sum_{i=1}^{P} \lambda_i}$$

where Φ_i is the relative percent variance of the ith principal component, and λ_i is the eigenvalue corresponding to the ith principal component.

Relative percent variance measures how much of the total sample variance is accounted for by each principal component. The cumulative percent variance of all eigenvalues is equal to 100 percent because the P eigenvalues always account for the total variance in the data set. Ideally, the cumulative percent variance of the first one to three principal components will be high (greater than 70 percent), indicating that the data structure was effectively summarized in few dimensions. The relative percent variance criterion can be useful in several ways: (1) it can evaluate the importance of each principal component; (2) it can determine how many principal components to retain; and (3) it can evaluate the effectiveness of the ordination as a whole in parsimoniously summarizing the data structure.

Although relative percent variance is intuitively appealing as a criterion, it has been criticized as being unfounded and unreliable. Besides being somewhat arbitrary, it is influenced by several factors and can be very misleading. In particular, it is influenced by the number of variables in the data set (Karr and Martin 1981). For example, with only three variables in the data set, the cumulative percent variance of the three components would be 100 percent, and yet the second and third components may be meaningless. With 20 variables in the data set, the cumulative percent variance of the first three components would have to be much less (unless the last 17 variables were 100 percent redundant), yet all three components may be very meaningful. Relative percent variance is also influenced by the number of observations: it declines with increasing sample size (Karr and Martin 1981). Hence, the use of this criterion to evaluate the effectiveness of the ordination must be employed with caution.

Alternatively, it has been suggested that the percent of "unexplained variance" remaining in the data set accounted for by each component is a more appropriate criterion than the relative percent variance criterion (Stauffer, Garton, and Steinhorst 1985). For example, with $P = 22$ eigenvalues, the percent of remaining variance accounted for by the fourth eigenvalue would be determined by how much of the sample variance is explained by the fourth component after removing the variance explained by the first three components. Specifically, *percent of remaining variance* (Ω) is defined as follows:

$$\Omega_i = \frac{\lambda_i}{\sum_{i=i}^{P} \lambda_i}$$

where Ω_i is the percent remaining variance of the ith component, and λ_i is the eigenvalue corresponding to the ith component.

However, we believe that percent of remaining variance can lead to erroneous conclusions regarding the importance of the principal components for the following reason. If the data structure is summarized very effectively by the component transformation and the first component accounts for a very large percent of the total variance (i.e., Φ_1 is very large), then the second and subsequent components will, by necessity, explain a very small percent of the total variance, since the sum must equal 100 percent. In an extreme case (to illustrate the point), the first component might explain 90 percent of the variance, the second 5 percent, the third 2 percent, and so on. Few researchers would consider the second component important or significant, since 5 percent of the total sample variance is a trivial amount. Yet, in this case, the second component accounts for 50 percent of the remaining variance. Consequently, evaluating the importance of the components based on percent of remaining variance can lead to spurious results; for this reason we prefer to use the relative percent variance criterion (Φ).

2.8.5 Significance Tests

There are a number of objective tests for evaluating the statistical significance of the principal components when the data come from a sample as opposed to an entire population. With a sample, we can ask what is the probability that the sampling process produced entities that show the computed variance-covariance structure, when in fact there are is no interesting structure in the population (i.e., Type I error). Generally we must assume independent random samples to ensure valid probability values.

Parametric statistical tests exist for determining the significance of each eigenvalue, but these tests are rarely employed because of the assumptions involved (Tatsuoka 1971). Alternatively, we can use a nonparametric test based on a resampling procedure to judge the significance of a principal component. Resampling procedures are conceptually simple, computer-intensive, nonparametric methods for determining the variability of statistics with unknown or poorly known distributions. As the name implies, these procedures involve "resampling" the original data. The function of resampling is to generate a distribution for the statistic that permits the calculation of dispersion estimates (i.e., the variability of the statistic under either the null or alternative hypothesis). The main advantage of these procedures over the usual parametric methods is that statistical inferences under resampling are based on the distributional properties of a "pseudosample" that is generated by resampling the data itself, and not the distributional properties of a sample drawn from an unknown underlying population. Consequently, the detailed specification of a probability model for the data is unnecessary, and no complex calculations or restrictive assumptions about the underlying population are required, since the statistical properties of a pseudosample are easy to determine by simulation, if they cannot be derived analytically. The only assumption required is that the observations are independent.

The most popular resampling procedures are the *jackknife* and the *bootstrap* (Efron 1979, 1982; Efron and Gong 1983). A less well-known but more attractive procedure is the *randomization test* (Edgington 1995; Manly 1997). Each of these procedures can be used to test the statistical significance of virtually any parameter estimated in PCA (or any other multivariate technique discussed in this book). In the following discussion, we present a general description of these procedures and discuss their application for testing the significance of each eigenvalue, but they are easily extended to other measures of component performance. Rather than testing each eigenvalue directly, however, we use these procedures to test the significance of the relative percent variance (Φ) accounted for by each eigenvalue.

Now each of the popular resampling procedures are described.

1. Jackknife Procedure. The premise of the jackknife procedure is to determine the effect of each sampling entity on a statistic by iteratively removing successive sampling entities and recalculating the desired statistic. Briefly, in the present context, the jackknife procedure begins by computing for each eigenvalue the relative percent variance from the original $N \times P$ data set. A single entity is then removed from the data and the Φ_i are recalculated using the remaining $N-1$ entities. The next entity is then removed from the data and the Φ_i are again calculated using the remaining $N-1$ entities. This procedure is repeated N times, removing successively each entity, until N "pseudoestimates" of the Φ_i have been obtained. Let $\Phi_{i(\text{all})}$ represent the Φ for the ith component calculated from the full sample and let $\Phi_{i(j)}$ represent the Φ for the ith component with the jth sample removed. Define a pseudovalue $\Phi^*_{i(j)}$ as

$$\Phi^*_{i(j)} = N\Phi_{i(\text{all})} - (N-1)\Phi_{i(j)}$$

where N is the sample size. The jackknife estimate of Φ_i is defined as the mean of the pseudovalues:

$$\Phi^*_i = \frac{\sum_{j=1}^{N} \Phi^*_{i(j)}}{N}$$

The standard deviation (also the standard error) of the jackknife estimate is given by

$$SE(\Phi^*_i) = \sqrt{\frac{\sum_{j=1}^{N} (\Phi^*_{i(j)} - \Phi^*_i)^2}{N(N-1)}}$$

To determine if the observed Φ_i differ significantly from those expected from the same data but without any "real" (i.e., nonrandom) structure, simply divide the jackknife estimate (Φ^*_i) by its estimated standard error $(SE(\Phi^*_i))$. This ratio (referred to as the jackknife error ratio) can be treated as a t-statistic with $N-1$ degrees of freedom (Mosteller and Tukey 1977).

2. Bootstrap Procedure. The premise of the bootstrap is that, through resampling of the original data, confidence intervals may be constructed based on the repeated recalculation of the statistic under investigation. Briefly, in the present context, the procedure involves drawing N random sample entities (i.e., rows of the data matrix), with replacement after each drawing, from the original $N \times P$ data set (call this a "bootstrap sample") and calculating the Φ_i for the bootstrap sample. A consequence of replacement after each drawing is that a sampling entity could be represented more than once or not at all in the bootstrap sample. By drawing a large number (M) of bootstrap samples from the original data set, a bootstrap estimate of the

2.8 Assessing the Importance of the Principal Components

standard error for Φ_i and a confidence interval can be calculated. In this case, let $\Phi^*_{i(j)}$ represent the jth bootstrap calculation of Φ_i. The bootstrap estimate of Φ_i is defined as the mean of the M bootstrap estimates:

$$\Phi^*_i = \frac{\sum_{j=1}^{M} \Phi^*_{i(j)}}{M}$$

The standard deviation (also the standard error) of the bootstrap estimate is given by

$$SE(\Phi^*_i) = \sqrt{\frac{\sum_{j=1}^{M} \Phi^*_{ij} - \Phi^{*2}_i}{M-1}}$$

As in the jackknife procedure, to determine if the observed Φ_i differ significantly from those expected from the same data but without any "real" (i.e., nonrandom) structure, simply divide the bootstrap estimate (Φ^*_i) by its estimated standard error ($SE(\Phi^*_i)$). This ratio (referred to as the bootstrap error ratio) can be treated as a t-statistic with $N-1$ degrees of freedom. Indeed, the jackknife and bootstrap procedures are very similar, although it has been shown that the bootstrap is superior to the jackknife (Efron and Gong 1983).

In a simulation experiment comparing a number of methods, Jackson (1993) found that bootstrapping was superior to a number of other statistical tests of component significance, although he did not compare it with the randomization test described below. He suggests a combined approach where bootstrap confidence intervals are constructed for both eigenvalues, as discussed above, and for eigenvector coefficients (or principal component weights). Eigenvector coefficients that are not significantly different than zero are discarded from each component. Only components having two or more coefficients that are significantly different than zero are considered meaningful. This is because a component with only one significant eigenvector coefficient represents only a single original variable, and therefore nothing is gained by the component analysis.

3. Randomization Test. The premise of the randomization procedure is that, through resampling of the original data, we can generate the actual distribution of the statistic under the null hypothesis and test the observed sample against this distribution (Edgington 1995). Briefly, in the present context, the procedure involves randomizing the data matrix (call this a "random permutation") and calculating the Φ_i for the random permutation. The randomization procedure differs from the bootstrap in how the randomization is done. In the bootstrap, we randomly draw N entire vectors corresponding to the original sample observations (i.e., rows of the data matrix) with replacement after each drawing. Thus, the variance structure of each sampling entity is left intact. In the randomization procedure, we randomize the vector of observed values for each variable (i.e., columns of the data matrix) such that every value in the original data matrix is represented in the random permutation set, but in random order. Thus, in the randomization procedure, each of the N sample observations in the random permutation includes a randomly selected value (without replacement) from the first variable, a randomly selected value (without replacement) from the second variable, and so forth for P variables. Thus, the random permutation matrix contains all the original $N \times P$ data values, but it is randomized within each variable to remove any "real" data structure. Furthermore, while each bootstrap sample can be drawn more than once because it is drawn from the same complete data set, each random permutation sample can be drawn only once because it is not replaced after selection.

We repeat this step a large number of times (ideally computing all possible random permutations of the data matrix) and rank order the Φ_i^* estimates. The resulting permutation distribution of Φ_i^* reflects the expected distribution under the null hypothesis—that is, when there is no "real" data structure (i.e., no interesting covariance structure determined by underlying ecological gradients). Thus, we can compare the magnitude of the original Φ_i to the distribution under the null hypothesis and determine the percentage of the distribution that is greater than or equal to the original estimate (for an upper, one-sided test). This percentage equals the probability (P value) of observing a Φ as large as this if the original sample were actually drawn from a population defined by the sample data characteristics, but without any interesting structure.

All three of these resampling procedures provide a measure of the statistical significance of the statistic under investigation. In the present context, they all objectively measure the importance of the principal components, as measured by the relative percent variance criterion. They differ, however, in the nature of the test procedure developed by the resampling procedure. By removing any real data structure from the random permutation samples, the randomization procedure calculates the expected distribution of the statistic (in this case, Φ_i) under the null hypothesis of no real data structure, and compares the observed value of the statistic to that distribution. In the present context, by comparing the observed Φ_i to the permutation distribution, we can determine directly the probability (P value) of observing a Φ_i as large as this if the original sample were actually drawn from a population defined by the sample data characteristics, but without any real structure (i.e., Type I error rate).

In contrast, by preserving the covariance structure of each observation vector in each jackknife pseudosample or bootstrap sample, the jackknife and bootstrap procedures calculate the expected distribution of the statistic under the alternative hypothesis of real data structure, and thus allow us to calculate a standard error or confidence interval about the observed value of the statistic. To calculate a P value, however, we must translate the distribution derived for the alternative hypothesis to the expected distribution under the null hypothesis. To do this, we merely subtract the observed value of the statistic from each pseudoestimate or bootstrap sample estimate. Thus, in the present context, by comparing the observed Φ_i to the translated jackknife or bootstrap distribution, we can determine the probability (P value) that the ith principal component defines a real gradient.

2.8 Assessing the Importance of the Principal Components   49

Regarding these points, the randomization procedure is more direct and intuitive because it resembles the standard statistical testing procedure; therefore, we prefer the randomization test over both the jackknife and bootstrap procedures.

It is important to note that even if a principal component is deemed statistically significant based on the Φ criterion, we may decide that it lacks substantive ecological importance for two reasons: first, it may not describe enough variance to meet the researchers ecological needs; second, it may not have a meaningful ecological interpretation as judged by the principal component loadings (see Section 2.9.1). Ultimately, the utility of each principal component must be grounded on ecological criteria. Nevertheless, objective statistical tests can be useful tools in helping to assess the importance of the components and in determining how many components to retain and interpret.

In practice, there are different philosophies on whether any of the heuristic or statistical measures of component performance should be used to evaluate PCA results. Some argue that principal components should be evaluated solely on whether they offer a meaningful ecological interpretation (Harner and Whitmore 1981; Gauch 1982). This is because neither mere percentage of variance accounted for nor statistical significance has proved to be a reliable indicator of the quality of results (Austin and Greig-Smith 1968; Robertson 1978; Karr and Martin 1981). In this school of thought, little or no emphasis is placed on the relative percent variance criterion, and ecological meaningfulness is the primary basis for determining the number of components to retain and the basis for evaluating success of PCA. Others argue that no more than three principal components, preferably less, should be retained and that relative percent variance and other objective measures should be used to help determine whether principal components represent real patterns and not simply sample bias. This argument is largely founded on simulation studies which reveal that rarely do the second, third, and later axes account for more variation than can be expected from ordinations on random data (Karr and Martin 1981; although see Stauffer et al. 1985).

In practice, most researchers seldom use a single criterion in determining how many principal components to retain. Instead, they initially use a criterion such as the latent root as a guideline for the maximum number of components to be considered, then relative percent variance, broken stick method, statistical tests based on resampling procedures, and the ecological interpretability of the components to select the final number of components. Most researchers agree that, in the end, the assessment of how many components to retain must be in terms of ecological utility.

Empirical Example. In the example, we computed the 22 eigenvalues of the correlation matrix (Fig. 2.7). The first eight eigenvalues were greater than one (although only six are shown in Figure 2.7). Thus, on the basis of the latent root criterion, we would retain eight components for interpretation. In the scree plot (Fig. 2.9) there is a major slope change after the fourth eigenvalue. Thus, on the basis of the scree test criterion, we would retain four components for interpretation. Based on the broken stick distribution, the first eight components account for more variation than might be expected from random data. According to the relative percent variance criterion ("proportion" row in Fig. 2.7), the first principal component accounts for 24.1 percent of the total variance in the data set; the second component accounts for an additional 15.5 percent. Seven components are required to account for more than 70 percent of the variance. On the basis of the relative percent variance criterion alone, we would conclude that PCA is not very effective in reducing the dimensionality of the data set; three components capture only half of the variation in floristics. Nevertheless, each of these components explains significantly more variation than would be expected if there were no real structure to this data set.

2.9 Interpreting the Principal Components

Once we have derived the principal components and assessed their importance, we can proceed to interpreting their meaning. We find this meaning by (1) examining the relationships between the individual variables and the principal components, and (2) examining the relative positions of the sampling entities in ordination space. Most of the methods are based in some manner on an assessment of the relative importance of the variables in the principal components. In this section, we review some of the more common methods of interpreting the principal components.

2.9.1 Principal Component Structure

As noted previously, when the correlation matrix or standardized covariance matrix is used in the eigenanalysis, the eigenvector coefficients are directly proportional to the correlations between the original variables and the principal components. However, when the raw covariance matrix is used in the eigenanalysis, the eigenvectors reflect the scales of the original variables, and thus do not necessarily reflect the relative importance of the original variables in the principal components. In this case, the product-moment correlations between the principal components and the individual variables must be determined separately. These simple bivariate correlations are not affected by differences in measurement scale or variance and therefore reflect the true relationship between each variable and the principal component. These correlations are referred to as principal component *loadings*, and the matrix of loadings is called the principal component *structure* (or structure matrix). SAS refers to this as the factor pattern or pattern matrix in the PROC FACTOR procedure, but this is not to be confused with the factor pattern in factor analysis (FA), which is something different than the factor structure. The principal component loadings are defined as follows:

$$s_{ij} = v_{i(j)}\sqrt{\lambda_i}$$

where s_{ij} is the correlation between the ith principal component and the jth variable, v_{ij} is the principal component weight of the jth variable in the ith principal component (i.e., eigenvector element), and λ_i is the eigenvalue associated with the ith principal component.

The component loadings tell us how closely a variable and a component are related. When the absolute magnitude of the loading is large (i.e., approaching a value of 1), the component is carrying nearly the same information as the variable. Conversely, when the loading is near zero, they have little in common. Hence, we can define a principal component on the basis of the loadings, simply by noting the variables with the largest loadings. The squared loadings indicate the percent of the variable's variance accounted for by that component. Loadings are intuitively appealing because most researchers are familiar with simple correlation coefficients and can visualize (in graphic terms) differences in the magnitude of the loading. The principal component structure provides a way to establish an ecological interpretation of each principal component and is perhaps the single most important piece of information resulting from PCA. Note that the principal component structure derived from the correlation matrix is not the same as that derived from the raw covariance matrix, because the components will vary.

2.9.2 Significance of Principal Component Loadings

An important decision in PCA is determining which principal component loadings are worth considering in the ecological interpretation of each component. There are a number of general rules that can be used (Hair, Anderson, and Tatham 1987).

Rule 1. The first rule is not based on any mathematical proposition; it is a rule of thumb frequently employed by researchers (Hair, Anderson, and Tatham 1987). Principal component loadings greater than 0.30 or less than –0.30 are considered significant. Loadings greater than 0.40 or less than –0.40 are considered more important, and loadings greater than 0.50 or less than –0.50 are considered very significant. Thus, the larger the absolute size of the loading, the more significant the loading is in interpreting the principal component structure. These guidelines are considered useful when the sample size is greater than or equal to 50.

Rule 2. A similar rule discussed by Tabachnik and Fidell (1989) suggests that loadings greater than 0.32 or less than –0.32 are poor, greater than 0.45 or less than –0.45 are fair, greater than 0.55 or less than –0.55 are good, greater than 0.63 or less than –0.63 are very good, and greater than 0.71 or less than –0.71 are excellent. These benchmarks are intuitively appealing because they are tied to the percentage of variance (square of loadings) accounted for by the component; specifically, these benchmarks identify variables that account, respectively, for more than 10 percent, 20 percent, 30 percent, 40 percent, and 50 percent of the variance in the component.

Rule 3. In determining a significance level for interpretation of loadings, an approach can be used that is similar to that of interpreting simple correlation coefficients. Specifically, loadings greater than or equal to ±0.19 and ±0.26 are recommended for the 5 percent and 1 percent significance levels, respectively, when the sample size is 100 (Hair, Anderson, and Tatham 1987). When the sample size is 200, ±0.14 and ±0.18 are recommended for the 5 percent and 1 percent levels, and when the sample size is greater than or equal to 300, loadings of ±0.11 and ±0.15 are recommended. Since it is difficult to assess the amount of error involved in factor analytic studies, it is probably safer to adopt the 1 percent level as the criterion for significance. Note, however, that significant correlation coefficients may not necessarily represent ecologically important variables.

The disadvantage of these rules is that the dimensions of the data set (i.e., number of sampling entities and variables) being analyzed and the specific component being examined are not considered. It has been shown, for example, that as one moves from the first component to later components, the acceptable level for a loading to be judged significant should increase. The following general guidelines can be stated (Hair, Anderson, and Tatham 1987): (1) the larger the sample size, the smaller the loading to be considered significant; (2) the larger the number of variables being analyzed, the smaller the loading to be considered significant; and (3) the larger the number of components, the larger the size of the loading on later factors to be considered significant for interpretation. Adjustment for the number of variables is particularly true as one moves from the first component extracted to later components. Specific rules for adjusting the significance level based on these considerations have been developed (Hair, Anderson, and Tatham 1977), but are too detailed to present here.

Rule 4. As we discussed previously, resampling procedures such as the jackknife, bootstrap, and randomization tests can be used to test the significance of virtually any statistical parameter (see Sec. 2.8.5 for a complete discussion of these procedures). Here, we can use these same procedures to determine which principal component loadings are significant. Simply substitute the loadings for the relative percent variance parameter (Φ) in the previous discussion. Because each principal component is a linear combination of the original variables, a component loading (i.e., correlation coefficient) of zero denotes the absence of association between a variable and a component. In each case, the null hypothesis tested is that the particular loading is approximately equal to zero. Thus, rejection of the null hypothesis implies that a particular loading probably represents a real association between the corresponding variable and principal component.

In all three of these resampling procedures, since multiple tests are performed (up to $P \times P$, where P = number of variables) the overall experimentwise error rate will, in general, be greater than the comparisonwise error rate of each individual test. If there is a large number of loadings being tested and a 5 percent significance level is applied to each test, then there is a real possibility that, even when the null hypothesis is true, some of the tests (approximately 5 percent of the them) will produce a P value that is less than 5 percent. One way of dealing with this multiple comparisons problem is to use the Bonferroni method, which adjusts the significance level of the individual tests by dividing the significance level by the total number of tests.

2.9.3 Interpreting the Principal Component Structure

Interpreting the complex interrelationships represented in the principal component structure is often a difficult task. However, there are a few simple procedures that can facilitate interpretation. Start with the first variable on the first component and move horizontally from left to right through the structure matrix (i.e., across components), looking for the highest absolute loading for that variable on any component; highlight the highest absolute loading if it is considered significant (see preceding section). Repeat this process for each variable until all variables have been highlighted once for their highest significant loading on a component.

Under ideal circumstances each variable will have only one loading on one component that is considered significant. This is a "simple structure solution" under which the interpretation of the meaning of each component is simplified considerably. In practice, however, several variables will have moderate-sized loadings, all of which are significant, and the job of interpreting the components is much more difficult. This is because a variable with several significant loadings must be considered in the interpretation of all the components on which it has a significant loading.

Next, because most component solutions do not result in a simple structure, you should, after highlighting the highest significant loading for each variable, continue to evaluate the structure matrix by highlighting all significant loadings for a variable on all the components, perhaps distinguishing these loadings from the single highest significant loading for each variable. Ultimately, you would like to minimize the number of significant loadings on each row of the structure matrix (i.e., the loadings associated with each variable) and to maximize the number of loadings with negligible values such that the variance associated with each variable is both concentrated on a few components and fully accounted for.

Once all the significant loadings have been highlighted, you should examine the structure matrix to identify variables that have not been highlighted and, as such, do not load on any component. If there are variables that do not load on any component, you have two options: (1) interpret the solution as it is and simply ignore those variables without a significant loading; or (2) critically evaluate each of the variables that do not load significantly on any component. If the variable(s) is of minor importance to the study's objective, you may decide to eliminate the variable(s) and derive a new component solution with the nonloading variables eliminated.

When a final component solution has been obtained, you can attempt to assign some meaning to the pattern of the loadings. Variables with higher loadings are considered more important in interpreting each component; they greatly influence the name or label selected to represent a component. Thus, you should examine all the highlighted variables for a particular component and, placing greater emphasis on those variables with higher loadings, attempt to assign a name or label to a component that accurately reflects, to the greatest extent possible, what the several variables loading on that component represent.

It is also important to note the sign (+ or −) of each significant loading. Positive correlations indicate a direct relationship between a variable and a component;

that is, larger values of the variable are associated with positive values of the component, and smaller values of the variable are associated with negative values of the component. Conversely, negative correlations indicate an inverse relationship between a variable and a component; that is, larger values of the variable are associated with negative values of the component, and vice versa. Components often contain both significant positive and significant negative loadings. Under this condition, larger values of the positive variables and smaller values of the negative variables are associated with the positive end of the gradient, while smaller values of the positive variables and larger values of the negative variables are associated with the negative end.

Finally, it is important to note that the meaning or label assigned to each component is not derived or assigned by the computer program, but rather is intuitively developed by the researcher based upon its appropriateness for representing the underlying dimensions of a particular component.

Empirical Example. In the example, we retained the first three principal components for interpretation based on the principal component structure (Fig. 2.10). Recall that since we used the correlation matrix in the eigenanalysis, the eigenvectors are directly proportional to the loadings; therefore, we could have based our ecological interpretation on the eigenvectors, and the conclusions would have been the same.

Several floristic variables are important in each principal component (Fig. 2.10). We based our interpretation of each component on those variables with loadings

Factor Pattern

	PC1	PC2	PC3
FORB	-0.39044	-0.47332	0.38298
GRASS	-0.26748	0.01359	0.32364
FERN	0.51597	0.60069	0.01890
SALAL	0.44672	0.47740	0.44371
GRAPE	0.45105	-0.19570	-0.40575
SALMON	-0.62496	-0.15342	0.33015
CURRANT	0.46136	-0.37994	-0.29764
HUCKLE	0.64443	-0.05148	-0.02180
THIMBLE	0.06692	-0.19626	0.14013
DEVIL	0.37390	-0.03166	-0.56994
VINE	0.02887	0.78862	-0.09514
ELDER	-0.14718	0.21745	-0.18469
HAZEL	-0.15745	0.49190	-0.48609
PLUM	0.48024	0.56095	0.53361
OCEAN	-0.03474	0.14220	0.21939
ALDER	-0.78552	0.20863	0.00897
MAPLE	-0.30308	0.38885	-0.41357
FIR	0.57516	0.61772	0.15166
HEMLOCK	0.37281	-0.48296	0.31080
CEDAR	0.59051	-0.45752	-0.21071
HARDWD	-0.81548	0.29845	-0.10201
CONIFER	0.88423	-0.04101	0.16149

FIGURE 2.10 Principal component structure (i.e., component loadings) for the first three principal components.

2.9 Interpreting the Principal Components 55

greater than 0.40 or less than −0.40, and placed most emphasis on those with loadings greater than 0.60 or less than −0.60 (Rule 1 of preceding section). The first component largely represents a gradient from hardwood-dominated (HARDWD, ALDER) sites to conifer-dominated (CONIFER, FIR, CEDAR) sites. Note the inverse relationship between HARDWD (−) and CONIFER (+), indicating that high values of CONIFER are associated with low values of HARDWD. The first component also reflects a gradient in several shrub species associated with hardwood sites (SALMON) and conifer sites (HUCKLE, FERN, PLUM, CURRANT, SALAL, and GRAPE). Currant and plum are indicative of steep gradient, constrained stream reaches; these are generally sites where upslope shrubs such as salal, Oregon grape, and huckleberry extend downslope to the stream. Conversely, salmonberry is indicative of low gradient, unconstrained stream reaches and is usually found in association with red alder patches. In summary, we label PC1 "hardwood-conifer gradient."

The second component largely represents a gradient in vine maple (VINE) cover and, to a lesser extent, Douglas-fir (FIR), sword fern (FERN), plum (PLUM), California hazel (HAZEL), salal (SALAL), Western hemlock (HEMLOCK), Western redcedar (CEDAR), and forb (FORB) cover. Vine maple is associated with Douglas-fir, sword fern, plum, hazel, and salal on one end of the gradient (positive end), while hemlock, cedar, and forb are associated with the other end of the gradient (negative end). Of particular importance in PC2 is the gradient in vine maple cover and the separation of sites with hemlock and cedar from sites with Douglas-fir. Hemlock and cedar are both shade-tolerant, long-lived trees that outcompete Douglas-fir during advanced stages of stand development, typical of sites that have not experienced catastrophic disturbance for several centuries. Conversely, vine maple and Douglas-fir are typical of more recently disturbed sites. Thus, we label PC2 "seral stage gradient."

The third principal component represents a gradient in several floristic variables, but most importantly represents a gradient in devil's club (DEVIL) cover. Devil's club is associated with recently disturbed wet soils. Thus, we label PC3 "devil's club gradient."

Note that GRASS, THIMBLE, ELDER, and OCEAN do not load significantly on any component. Apparently, the variation in these variables is not captured by the first three components (final communality estimates support this conclusion; see Section 2.9.4). Perhaps the variation in these variables is associated with later components. Regardless, it is apparent that these variables do not contribute significantly to any dominant vegetation gradient present in the sample data set. A new component solution could be derived after eliminating these variables from the data set.

We conclude that the first three principal components describe readily interpretable and ecologically consistent gradients. Based on these findings, we might hypothesize that local variation in riparian vegetation is largely influenced by geomorphology, natural successional processes, and soil moisture/disturbance conditions. Note that these findings cannot be used to infer causation, but rather to suggest hypotheses which may form the basis for subsequent experimentation. Thus, the PCA results are deemed useful, despite the fact that 50 percent of the variance in the data set remains unexplained.

2.9.4 Communality

Another useful and often overlooked measure of PCA performance that can aid interpretation is the final communality estimate for each variable. The *communality* of a variable is equal to the squared multiple correlation for predicting the variable from the principal components. Also, each variable's final communality is equal to the sum of the squared loadings from the retained principal components. This is equivalent to the sum of the squared loadings on each principal component, as follows:

$$c_j = \sum_{i=1}^{P} s_{ij}^2$$

where c_j is the communality of the jth variable and s_{ij} is the loading (or correlation) between the ith principal component and the jth variable. In other words, communality refers to the proportion of a variable's variance that is accounted for by the principal components. In PCA, prior communalities are equal to one because 100 percent of the variation in each variable is accounted for by the P principal components. (Note: This represents an important distinction between PCA and factor analysis; see below.) Final communality estimates (i.e., after the components are extracted) are equal to the squared multiple correlations for predicting the variables from the retained principal components; retained principal components are those components (usually the first one to three) that are extracted and retained for interpretation. Final communality estimates indicate how well the original variables are accounted for by the retained principal components. For example, after retaining the first two principal components, a high communality estimate ($c_j >$ 0.7) for a particular variable indicates that most of the variation in the variable is accounted for by the first two principal components. Note that final communality estimates increase from zero to one ($0 < c_j < 1$) as the number of retained principal components increases from zero to P, and that the average final communality estimate is always equal to the cumulative percent variance of the retained principal components.

Although final communality estimates are very useful in evaluating the effectiveness of PCA, they are often overlooked and rarely reported with the PCA results. We recommend that you include final communality estimates in reported results.

Empirical Example. In the example, final communality estimates were calculated based on retention of the first three principal components. Figure 2.11 indicates that these three components account for the variation in the 22 floristic variables in widely varying degrees. Final communality estimates range from 0.069 for OCEAN to 0.809 for CONIFER (rounded values) and average 0.494. Hence, some of the variables are very well accounted for by the retained components, while others are not. Perhaps most importantly, the final communalities for HARDWOOD and CONIFER are quite high (greater than 0.76), indicating that these two variables, which are associated with the dominant gradient in the data set, are very well accounted for by the retained components. Note that the variables with low final communalities are those variables that do not load significantly on any of the

```
       Final Communality Estimates: Total = 10.866351

    FORB       GRASS        FERN       SALAL       GRAPE      SALMON     CURRANT      HUCKLE
0.523146    0.176469    0.627411    0.624347    0.406375    0.523107    0.445795    0.418417

 THIMBLE       DEVIL        VINE       ELDER       HAZEL        PLUM       OCEAN       ALDER
0.062635    0.465637    0.631807    0.103056    0.503044    0.830037    0.069557    0.660649

   MAPLE         FIR     HEMLOCK       CEDAR      HARDWD     CONIFER
0.414108    0.735383    0.468836    0.602422    0.764482    0.809631
```

FIGURE 2.11 Final communality estimates based on retention of the first three principal components.

retained principal components. This is not too surprising, because the final communality of a variable is equal to the sum of the squared loadings on the retained principal components. Thus, if the loadings are small and insignificant on all of the retained components, the sum of their squares will be low as well.

2.9.5 *Principal Component Scores and Associated Plots*

A final way of interpreting the principal components involves simple graphical representations of the component scores. Each sampling entity in the data set has a score on each principal component (or a location on each principal component axis) that is derived by multiplying the observed values for each variable (in standardized form) by the corresponding standardized scoring coefficients (i.e., standardized principal component weights) and summing the products. Computationally, the principal component scores are calculated as follows:

$$z_{ij} = a_{i1}x^*_{j1} + a_{i2}x^*_{j2} + \cdots + a_{ip}x^*_{jp}$$

where z_{ij} is the standardized score for the ith principal component and jth sample, a_{ik} is the standardized scoring coefficient for the ith principal component and kth variable, and x^*_{jk} is the standardized value for jth sample and the kth variable.

Principal component scores are useful for graphically displaying the sampling entities in ordination space. Scatter plots of principal component scores for each pair of retained principal components graphically illustrate the relationships among entities, since entities in close proximity in ordination space are ecologically similar with respect to the environmental gradients defined by the principal components. Scatter plots are also useful in evaluating the linearity assumption, since pronounced nonlinearities often show up as arched or curvilinear configurations of entities in two-dimensional ordination space.

Principal component scores also represent the values of the new uncorrelated variables (components) that can serve as the input data for subsequent analysis by other procedures such as cluster analysis, discriminant analysis, or multiple regression.

OBS	PC1	PC2	PC3
1	-0.73163	0.48602	-0.74422
2	-1.00046	0.27023	-0.33100
3	1.16265	1.84252	1.39165
4	3.03046	3.23863	3.01008
5	0.78553	1.65787	0.21373
.	.	.	.
.	.	.	.
.	.	.	.
48	-0.09459	0.01037	-0.12769

FIGURE 2.12 Principal component scores for the first three principal components. Only a portion of the scores is shown.

Empirical Example. In the example, we computed the principal component scores for the first three components (Fig. 2.12) and constructed scatter plots of all three pairwise combinations; although only the plot of PC1 versus PC2 is shown (Fig. 2.5). The scatter plot of PC1 versus PC2 indicates several things. First, the plots along stream 2 form a tight cluster and therefore are very similar with respect to the gradients defined by PC1 and PC2. Specifically, these sites occupy a position near the mean of PC1, indicating that they contain a mixture of hardwoods and conifers, and they occupy a position on the negative end of PC2, indicating that they contain a preponderance of hemlock, cedar, and forb. Thus, these sites are probably characteristic of well-developed, late-seral stand conditions. The tightness of the cluster indicates that the plots are relatively similar in composition with respect to these two dominant gradients. Thus, we can conclude that there is comparatively little local variability in floristic composition along this stream, perhaps due to the lack of fine-scale disturbance events over the past century or more.

A second pattern evident in the scatter plot is the separation of sites along streams 1 and 6 from those along streams 2, 3, and 4 with respect to the gradient defined by PC2. Thus, with respect to the seral stage gradient defined by PC2, there appears to be an important source of spatial variability at the subwatershed scale. This scale of pattern in streamside vegetation is not surprising given the natural disturbance regime in this landscape. Infrequent, coarse-scale disturbance events (e.g., fire, debris flows) often affect entire watersheds. Unfortunately, these findings also highlight the fact that this data set does not consist of truly independent samples, and therefore violates one of the assumptions of the PCA model.

2.10 Rotating the Principal Components

Methods exist for rotating the established principal component axes about the centroid to improve component interpretation. The term *rotation* means exactly what it implies: the reference axes of the components are turned about the origin (centroid) until some other position has been reached. The simplest case is an *orthogonal* rotation, in which the angle between the axes is maintained at 90 degrees. A number of orthogonal procedures exist, including *varimax* rotation, *quartimax* rotation, and *equimax* rotation, of which the first has been found to be superior under most circumstances and is most frequently used (Bhattacharyya 1981; Hair, Anderson, and Tatham 1987).

It is possible to rotate the axes and not retain the 90-degree angle between the reference axes. This rotational procedure is referred to as an *oblique* rotation. None of the oblique solutions have been demonstrated to be analytically superior to the orthogonal solutions, therefore we will not comment on them further.

From a geometrical perspective, if the multidimensional cloud of data points is distributed multivariate normal (i.e., hyperellipsoidal with normally varying density around the centroid), then the unrotated component axes will be best. In this case, the unrotated components will provide the set of orthogonal axes that best explains the maximum variance in the original data (i.e., does the best possible job of describing the variance-covariance structure). However, if the multidimensional cloud of data points is not multivariate normal, as is almost always the case, and is shaped such that the second, third, and perhaps subsequent dimensions of the data cloud are not best described by axes perpendicular to the first, then rotating the set of orthogonal axes to better explain the overall variance-covariance structure of the data (i.e., better describe the shape of the data cloud) will often prove successful. This is possible because the second and subsequent unrotated component axes are constrained by orthogonality with the preceding axes. Thus, the orientation of the second and later axes are in part dictated by the orientation of the first axis. If the multidimensional cloud of data points can be better defined by a different set of orthogonal axes at the expense of the first axis, then rotation of the component axes will prove beneficial.

From a practical standpoint, rotations generally serve to enhance interpretation of the principal component structure. The ultimate effect of rotating the structure matrix is to redistribute the variance from earlier components to later ones to achieve a simpler, theoretically more meaningful, principal component structure. It does this by increasing the principal component loadings of important variables and decreasing the principal component loadings of unimportant variables. The various rotation methods approach this task from different perspectives. Varimax rotations, for example, increase the absolute values of large loadings and reduce the absolute values of small loadings within columns of the structure matrix (i.e., components). Thus, varimax rotation simplifies the structure within a component and is useful for improving the interpretation of components in terms of well-understood variables because it increases the distinction between significant loading and nonloading variables on each component. Quartimax rotations, in contrast, increase large loadings and reduce small loadings within rows of the structure matrix (i.e., variables). Thus, quartimax rotation is useful for interpreting variables in terms of well-understood components because each variable loads more distinctly on fewer components.

Furthermore, rotations always reduce the eigenvalue (variance) of the first component, since by definition the first unrotated component explains the maximum variance in the data set. However, rotations always maintain the cumulative percent variance (or total variance accounted for by the retained components) because increases in the variances of later axes are equally offset by the decline in the first axis.

Empirical Example. In the example, we rotated the first three principal components using the varimax method (Fig. 2.13). Note that the rotated factors account for the same amount of sample variance, as indicated by the sum of the rotated eigenvalues (compare to Figure 2.6). The rotated structure matrix provides a somewhat simpler solution for the first component (i.e., fewer variables with large loadings), but not for the second component. Moreover, the rotation does not offer a more interpretable set of ecological gradients. Thus, we deem the rotation ineffective.

Rotated Factor Pattern

	PC1	PC2	PC3
FORB	-0.40492	-0.42967	-0.41782
GRASS	-0.02092	-0.40993	-0.08937
FERN	0.73127	0.21825	0.21219
SALAL	0.77739	-0.09065	-0.10862
GRAPE	-0.02632	0.63453	-0.05530
SALMON	-0.34880	-0.62909	-0.07546
CURRANT	-0.10249	0.60911	-0.25352
HUCKLE	0.34531	0.47925	-0.26363
THIMBLE	-0.03720	-0.00462	-0.24745
DEVIL	-0.02716	0.65530	0.18838
VINE	0.51943	-0.08844	0.59513
ELDER	-0.01487	-0.02635	0.31959
HAZEL	0.04468	0.11001	0.69925
PLUM	0.89142	-0.14619	-0.11843
OCEAN	0.16556	-0.20399	-0.02314
ALDER	-0.32805	-0.60429	0.43343
MAPLE	-0.08452	-0.01884	0.63766
FIR	0.83262	0.16637	0.12023
HEMLOCK	0.02041	0.15803	-0.66592
CEDAR	-0.04233	0.65818	-0.40918
HARDWD	-0.32964	-0.57003	0.57523
CONIFER	0.57182	0.52166	-0.45883

Variance explained by each factor

PC1	PC2	PC3
3.897633	3.769050	3.199668

FIGURE 2.13 Rotated principal component structure (i.e., loadings) using the varimax rotation method for the first three components.

2.11 Limitations of Principal Components Analysis

Most of the limitations of PCA have already been discussed. Nevertheless, we briefly summarize some of these limitations again to emphasize the importance of their consideration when interpreting the results of PCA.

- In general, PCA produces severely distorted ordinations for data sets with large community patterns or long environmental gradients. Other ordination methods (e.g., reciprocal averaging, detrended correspondence analysis, and canonical correspondence analysis) are usually superior under these conditions. Principal components analysis is most effective when applied to narrow environmental gradients where the sample distribution may approach linearity (Beals 1973; Kessell and Whittaker 1976; Fasham 1977; Gauch 1982).
- Inherent to all ordination techniques, PCA is only capable of detecting gradients that are intrinsic to the data set. Sampling entities may lie on other more important gradients not measured using the selected variables, and these dominant, undetected gradients may distort or confuse any relationships that are intrinsic to the data (Beals 1973; Nichols 1977; Gauch 1982).
- By only interpreting the first few principal components, one may overlook a later axis that explains most of the variation in some variable. Consequently, the information in this variable will be lost. Final communality estimates indicate how much of the information in each variable is accounted for by the retained principal components and how well each variable is "captured" by the retained principal components.
- It is sometimes argued that there is little justification for selecting a linear combination of variables (i.e., principal components) simply because doing so maximizes the variance within the total set; that is, it is a "best" summary (Johnson 1981). One can increase the percentage of variance explained by the first principal component merely by adding redundant variables to the data set. As more and more of these are included, the PCA appears better and better, but in fact only noise is being added to the system, and interpretation becomes increasingly awkward. A large percent of variance explained may reflect only ignorance in the selection of variables.

2.12 R-Factor Versus Q-Factor Ordination

Thus far, in our discussion of PCA we have been considering data for which rows represent sampling entities and columns represent variables. Specifically, we create components out of the environmental variables (columns) and position sample entities (rows) in the reduced ordination space defined by the principal components. We computed the PCA on the correlation or covariance matrix of the environmental variables. This is the most common type of PCA and is referred to as *R-factor analysis*.

However, PCA can be applied to the correlation or covariance matrix of the sampling entities (rows). This type of analysis is referred to as *Q-factor analysis* and can be used if the objective is to combine or condense large numbers of sample entities (rows) into distinctly different groups within a larger population, similar to cluster analysis. The result of a Q-factor analysis is a principal component structure that can be used to identify similar sample entities. For example, the component structure might tell you that observations 1, 8, and 34 are similar because they exhibited a high loading on the same component. Similarly, observations 2, 21, and 30 would perhaps load together on another component. From the results of a Q-factor analysis, we could identify groups or clusters of entities demonstrating a similar response pattern of the variables included in the analysis.

In this respect, Q-factor analysis is quite similar to cluster analysis because both approaches compare a series of observations to a number of variables and place observations into several groups. The difference is that the resulting groups for a Q-factor analysis would be based on the intercorrelations between the means and standard deviations of the observations, while in a typical cluster analysis approach, groupings would be devised based on distance between the observations' scores on the variables being analyzed. For a number of reasons, Q-factor analysis is quite inferior to cluster analysis when the objective is to form groups of observations based on a number of variables. For more discussion of this, the interested reader is referred to Stewart (1981).

Indeed, the only appropriate use of Q-factor analysis is as the transpose of R-factor analysis. When more variables (columns) than samples (rows) are present, Q-factor analysis will result in a more stable solution than R-factor analysis. The reason is that the standard error of a correlation is a function of the sample size. Further, statistical independence of correlation coefficients in an R-factor analysis can be obtained only when there are fewer variables than samples. Indeed, most computer software packages will not even provide a solution to an R-factor analysis if the situtation is reversed. Thus, when faced with a circumstance involving more variables (columns) than samples (rows), a researcher can compute components by using the Q-factor approach simply by transposing the data matrix and conducting the analysis as before. The solution will be more reliable than the one obtained with the R-factor approach. However, in general the Q-factor approach has little utility in wildlife research because we would not encourage the use of ordination when we have more variables than samples.

The purpose of mentioning Q-factor analysis is to point out that PCA (and factor analysis) can be applied to a correlation or covariance matrix of either the rows or columns in a two-way data matrix. In other words, rows can be ordinated in column space (R-factor analysis) or columns can be ordinated in row space (Q-factor analysis) (although we are almost always interested in the former). This is an important principal utilized in the ordination technique referred to as *reciprocal averaging* or *correspondence analysis*, which we discuss below.

2.13 Other Ordination Techniques

It is important to remember that ordination refers to a family of techniques that share a goal of organizing entities along a continuum and extracting dominant gradients of variation. Principal components analysis is without a doubt the most well understood and most widely used ordination method, but this does not mean that it is the best choice of ordination methods in all, or even most, cases. If the objective criterion for ordering sampling entities along a continuum is the single dimension that maximizes the variation along that axis, then PCA is the best method. However, for many ecological applications PCA ranks poorly when compared with other ordination techniques. Because of this, several more robust techniques have been increasing in popularity, and for some applications they are used almost exclusively over principal components analysis.

In this section, we limit our comparisons to six ordination techniques: (1) polar ordination, (2) factor analysis, (3) nonmetric multidimensional scaling, (4) reciprocal averaging, (5) detrended correspondence analysis, and (6) canonical correspondence analysis. The first five are unconstrained ordination techniques, while the last is a constrained ordination technique. These techniques each differ dramatically from PCA in at least one respect. While it is beyond the scope of this book to present each of these techniques in detail, we believe that with a solid understanding of PCA, you will have an excellent framework for learning these other techniques. Consequently, in this section we provide a brief overview of each technique to highlight the major differences with PCA and identify the conditions under which each technique offers potential advantages over PCA.

2.13.1 Polar Ordination

Polar ordination (PO), also known as *Bray and Curtis ordination*, is a relatively simple and intuitively straightforward technique (see Bray and Curtis 1957 and Gauch 1982 for a detailed description). Polar ordination begins with the computation of a dissimilarity matrix (see Chapter 3, Section 3.7.1.4 for a discussion of dissimilarity measures) from the original data matrix; percent dissimilarity is generally superior to Euclidean distance and coefficient of community for samples-by-species data sets (Gauch 1982). Next, two observations are chosen to serve as "poles." Ideally, these observations represent opposite ends of an environmental gradient and thus are likely to have the largest, or close to the largest, dissimilarity value. Finally, ordination scores are computed by projecting observations onto the axis formed by connecting the two endpoints. Observations of similar environmental makeup (or, e.g., species composition) will ordinate near each other. The distance (E) off the axis of a sample entity measures the extent to which the ordination fails to define the location of the entity's position in the multidimensional dissimilarity space and indicates the presence of additional gradients (or stochastic variation). Large values of E indicate that much of the data structure is not captured by the PO axis and additional axes may be useful. Techniques for extracting additional PO axes have been suggested (Bray and Curtis 1957).

Polar ordination has several appealing properties. First, and most important, PO allows the investigator to subjectively determine the ordination gradient. Consequently, environmental interpretation of the results is relatively direct. Second, PO can be very useful when the investigator is specifically interested in the organization and relationship among entities on a predetermined, well-specified gradient. For example, if animal species data were collected at observations in streamside environments located sequentially along a transect beginning at the headwaters (one pole) and ending at the estuary (the other pole), then PO could be used to evaluate animal community changes along the intrariparian gradient. Third, PO does not have some of the restrictive assumptions attached to most eigenvector ordination techniques (all others considered here). Specifically, PO makes no assumption about random sampling and does not require a multivariate normal distribution of points. As a result, PO is somewhat robust to failure of the linearity assumption. Last, unlike PCA and reciprocal averaging, PO is unaffected by clusters of observations and outliers, as long as outliers are not selected as endpoints (Gauch et al. 1977).

Polar ordination is not without limitations (Beals 1973; Gauch 1973, 1982). First, PO is criticized heavily for its subjectivity in assigning endpoints. This emphasizes the need to use PO only when the endpoints are obvious and usually predetermined. Second, PO is subject to serious distortions leading to erroneous conclusions if outliers are inadvertently selected as endpoints. Third, unlike PCA, the second and later PO axes are only approximately orthogonal to the first axis, and thus do not represent true independent relationships. Last, like PCA, later axes can under certain conditions suffer from the "arch effect" and "compression effect" (see Gauch 1982 for a discussion of these phenomena), which often limits the use of PO to situations in which only the first polar axis is of interest.

2.13.2 Factor Analysis

Principal components analysis is often confused with *factor analysis* (FA). In both PCA and FA, the objective is to parsimoniously summarize the variance-covariance structure of the original variables. This is accomplished by generating a few components or factors (i.e., linear combinations of the variables) that extract most of the variation in the original variables. The resulting component structure or factor structure is used to interpret the major gradients of variation in the data set.

One reason that PCA and FA are often confused is that some statistical software packages offer the PCA solution as a default to the FA procedure. Actually, PCA and FA differ dramatically with respect to how they summarize the covariance structure of a data set. In PCA, 100 percent of the variation in the original variables is accounted for by the principal components (i.e., prior communalities are equal to one), when in fact, some of the variation is "unique" (inherent) to the measured variables (i.e., unique variation is not shared by one or more other variables and cannot be accounted for by the common factors or components). Models that do provide explicitly for a separation of shared and unique variance lead to the statistical techniques known collectively as factor analysis. All FA models separate the unique variance from common variance and make the assumption that the intercorrelations among the P original variables are generated by some smaller number of hypothetical (latent) variables (common factors). Under this assumption, the unique part of a variable is factored out and does not contribute to relationships among the variables. The basic model is as follows:

$$Z_i = a_{i1}F_1 + a_{i2}F_2 + \cdots + a_{im}F_m + E_i$$

where Z_i is the standardized value for the ith variable, F_j is the jth common factor, E_i is the unique variance of the ith variable, and a_{ij} is the factor weight of the ith variable on the jth common factor.

At first glance this would appear to be a much less parsimonious model than the PCA model since it involves $P + M$ latent variables rather than just P variables (as in PCA). However, the sole purpose of the P unique factors is to permit elimination of unique variance from the description of the relationships among the original variables, for which description only the M (usually $\ll P$) common factors are employed. Note that the matrix of factor weights (a_{ij}) in the present context is not the same kind of matrix as discussed in PCA. In PCA, the emphasis traditionally has been on describing the principal components in terms of the original variables, so that the principal component weights represent the coefficients used to generate component scores from the original variables. In FA, on the other hand, the emphasis traditionally has been on describing the original variables in terms of the latent variables, so that the component weights represent the coefficients used to generate scores on the original variables from scores on the latent variables. Actually, the emphasis in FA, like in PCA, is usually on reproducing the intercorrelations among the original variables. However, the solution to this problem is much less straightforward in FA.

The major advantage of FA is the ability to reproduce the original pattern of intercorrelations from a small number of common factors without systematic errors, such as occur in PCA when not all components are retained. It is tempting to infer that the relationships revealed by FA are more "reliable" than those contained in a PCA of the same data, because the error variance has been removed along with all the other nonshared variance. However, from an ecological perspective, the most important criteria in evaluating an ordination technique should be the straightforwardness and interpretability of the results. In FA, the straightforward relationship between values of the factors and their sources on the original variables is lost. Consequently, factor interpretation is usually more difficult than in PCA. In general, FA leads to more readily interpretable factors only under certain conditions (i.e., if strong evidence exists for differences among the original variables in communalities). Also, in FA the uniqueness of the solution is lost. There are many different ways of determining the factor structure. Consequently, a given factor structure simply represents a description of the original intercorrelations in terms of a particular frame of reference. Also, the unique variance in any data set is unknown; it must be estimated from the data. Estimating the unique variance is perhaps the most difficult and ambiguous task in FA, and it is the source of the major differences between the various methods of factoring. For these reasons, FA has not been used much in wildlife research.

2.13.3 Nonmetric Multidimensional Scaling

Nonmetric multidimensional scaling (NMMDS) is actually a family of related ordination techniques that use only rank order information in the dissimilarity matrix (see Fasham 1977 and Gauch 1982 for a more detailed description). The intention behind NMMDS is to replace the strong and problematic assumption of linearity with a weaker and, it is hoped, less problematic assumption of monotonicity. Like PO, NMMDS operates on a sample dissimilarity matrix rather than the original data matrix. The common NMMDS techniques attempt to locate sample entities in a low-dimensional ordination space such that the intersample distances in the ordination have the same rank order as do the intersample dissimilarities in the dissimilarity matrix. Thus, the NMMDS algorithm uses only rank order of the intersample dissimilarities and not their magnitudes.

The major strength of NMMDS is, of course, that it makes no assumption of linearity. In addition, NMMDS appears to be less susceptible to arch and compression effects than PCA, and it is not affected by sample clusters, irregularly spaced observations along the underlying gradients, outliers, and a moderate level of noise in the data.

Of course, NMMDS is not without its limitations. First, in spite of the innocuous assumptions, it cannot be relied on to extract the correct configuration, even when applied to simulated data. Part of the difficulty lies in the first stage of the analysis, for which the data are summarized in a dissimilarity matrix similar to polar ordination. Dissimilarity matrices contain no explicit information on the variables (only sample dissimilarities) and are therefore impoverished compared to the original samples-by-variables data matrix. Second, the choice of the initial configuration (i.e., the trial vector) used in NMMDS algorithms gives rise to one of the main problems of the method. If the initial configuration is not appropriately chosen, then it is possible for the algorithm to become trapped in a local minimum, thereby missing the global minimum and producing spurious ordination scores. Because the choice of initial configuration is somewhat arbitrary, the solution is never unique. Third, NMMDS assumes that the dimensionality of the data is known a priori. In practice, the investigator must rely on his ecological knowledge to make a judgement; yet, NMMDS is not robust enough to withstand the incorrect assignment of dimensionality. In most ecological investigations employing ordination, the dimensionality is unknown a priori.

Overall, under most conditions NMMDS offers little or no advantage over PCA and the other ordination techniques described below, and as such, has not, to our knowledge, been employed in wildlife studies. We mention it here because of its rather different approach and underlying assumptions.

2.13.4 Reciprocal Averaging

Reciprocal averaging (RA), also known as *correspondence analysis*, is a dual ordination procedure in which samples (rows) and variables (columns) are ordinated simultaneously on separate but complementary axes. Reciprocal averaging is an eigenanalysis problem similar to PCA (see Hill 1973, 1974; and Gauch 1982 for a detailed description), where the eigenvalues reflect the degree of correspondence between variables and sample scores. A high eigenvalue indicates a long and important gradient. Computationally, RA also can be solved using a simple weighted averaging procedure. The procedure involves a process of repeated cross-calibration between row scores and column scores that results in a unique simultaneous ordination of both rows and columns. Conceptually, it is equivalent to simultaneous Q- and R-factor ordinations. In RA, row scores are weighted averages of column scores, and reciprocally, column scores are weighted averages of row scores, hence the term "reciprocal averaging."

The first step is to assign arbitrary column (variable) ordination scores. Next, weighted averages are used to obtain sample ordination scores from the variable scores. Then, the second iteration yields new variable scores by computing weighted averages of the sample scores. The new variable scores are then rescaled and a new set of sample scores are produced from weighted averaging of the variable scores. This iterative process is repeated until the scores stabilize. Ultimately, the ordination scores of the samples and variables are derived such that the correlation between sample and variable scores is maximized. Second and higher axes are derived in the same manner, such that the sample-variable correlation is maximized under the constraint that these axes be orthogonal. The result is a unique solution that is not affected in any way by the values initially chosen to start the process.

Reciprocal averaging is closely related to PCA, both computationally and geometrically. Like PCA, RA can be viewed geometrically as the derivation of new axes, which maximally account for the structure of a multidimensional cloud of points, making possible the reduction of the dimensionality. The general intentions of PCA and RA, therefore, are identical: a multidimensional cloud of points is to be projected efficiently onto fewer dimensions. The details of these two analyses differ, however. From a practical viewpoint, the most important question in comparing these two techniques is "How well do RA and PCA perform under various circumstances?"

The feature that most distinguishes RA from other ordination techniques is the process of dual ordination of samples and variables. The RA ordination diagram, called a "joint plot," is very useful in visualizing the relationships between sampling entities and variables. For example, in the study of animal communities, ecologists often collect information on the presence or abundance of a set species across many sites (i.e., samples-by-species matrix). In this case, the row ordination describes the major gradients in species composition among sites, and the column ordination provides the locations of theoretical maxima for each species in ordination space. Both of these ordinations are useful by themselves, and each are necessary to fully understand and interpret the other. Similarly, in the study of behavioral ecology, ecologists often collect information on animal behaviors for a suite of species (i.e., in a species-by-behavior matrix). In this case, both species and behavioral categories are analyzed simultaneously. The dispersion of species along gradients in this case describes the behavioral similarity among species. Conversely, the locations of behavioral categories in ordination space relative to species points describe the gradients of behavioral characters across the sampled taxa. In this case as well, both row and column ordinations are needed to fully understand the relationships among the data.

The RA ordination solution has a number of other desirable properties. For most suitable community data sets, RA has been shown to be equal or superior to PCA. However, like PO and PCA, RA suffers from arch and compression effects under certain conditions, although RA appears to be less vulnerable than both PO and PCA. Moreover, RA appears to be better than PCA at handling long ecological gradients.

2.13.5 Detrended Correspondence Analysis

Detrended correspondence analysis (DCA) (Hill and Gauch 1980) is an improved RA technique. It is essentially an RA ordination with two additional steps, detrending and rescaling, added to remove RA's two major faults, the arch effect and axis compression. The orthogonality criterion for second and higher axes in RA (which is where arch and compression effects cause distortions), is replaced with the stronger criterion that the second and higher axes have no systematic relationship of any kind to lower axes. "Detrending" to eliminate the arch effect is accomplished by dividing the first axis into a number of equal segments and within each segment adjusting the ordination scores to a mean of zero. Detrending is applied to the ordination scores at each iteration, except that, once convergence is reached, the final ordination scores are determined by weighted averages without detrending. After the solution has stabilized, the axes may be rescaled to unit within-sample variance. This expands or contracts axis segments to ensure that change along the axis occurs at uniform rates; that is, it makes sure that equal distances in the ordination correspond to equal ecological differences. Rescaling has been severely criticized as being arbitrary. We suggest that in most cases the final axes be interpreted without rescaling.

Detrended correspondence analysis has several appealing properties (Hill and Gauch 1980; TerBraak and Prentice 1988). Like RA, DCA produces a simultaneous ordination of rows and columns (e.g., samples and species) which maximizes the correlation between sample and variable scores. Likewise, DCA is capable of handling large, complex data sets and uncovering extremely long ecological gradients. In contrast to PCA and RA, DCA is not subject to arch and compression effects, and performs particularly well when the data have nonlinear and unimodal distributions. These are exactly the conditions where PCA fails dramatically. Ecological data more often than not are nonlinear and unimodal and often approximate the Gaussian model. For data sets that contain this structure, DCA has proven to be better than nearly all other unconstrained ordination techniques.

Detrended correspondence analysis' most important limitation is its sensitivity to outliers and discontinuities in the data (Palmer 1993). Also, the sample ordination is generally more robust than the variable ordination. That is, the location of sample points in the ordination is usually more reliable than the location of variable points. In particular, the ordination of variables whose maxima are outside the range of the sample observations is unreliable. In addition, DCA performs poorly with skewed variable distributions and is occasionally unstable (Palmer 1993). It may compress one end of a gradient into a "tongue" and it will smash any real complex structure that is present in second and higher axes. It has also been criticized because of its use of an arbitrary, ad hoc standardization of the second and successive axes.

Detrended correspondence analysis has become very popular in ecology, and, under most conditions, is superior to PCA and RA. Its use, particularly in conjunction with canonical correspondence analysis (see Section 2.13.6), will undoubtedly further increase as it becomes more widely understood and readily available on standard statistical software. Currently, for example, DCA cannot be conducted using SAS.

2.13.6 Canonical Correspondence Analysis

All of the techniques we have described thus far are *unconstrained* ordination techniques. They are conducted on a single set of interdependent variables and seek to extract the major patterns among those variables irrespective of any variables outside that set. They do not find relationships between these patterns and causal factors; they do not inform as to the possible mechanisms driving the patterns; they simply describe the major patterns inherent to the data set. This, of course, can be very informative. But, unconstrained ordination techniques tell little about what those patterns mean and nothing about what causes them. The axes in PCA, RA, and DCA are the major independent gradients in the data, yet it is often difficult to understand their ecological meaning. To interpret the ecological meaning of the axes, they must be associated with other ecological factors in a separate, secondary analysis. Often it is more advantageous to directly ordinate the first set of variables on axes that are combinations of the second set of variables, particularly when the first set of variables consists of species scores (e.g., abundances) and the second set consists of environmental variables (e.g., habitat characteristics). This is called *constrained* ordination.

Canonical correspondence analysis (CCA) is the best multivariate constrained ordination technique developed to date. The technique is essentially a constrained reciprocal averaging ordination. It is a hybrid of ordination and multiple regression. Canonical correspondence analysis' calculations involve the addition of a multiple least squares regression of species onto environmental variables to the weighted averages procedures of reciprocal averaging. Samples and variables are ordinated in CCA in exactly the same way as in RA, with the added limitation that the axes must be linear combinations of the explanatory variables included in the second matrix. This is very different than PCA or DCA, where the axes are major gradients within the species data themselves, irrespective of any ecological explanatory variables. Canonical correspondence analysis extracts the major gradients in the data that can be accounted for by the measured explanatory variables; it performs ideally when the variables in the first matrix (i.e., the dependent variables, usually species abundances) have Gaussian response surfaces with respect to combinations of the variables in the second matrix (i.e., the independent variables, usually environmental factors).

Like RA and DCA, CCA is a dual ordination of both samples and species. In the CCA ordination diagram, called a "triplot," the distribution of species and sample points jointly represent the dominant ecological relationships in so far as they can be explained by the explanatory variables (Fig. 2.14). One of the major advantages of CCA over unconstrained ordination techniques, such as DCA, is that the axes are linear combinations of the independent variables that explain the most variance in the dependent variables. This gives an immediate understanding of what the axes mean and how the independent variables are related to the distribution of the dependent variables.

The explanatory variables from the second matrix are plotted in the CCA triplots as arrows emanating from the grand mean of all explanatory variables (TerBraak 1986). The arrows are pure ecological gradients. The directions of the arrows in the ordination space indicate the direction of maximum change in each structuring variable. The length of an arrow is equal to the rate of change of the weighted averages along that axis and measures how much species distributions change along that explanatory variable. In general, the length of an arrow indicates the importance of an environmental variable. The direction of the arrows relative to the axes visually indicates what the axes represent. The cosine of the angle between an arrow and an axis is the correlation coefficient between that variable and that axis. Similarly, the angle between arrows indicates the correlations between environmental variables. Species points and arrows jointly reflect species distribution on each environmental variable. Arrows and axes jointly reflect the composition of the major ecological gradients in the data.

The position of sample points relative to the arrows indicates the environmental conditions at each sample point. In addition, the location of the species points relative to the arrows indicates the characteristics of the ecological optima of each species (Palmer 1993). Therefore, the abundance, or probability of occurrence, of a species will decrease with distance from its species point. Furthermore, since each sample point lies at the centroid of the species occurring at that site, the diagram indicates which species are likely to be present at any given site.

Empirical Example. The empirical example was taken from a study on bird-habitat relationships in mature, unmanaged forest stands in the central Oregon Coast Range (McGarigal and McComb, unpublished data). We conducted a CCA (canonical correspondence analysis) with 22 vegetation structure variables as the independent data set, and bird species abundances as the dependent set. In the CCA triplot, the inverted triangles indicate the location of individual plots in the axis 1-, axis 2-space (Fig. 2.14). The solid circles indicate the locations of predicted max-

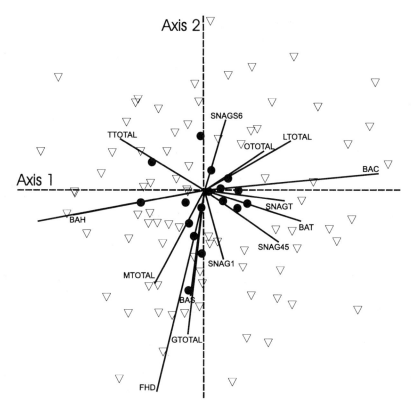

FIGURE 2.14 Example CCA triplot. Inverted triangles represent plots, circles represent species, and lines represent independent variables included in the second data matrix.

imum abundance and optimum environmental conditions for each bird species in the axis 1-, axis 2-space. The lines emanating from the center of the diagram represent the environmental factors. Their directions relative to each other and to the axes, species points, and plots indicate the coordinated gradients in this ecological space and the distribution of species maxima with respect to them. The CCA triplot succinctly portrays the complex relationships between bird species abundances and the sampled environmental factors. Our primary objective is to provide a real-data example of a CCA triplot. Unfortunately, it is not in the scope of this book to provide a detailed interpretation of the CCA results.

In CCA, the final regression coefficients are called *canonical correlation coefficients*, and the multiple correlation coefficient of the final regression is called the *species-environment correlation* (TerBraak 1986). The canonical coefficients define the axes as linear combinations of the explanatory variables. The species-environment correlation measures how well the extracted variation can be explained by the explanatory variables. Correlations between explanatory variables and the ordination axes are called *intraset* correlations. They tell us the relative magnitudes and importance of each variable in structuring the ordination. *Interset* correlations are correlations between environmental variables and species scores. They tell us which independent variables are most important in determining the value of the dependent variables.

In simulation experiments involving noisy and skewed data sets, CCA consistently performed better than DCA (Palmer 1993); CCA is robust to skewed species distributions and high noise levels. However, when the dependent variables are highly skewed it may be preferable to transform them by taking square roots or logarithms. This will prevent a few high abundance values from dominating a variable's contribution. Canonical correspondence analysis performs well with data from complex sampling designs. It will not create an arch effect, and, unlike DCA, it will not create a misleading tongue compression of one of the gradients. Because an artificial arch does not appear in CCA, detrending is not necessary. In fact, detrending can be harmful because it will remove any real arch (Palmer 1993). Canonical correspondence analysis also performs very well with nonorthogonal and collinear gradients. Unlike canonical correlation analysis (CANCOR, Chapter 5), CCA is not hampered by multicollinearity or high correlations between dependent or independent variables.

Canonical correspondence analysis is a very powerful and robust ordination technique, but like all techniques, it has limitations. To use CCA results for statistical inferences, a number of assumptions must be strictly met. The first is that observations are independently and randomly selected. This is seldom the case in ecological studies. Another major limitation is that the independent variables must be measured without error and be constant within a site. This is a difficult requirement to meet and is a serious limitation of multiple regression techniques in general. Within-site variation, similarly, is a problem for constrained ordination techniques of all kinds and is not a limitation unique to CCA. Unlike the data in pure regression, however, the data in CCA do not need to come from a normal distribution, because CCA is robust to variations in the distribution of the variables.

Comparison of constrained and unconstrained ordination results can greatly increase the usefulness of information obtained from an analysis. Unconstrained ordination techniques, like PCA or DCA, extract the largest gradients in the species data without regard to secondary ecological variables, while constrained ordination extracts the largest gradients in the species data explainable by the measured explanatory variables. When the analyses produce similar results, we assume that the measured ecological variables account for the main variation in the dependent variables. When the CCA accounts for substantially less variance than DCA, and the CCA species-environment correlation is high, we infer that the measured variables are important but that other unaccounted factors are also important. If the species-environment correlation is low in CCA, then we infer that the measured environmental variables are not important in structuring the system. We believe that comparing unconstrained and constrained ordinations is valuable whenever a constrained ordination is conducted and recommend that DCA be used whenever CCA is performed. The comparison of the results will provide information beyond what either analysis alone can provide. In particular, it is important to know how closely the most important gradients in the dependent variables (explainable by the measured independent variables) match the largest overall gradients in the dependent variables.

Bibliography

Procedures

Anderson, A.J.B. 1971. Ordination methods in ecology. *Journal of Ecology* 59:713–726.
Anscombe, F.J. 1960. Rejection of outliers. *Technometrics* 2:123–146.
Austin, M.P. 1976. On non-linear species response models in ordination. *Vegetatio* 33:33–41.
Austin, M.P., and Greig-Smith, P. 1968. The application of quantitative methods to vegetation survey. II. Some methodological problems of data from rain forest. *Journal of Ecology* 56:827–844.
Austin, M.P., and Noy-Meir, I. 1971. The problem of nonlinearity in ordination: experiments with two-gradient models. *Journal of Ecology* 59:763–773.
Austin, M.P., and Orloci, L. 1966. Geometric models in ecology. II: An evaluation of some ordination techniques. *Journal of Ecology* 54:217–227.
Beals, E.W. 1973. Ordination: mathematical elegance and ecological naivete. *Journal of Ecology* 61:23–35.
Bhattacharyya, H. 1981. Theory and methods of factor analysis and principal components. In *The Use of Multivariate Statistics in Studies on Wildlife Habitat,* ed. D.E. Capen, pp 72–79. U.S. For. Serv. Gen. Tech. Report RM–87.
Blalock, H.M., Jr. 1979. *Social Statistics.* New York: McGraw-Hill.
Bliss, C.I., Cochran, W.G., and Tukey, J.W. 1956. A rejection criterion based upon the range. *Biometrika* 43:418–422.
Bray, J. R., and Curtis, J. T. 1957. An ordination of the upland forest communities of southern Wisconsin. *Ecological Monographs* 27:325–349.
Cattell, R.B. 1966. The scree test for the number of factors. *Multivariate Behavioral Research* 1:245–276.
Cliff, N. 1988. The eigenvalues-greater-than-one rule and the reliability of components. *Psychological Bulletin* 103:276–279.
Dale, M.B. 1975. On objectives of methods of ordination. *Vegetatio* 30:15–32.
Edgington, E.S. 1995. *Randomization Tests*, 3rd ed. New York: Marcel Dekker.
Efron, B. 1979. Computers and the theory of statistics: thinking the unthinkable. *Society for Industrial and Applied Mathematics (SIAM) Review* 21:460–480.
Efron, B. 1982. The jackknife, the bootstrap and other resampling plans. *Society for Industrial and Applied Mathematics (SIAM),* Monograph #38, CBMS-N8F.
Efron, B, and Gong, G. 1983. A leisurely look at the bootstrap, the jackknife, and cross-validation. *American Statistician* 37:36–48.
Fasham, M.J.R. 1977. A comparison of nonmetric multidimensional scaling, principal components and reciprocal averaging for the ordination of simulated coenoclines and coenoplanes. *Ecology* 58:551-561.

Frontier, S. 1976. Etude de la decroissance des valeurs propres dans une analyze en composantes principales: comparison avec le modele de baton brise. *Journal of Experimental Marine Biology and Ecology* 25:67–75.

Gauch, H.G. 1973. A quantitative evaluation of the Bray–Curtis ordination technique. *Ecology* 54:829–836.

Gauch, H.G. 1973. The relationship between sample similarity and ecological distance. *Ecology* 54:618–622.

Gauch, H.G. 1982. *Multivariate Analysis in Community Ecology.* Cambridge: Cambridge University Press.

Gauch, H.G. 1982. Noise reduction by eigenvector ordinations. *Ecology* 63:1643–1649.

Gauch, H.G., and Whittaker, R.H. 1972. Comparison of ordination techniques. *Ecology* 53:868–875.

Gauch, H.G., Whittaker, R.H., and Wentworth, T.R. 1977. A comparative study of reciprocal averaging and other ordination techniques. *Journal of Ecology* 65:157–174.

Grubbs, F.E. 1950. Sample criteria for testing outlying observations. *Annals of Mathematical Statistics* 21:27–28.

Guttman, L. 1954. Some necessary conditions for common factor analysis. *Psychometrika* 19:149–161.

Hair, J.F., Jr., Anderson, R.E., and Tatham, R.L. 1987. *Multivariate Data Analysis,* 2nd ed. New York: Macmillan.

Harman, H.H. 1968. *Modern Factor Analysis*, 2nd ed. Chicago: University of Chicago Press.

Harris, R.J. 1975. *A Primer of Multivariate Statistics.* New York: Academic Press.

Hill, M.O. 1973. Reciprocal averaging: an eigenvector method of ordination. *Journal of Ecology* 61:237–249.

Hill, M.O. 1974. Correspondence analysis: a neglected multivariate method. *Journal of the Royal Statistical Society, Ser. C* 23:340–354.

Hill, M.O., and Gauch, H.G. 1980. Detrended correspondence analysis, an improved ordination technique. *Vegetatio* 42:47–58.

Horn, J.L. 1965. A rationale and test for the number of factors in factor analysis. *Psychometrika* 30:179–185.

Jackson, D.A. 1993. Stopping rules in PCA: a comparison of heuristical and statistical approaches. *Ecology* 74(8):2205–2214.

Johnson, D.H. 1981. The use and misuse of statistics in wildlife habitat studies. In *The Use of Multivariate Statistics in Studies on Wildlife Habitat,* ed. D.E. Capen, pp 11–19. U.S. Forest Service Gen. Tech. Report RM–87.

Johnson, D.H. 1981. How to measure—a statistical perspective. In *The Use of Multivariate Statistics in Studies on Wildlife Habitat*, ed. D.E. Capen, pp 53–58. U.S. Forest Service Gen. Tech. Report RM–87.

Karr, J.R., and Martin, T.E. 1981. Random numbers and principal components: further searches for the unicorn? In *The Use of Multivariate Statistics in Studies on Wildlife Habitat*, ed. D.E. Capen, pp 20–25. U.S. Forest Service Gen. Tech. Report RM–87.

Kessell, S.R., and Whittaker, R.H. 1976. Comparison of three ordination techniques. *Vegetatio* 32:21–29.

Kruskal, W.H. 1960. Some remarks on wild observations. *Technometrics* 2:1–3.

Ludwig, J.A., and Reynolds, J.F. 1988. *Statistical Ecology: A Primer on Methods and Computing.* New York: Wiley and Sons.

Marcus, M., and Minc, H. 1968. *Elementary Linear Algebra.* New York: Macmillan.

Manly, B.F.I. 1997. *Randomization, Bootstrap and Monte Carlo Methods in Biology*, 2nd ed. London: Chapman and Hall.

McCune, B. 1997. Influence of noisy environmental data on canonical correspondence analysis. *Ecology* 78(8):2617–2623.

Meot, A., Legendre, P., and Borcard, D. 1998. Partialling out the spatial component of variation: questions and propositions in the linear modelling framework. *Environmental and Ecological Statistics* 5(1):1–27.

Miles, D.B. 1990. A comparison of three multivariate statistical techniques for the analysis of avian foraging data. In *Studies in Avian Biol. No. 13—Avian Foraging: Theory, Methodology, and Applications*, ed. M.L. Morrison, C.J. Ralph, J. Verner, and J.R. Jehl, Jr., pp 309–317.

Morrison, D.F. 1967. *Multivariate Statistical Methods*, 2nd ed. New York: McGraw-Hill.

Mosteller, F., and Tukey, W. 1977. *Data Analysis and Regression: A Second Course in Statistics*. Reading: Addison-Wesley.

Nichols, S. 1977. On the interpretation of principal components analysis in ecological contexts. *Vegetatio* 34:191–197.

Noy-Meir, I. 1973. Data transformations in ecological ordination. I. Some advantages of noncentering. *Journal of Ecology* 61:329–341.

Noy-Meir, I., and Austin, M.P. 1970. Principal component ordination and simulated vegetational data. *Ecology* 61:551-552.

Noy-Meir, I., Walker, D., and Williams, W.T. 1975. Data transformations in ecological ordination. II. On the meaning of data standardization. *Journal of Ecology* 63:779–800.

Okland, R.H. 1996. Are ordination and constrained ordination alternative or complementary strategies in general ecological studies? *Journal of Vegetation Science* 7(2):289–292.

Oksanen, J., and Minchin, P.R. 1997. Instability of ordination results under changes in input data order—explanations and remedies. *Journal of Vegetation Science* 8(3):447–454.

Orloci, L. 1966. Geometric models in ecology. I. The theory and application of some ordination methods. *Journal of Ecology* 54:193–216.

Palmer, M.J. 1993. Putting things in even better order: the advantages of canonical correspondence analysis. *Ecology* 74(8):2215–2230.

Philippi, T.E., Dixon, P.M., and Taylor, B.E. 1998. Detecting trends in species composition. *Ecological Applications* 8(2):300–308.

Robertson, P.A. 1978. Comparison of techniques for ordinating and classifying old-growth floodplain forests in southern Illinois. *Vegetatio* 37:43–51.

Stauffer, D.F., Garton, E.O., and Steinhorst, R.K. 1985. A comparison of principal components from real and random data. *Ecology* 66:1693–1698.

Stewart, D.W. 1981. The application and misapplication of factor analysis in marketing research. *Journal of Marketing Research* 18:51–62.

Swan, J.M.A. 1970. An examination of some ordination problems by use of simulated vegetational data. *Ecology* 51:89–102.

Tabachnick, B.G., and Fidell, L.S. 1989. *Using Multivariate Statistics*. New York: Harper and Row.

Tatsuoka, M.M. 1971. *Multivariate Analysis: Techniques for Educational and Psychological Research*. New York: Wiley and Sons.

TerBraak, C.J.F. 1986. Canonical correspondence analysis: a new eigenvector technique for multivariate direct gradient analysis. *Ecology* 67(5):1167–1178.

TerBraak, C.J.F. 1987. The analysis of vegetation-environment relationships by canonical correspondence analysis. *Vegetatio* 69:69–77.

TerBraak, C.J.F., and Prentice, C.I. 1988. A theory of gradient analysis. *Advances in Ecological Research* 18:272–313.

Wartenberg, D., Ferson, S., and Rohlf, F.J. 1987. Putting things in order: a critique of detrended correspondence analysis. *American Naturalist* 129:434–448.

Whittaker, R.H., and Gauch, H.G. 1978. Evaluation of ordination techniques. In *Ordination of Plant Communities,* ed. R.H. Whittaker, pp 277–336. The Hague: W. Junk.

Wilson, M.W. 1981. A statistical test of the accuracy and consistency of ordinations. *Ecology* 62:8–12.

Applications

Birch, J.M. 1997. Comparing wing shape of bats: the merits of principal-components analysis and relative-warp analysis. *Journal of Mammalogy.*78(4):1187–1198.

Blair, R.B., and Launer, A.E. 1997. Butterfly diversity and human land use: species assemblages along an urban gradient. *Biological Conservation* 80(1):113–125.

Bolger, D.T., Scott, T.A., and Rotenberry, J.T. 1997. Breeding bird abundance in an urbanizing landscape in coastal southern California. *Conservation Biology* 11(2):406–421.

Bray, J.R., and Curtis, J.T. 1957. An ordination of the upland forest communities of southern Wisconsin. *Ecological Monographs* 27:325–349.

Burke, A. 1997. The impact of large herbivores on floral composition and vegetation structure in the Naukluft mountains, Namibia. *Biodiversity & Conservation.* 6(9):1203–1217.

Carey, A.B. 1981. Multivariate analysis of niche, habitat and ecotope. In *The Use of Multivariate Statistics in Studies on Wildlife Habitat,* ed. D.E. Capen, pp 104–114. U.S. Forest Service Gen. Tech. Report RM–87.

Clark, T.E., and Samways, M.J. 1996. Dragonflies (Odonata) as indicators of biotope quality in the Kruger National Park, South Africa. *Journal of Applied Ecology.* 33(5):1001–1012.

Conner, R.N., and Adkisson, C.S. 1977. Principal components analysis of woodpecker nesting habitat. *Forest Science* 22:122–127.

Gibson, A.R., Baker, A.J., and Moeed, A. 1984. Morphometric variation in introduced populations of the common myna (Acridotheres tristis): an application of the jackknife to principal components analysis. *Systematic Zoology* 33:408–421.

Ellis, B.A., Mills, J.N., Childs, J.E., Muzzini, M.C., Mckee, K.T., Enria, D.A., and Glass, G.E. 1997. Structure and floristics of habitats associated with five rodent species in an agroecosystem in central Argentina. *Journal of Zoology* 243(3):437–460.

Fuller, R.J., Trevelyan, R.J., and Hudson, R.W. 1997. Landscape composition models for breeding bird populations in lowland English farmland over a 20-year period. *Ecography* 20(3):295–307.

Grigal, D.F., and Goldstein, R.A. 1971. An integrated ordination-classification analysis of an intensively sampled oak-hickory forest. *Journal of Ecology* 59:481–492.

Hamer, K.C., Hill, J.K., Lace, L.A., and Langan, A.M. 1997. Ecological and biogeographical effects of forest disturbance on tropical butterflies of Sumba, Indonesia. *Journal of Biogeography* 24(1):67–75.

Hanowski, J.M., Niemi, G.J., and Christian, D.C. 1997. Influence of within-plantation heterogeneity and surrounding landscape composition on avian communities in hybrid poplar plantations. *Conservation Biology* 11(4):936–944.

Harner, E.J., and Whitmore, R.C. 1981. Robust principal component and discriminant analysis of two grassland bird species' habitat. In *The Use of Multivariate Statistics in Studies on Wildlife Habitat,* ed. D.E. Capen, pp 209–221. U.S. Forest Service Gen. Tech. Report RM–87.

Harshbarger, T.J., and Bhattacharyya, H. 1981. An application of factor analysis in an aquatic habitat study. In *The Use of Multivariate Statistics in Studies on Wildlife Habitat*, ed. D.E. Capen, pp 180–185. U.S. Forest Service Gen. Tech. Report RM–87.

Hawrot, R.Y., and Niemi, G.J. 1996. Effects of edge type and patch shape on avian communities in a mixed conifer-hardwood forest. *Auk* 113(3):586–598.

James, F.C. 1971. Ordinations of habitat relationships among breeding birds. *Wilson Bulletin* 83:215–236.

Kirk, D.A., Diamond, A.W., Hobson, K.A., and Smith, A.R. 1996. Breeding bird communities of the western and northern Canadian boreal forest: relationship to forest type. *Canadian Journal of Zoology* 74(9):1749–1770.

Lima, A.P., and Magnusson, W.E. 1998. Partitioning seasonal time: interactions among size, foraging activity and diet in leaf-litter frogs. *Oecologia* 116(1–2):259–266.

Madhsudan, M.D., and Johnsingh, A.J.T. 1998. Analysis of habitat-use using ordination: the Nilgiri Thar in southern India. *Current Science* 74(11):1000–1003.

Martinsmith, K.M. 1998. Relationships between fishes and habitat in rainforest streams in Sabah, Malaysia. *Journal of Fish Biology* 52(3):458–482.

Maurer, B.A., McArthur, L.B., and Whitmore, R.C. 1981. Habitat associations of birds breeding in clearcut deciduous forests in West Virginia. In *The Use of Multivariate Statistics in Studies on Wildlife Habitat*, ed. D.E. Capen, pp 167–172. U.S. Forest Service Gen. Tech. Report RM–87.

Moser, E.B., Barrow, W.C. Jr., and Hamilton, R.B. 1990. An exploratory use of correspondence analysis to study relationships between avian foraging behavior and habitat. In *The Use of Multivariate Statistics in Studies on Wildlife Habitat*, ed. D.E. Capen, pp 309–317. U.S. Forest Service Gen. Tech. Report RM–87.

Niemi, G.J., and Hanowski, J.M. 1984. Relationships of breeding birds to habitat characteristics in logged areas. *Journal of Wildlife Management* 48:438–443.

Murkin, H.R., Murkin, E.J., and Ball, J.P. 1997. Avian habitat selection and prairie wetland dynamics: a ten year experiment. *Ecological Applications* 7(4):1144–1159.

Pappas, J.L., and Stoermer, E.F. 1997. Multivariate measure of niche overlap using canonical correspondence analysis. *Ecoscience* 4(2):240–245.

Pearman, P.B. 1997. Correlates of amphibian diversity in an altered landscape of Amazonian Ecuador. *Conservation Biology* 11(5):1211–1225.

Rotenberry, J., and Wiens, J.A. 1981. A synthetic approach to principal components analysis of bird/habitat relationships. In *The Use of Multivariate Statistics in Studies on Wildlife Habitat*, ed. by D.E. Capen, pp 197–208. U.S. Forest Service Gen. Tech. Report RM–87.

Schaeffer, J.A., Stevens, S.D., and Messier, F. 1996. Comparative winter habitat use and associations among herbivores in the high Arctic. *Arctic* 49(4):387–391.

Sczerzenie, P.J. 1981. Principal components analysis of deer harvest-land use relationships in Massachusetts. In *The Use of Multivariate Statistics in Studies on Wildlife Habitat*, ed. by D.E. Capen, pp 173–179. U.S. Forest Service Gen. Tech. Report RM–87.

Shank, C.C. 1982. Age-sex differences in the diets of wintering Rocky Mountain bighorn sheep. *Ecology* 63:627–633.

Smith, K.G. 1977. Distribution of summer birds along a forest moisture gradient in an Ozark watershed. *Ecology* 58:810–819.

Velazquez, A., and Heil, G.W. 1996. Habitat suitability for the conservation of the volcano rabbit (Romerolagus diazi). *Journal of Applied Ecology* 33(3):543–554.

Whitmore, R.C. 1975. Habitat ordination of passerine birds of the Virgin River Valley, southwestern Utah. *Wilson Bulletin* 87:65–74.

Whitmore, R.C. 1977. Habitat partitioning in a community of passerine birds. *Wilson Bulletin* 89:253–265.

Yu, C.C., Quinn, J.T., Dufournaud, C.M., Harrington, J.J., Rogers, P.P., and Lohani, B.N. 1998. Effective dimensionality of environmental indicators: a principal components analysis with bootstrap confidence intervals. *Journal of Environmental Management* 53(1):101–119.

Appendix 2.1

SAS program statements used to conduct principal components analysis on the empirical data set presented in Figure 2.2. Everything given in lower case varies among applications with regard to characteristics of the specific data set (e.g., variable names) and the personal preferences of the user (e.g., naming output files).

The following header information creates a library named "km" for SAS files in the specified directory and provides page formatting information:

```
LIBNAME km 'd:\stats\pca';
OPTIONS PS=60 LS=80 REPLACE OBS=MAX;
```

The following macro is used to compute descriptive univariate statistics and plots for each of the original variables, including the raw variables and three different transformations, in order to assess univariate normality:

```
%MACRO RES(Y);
  DATA A; SET km.pca;
    TITLE 'Assessing Normality of the Raw Variable';
    PROC UNIVARIATE NORMAL PLOT;
      VAR &Y; RUN;
  DATA A; SET km.pca;
    TITLE 'Assessing Normality of the Arcsine Square Root Transformed Variable';
    &Y=SQRT(&Y);
    PROC UNIVARIATE NORMAL PLOT;
      VAR &Y; RUN;
  DATA A; SET km.pca;
    TITLE 'Assessing Normality of the Logit Transformed Variable';
```

The following line is needed to transform percent to proportion:

```
    &Y=&Y/100;
    &Y=LOG((&Y+0.5)/(1.5-&Y));
    PROC UNIVARIATE NORMAL PLOT;
      VAR &Y; RUN;
  DATA A; SET km.pca;
    TITLE 'Assessing Normality of the Rank Transformed Variable';
    PROC RANK;
      VAR &Y;
    PROC UNIVARIATE NORMAL PLOT;
      VAR &Y; RUN;
%MEND;
```

The following line should be repeated for each variable in the data set, substituting the appropriate variable names:

```
%RES(forb);
```

The following procedure is used to inspect the standardized data for possible outliers:

```
DATA A; SET km.pca;
   PROC STANDARD OUT=standard MEAN=0 STD=1;
      VAR forb-conifer; RUN;
DATA B; SET standard;
   ARRAY XX forb-conifer;
      DO OVER XX;
         IF XX GT 2.5 THEN OUTLIER=1;
         ELSE XX='.';
      END; RUN;
DATA C; SET B;
   IF OUTLIER=1;
   TITLE 'List of Potential Outlier Observations';
   PROC PRINT;
      VAR Id forb-conifer; RUN;
```

The following procedure is used to compute the initial PCA on the raw variables and inspect the first three principal components for outliers, multivariate normality, and linearity assumptions:

```
DATA A; SET km.pca;
   TITLE 'PCA on Vegetation Composition Variables';
   PROC FACTOR NFACTOR=3 OUT=scores;
      VAR forb-conifer; RUN;
   TITLE 'Assessing Normality of the First 3 Principal Components';
   PROC UNIVARIATE DATA=scores NORMAL PLOT;
      VAR FACTOR1-FACTOR3; RUN;
   TITLE 'Scatter Plots of the First 3 Principal Components';
   PROC PLOT DATA=scores;
      PLOT (FACTOR1 FACTOR2 FACTOR3)=ID; RUN;
```

The following procedure is used to compute the full PCA on all 22 raw variables, including output of the correlation matrix, scree plot, eigenvectors, standardized scoring coefficients, and a list of the principal component scores for the first three components:

```
DATA A; SET km.pca;
   TITLE 'PCA on Vegetation Composition Variables';
   PROC FACTOR NFACTOR=22 CORR SCREE EIGENVECTORS SCORE OUT=scores;
      VAR forb-conifer; RUN;
      TITLE 'List of Principal Component Scores for the First 3
   Components';
   PROC PRINT DATA=scores;
      VAR FACTOR1-FACTOR3; RUN;
```

The following procedure is used to compute varimax and quartimax rotations on the first three principal component axes:

```
DATA A; SET km.pca;
   TITLE 'Varimax Rotation of the First 3 Principal Components';
```

```
PROC FACTOR ROTATE=VARIMAX NFACTOR=3;
   VAR forb-conifer; RUN;
TITLE 'Quartimax Rotation of the First 3 Principal Components';
PROC FACTOR ROTATE=QUARTIMAX NFACTOR=3;
   VAR forb-conifer; RUN;
```

3
Cluster Analysis

3.1 Objectives

By the end of this chapter, you should be able to do the following:

- Define *cluster analysis* (CA) and give examples of how it can be used to help solve a research problem.
- List seven important characteristics of CA.
- Classify clustering techniques using five criteria:
  — Exclusive versus nonexclusive
  — Sequential versus simultaneous
  — Hierarchical versus nonhierarchical
  — Agglomerative versus divisive
  — Polythetic versus monothetic
- Describe the key features of nonhierarchical clustering techniques and explain the conditions under which they are best used.
- Define *resemblance* and, with examples, discuss three classes of resemblance measures used in *polythetic agglomerative hierarchical clustering* (PAHC):
  — Distance coefficients
  — Association coefficients
  — Correlation coefficients
- Compare and contrast six different fusion strategies used in PAHC:
  — Single linkage
  — Complete linkage
  — Centroid linkage
  — Median linkage
  — Average linkage
  — Ward's minimum variance linkage

- Explain the hierarchical relationships among entities by examining and interpreting a dendrogram and icicle plot.
- Define cophenetic correlation and use it to evaluate the results of a CA.
- Explain under what conditions you might standardize your data.
- Discuss alternative methods for deciding on the number of important clusters and describing differences among the resulting clusters.
- Describe the distinguishing features of divisive techniques.
- Describe two methods for validating the results of CA.
- Discuss how CA and ordination techniques can be used in conjunction to complement each other.
- List five limitations of CA.

3.2 Conceptual Overview

Ecologists are often interested in discovering ways to clarify relationships in complex ecological systems. Often this involves finding ways to combine similar entities into assemblages or groups in order to reduce the dimensionality of the system. Sometimes the purpose might be to reduce the number of sampling entities to a manageable size so that other procedures, such as ordination, can be used effectively. In other cases, the purpose might be to eliminate redundant information from the data set in order to provide a succinct summary of the data structure. In ecological studies, for example, the sample often contains redundant entities (e.g., species, sites)—entities that depict the same basic response pattern. This happens because several species may respond in a similar manner to environmental gradients or management actions. Including the unique response of every individual entity can obscure the dominant underlying response patterns. In other cases, the purpose of the analysis might be much more pragmatic. Wildlife managers, for example, are often confronted with finding ways to manage hundreds (if not thousands) of species simultaneously. If every species exhibits a unique distribution and response to environmental factors and management actions, then finding management solutions will be exceedingly difficult. If, on the other hand, groups of species exhibit similar distributions along environmental gradients, then it might be possible to aggregate species into a manageable number of groups for purposes of management.

Cluster analysis (CA) refers to a large family of techniques, each of which attempts to organize entities (i.e., sampling units) into discrete classes or groups (Table 3.1). This is in contrast to ordination, which attempts to organize entities along a continuum (see Chapter 2). With a single variable, entities are simply arranged in rank order on the measured variable and then combined into groups based on their relative location along this axis. Unfortunately, the process of creating groups, even on the basis of a single variable, is not as straightforward as it may seem. For example, does one begin with each entity in a separate group and then aggregate similar entities, or does one begin with all entities in one large group and then disaggregate? How is similarity quantified between a single entity and a group of entities, or between two groups of entities, for the purpose of deciding which two

3.2 Conceptual Overview

TABLE 3.1 Important characteristics of cluster analysis.

A family of techniques with similar goals.

Organizes sampling entities (e.g., species, sites, observations) into discrete classes or groups, such that within-group similarity is maximized and among-group similarity is minimized according to some objective criteria.

Operates on data sets for which prespecified, well-defined groups DO NOT exist; characteristics of the data are used to assign entities into artificial groups.

Summarizes data redundancy by reducing the information on the whole set of N entities (for example) to information about G groups (for example) of nearly similar entities (where hopefully G is very much smaller than N).

Identifies outliers by leaving them solitary or in small clusters, which may then be omitted from further analyses.

Eliminates noise from a multivariate data set by clustering nearly similar entities without requiring exact similarity.

Assesses relationships within a single set of variables; no attempt is made to define the relationship between a set of independent variables and one or more dependent variables.

---

entities or groups should be aggregated or disaggregated next? Is aggregation based on the average similarity among entities from the respective groups, or the maximum or minimum similarity between any two entities from the respective groups? Answers to these and other questions can lead to quite a variety of approaches, even for a single variable.

With multiple variables, the task is even more difficult, because there is a different clustering for each variable. In such cases, methods that produce meaningful groupings based on combinations of variables must be used. With ecological data, the idea is to develop groupings that match ecological or environmental classes. Unfortunately, as the number of variables increases, the number of possible groupings based on combinations of variables quickly becomes overwhelming. Luckily, mathematicians have devised clever ways to measure the similarity between entities based on multiple variables simultaneously. Given the variety of ways to measure multivariate similarity and the variety of methods for aggregating or disaggregating entities, there exists a plethora of clustering algorithms for the ecologist to choose from.

The fundamental principle behind the use of CA in ecology is that much of the variability in a multivariate ecological data set is often concentrated on relatively few groups of entities, and these major groupings are usually highly related to certain ecological or environmental factors. Ecological phenomena are not always continuously distributed along gradients of change. Often they are grouped into a hierarchical structure of discontinuous clusters. Clustering techniques use the redundancy in ecological data sets to extract and describe the major discrete groups in these sets.

Consider the following hypothetical data set involving the abundances of four bird species at six sites:

	Species			
Sites	A	B	C	D
1	1	9	12	1
2	1	8	11	1
3	1	6	10	10
4	10	1	9	10
5	10	2	8	10
6	10	0	7	1

Suppose that researchers want to group these six sites into two clusters based on their similarity in species composition. Based on species A alone, the sites can be organized into two perfectly discrete groups (sites 1–3 and 4–6). With the addition of species B and C, the same two groups are readily apparent, although the separation is not perfect. In particular, species C suggests a continuous rather than discrete ordering of sites. Finally, with the addition of species D, the groups are less immediately apparent. In fact, on the basis of species D alone, we would form two different groups (sites 1, 2, 6 versus 3, 4, 5). Nevertheless, overall it seems reasonable to form the two previous groups based on similarities in the abundances of species A to C.

For most purposes, researchers might be satisfied condensing the original data matrix into these two groups of fairly similar sites. If these two groups can be related to certain ecological or environmental factors, then we can use them to parsimoniously explain the patterns in this multivariate data set. Of course, real ecological data sets usually contain a more complex structure; as a result, rarely are the natural groupings so readily apparent. Moreover, rarely would a single grouping of sites perfectly describe the distribution patterns of several species. Usually, finding the groupings of entities that best describe the overall structure of a complex multivariate data set is a complex problem.

Cluster analysis is known by a wide variety of other terms (see Chapter 1, Table 1.2). Most of these terms refer or apply to a specific discipline or to a specific clustering approach. The diverse nomenclature is due in part to the importance of the methods in widely diverse disciplines. Although the names differ across disciplines, they all have the common feature of clustering entities according to natural relationships. Within the field of wildlife research, the terms "cluster analysis" and "classification" are used almost exclusively. In this book, we use "cluster analysis" exclusively to reference the family of techniques discussed in this chapter and reserve "classification" to refer to a subset of discriminant analysis (DA) involved in the development of predictive or classification functions.

Like ordination, the CA family includes several techniques that vary in a number of important ways. Unlike many multivariate techniques, CA assesses relationships within a single set of interdependent variables, regardless of any relationships these variables may have to variables outside the set. Like PCA, CA does not attempt to define the relationship between a set of independent variables and one or more dependent variables; CA leaves this to subsequent analyses.

In addition, in contrast to DA, CA operates on data sets for which prespecified, well-defined groups do *not* exist, but are suspected. Characteristics of the data are used to assign entities into groups, natural or manufactured. In other words, in DA we begin with a priori, well-defined groups in an attempt to identify the variables that best distinguish the groups; whereas, in CA, we begin with an undifferentiated collection of entities and attempt to form subgroups which differ on selected variables. In DA, we essentially ask how the given groups differ, whereas, in CA, we ask whether a given group can be partitioned into subgroups that indeed differ. Consequently, CA often serves as a precursor to DA when prespecified groups do not exist; that is, CA is used to define groups, and subsequently, DA is used to describe how the groups differ.

The main purpose of CA in ecological studies is to organize entities into classes or groups such that within-group similarity is maximized (i.e., intragroup distances are minimized) and among-group similarity is minimized (intergroup distances are maximized) according to some objective criteria. As such, CA summarizes data redundancy (i.e., identical information shared by two or more entities) among sampling entities by reducing the information on the whole set of N entities (for example) to information about G groups (for example) of nearly similar entities (where it is hoped G is very much smaller than N). Consequently, CA can be used to gain insight into the organization of ecological systems by depicting how readily entities can be partitioned into discrete, discontinuous classes. Further, CA can facilitate the management of ecological systems by establishing groups of entities (e.g., species, habitats, etc.) with similar ecological relationships and, hopefully, similar responses to management activities.

It is beyond the scope of this book to cover all, or even a majority, of the multitude of techniques that fall under the general heading of CA. Instead, we focus on the techniques most commonly used in ecological research and those with the most potential for application in wildlife research.

3.3 The Definition of Cluster

Several attempts have been made to define "cluster," although in most published ecological applications of CA, a formal definition has not been provided. Most proposed definitions are vague and circular in nature in the sense that terms such as "similarity," "distance," and so on are used in the definition, but are themselves undefined (Everitt 1977). There is no universal agreement on what constitutes a cluster. In fact, it is probably true that no single definition is sufficient for the many and varied applications of CA. The ultimate criterion for determining what constitutes a meaningful cluster is the intuition of the user.

The most intuitive description of what constitutes a cluster is one in which sampling entities are considered as points in a multidimensional space, where each of the P variables represents one of the axes of this space. Each sample point has a position on each axis and therefore occupies a unique location in this P-dimensional space. Collectively, all the sample points form a cloud of points in P-dimensional space. Within this data cloud, clusters may be described as continuous regions containing a relatively high density of points, separated from other such regions by space containing a relatively low density of points (Everitt 1977). For example, in Figure 1.4, each species is represented by a probability surface in three-dimensional niche space. The data points for each species occupy a dense region of the data cloud and form a distinct cluster. Clusters described in this way are sometimes referred to as "natural clusters." This description matches the way we detect clusters visually in two or three dimensions. It is impossible to visualize clusters in this manner when there are more than three dimensions, but they are conceptually the same.

An advantage of defining clusters in this way is that it does not restrict a cluster's shape. A major problem in CA is that most clustering procedures look for clusters of a particular shape and are usually biased toward finding spherical clusters (Everitt 1977). However, there is usually no a priori reason for believing that any clusters present in the data are of one particular shape, and by using the "wrong" clustering technique, we may impose a particular structure on the data, rather than find the actual structure present (Everitt 1977).

Several sets of two-dimensional data are shown in Figure 3.1. The clusters present are fairly obvious. Except in Figure 3.1c, where there is an overlap between the two apparent clusters, most of the clusters are of the natural cluster variety; they are distinguished by regions of dense points separated from other such regions, rather than conforming to any particular shape. Most clustering techniques would have no difficulty in finding the spherical clusters present in Figure 3.1a. However, many would likely have difficulty in recovering the natural clusters depicted in the other figures, and, indeed, many would likely produce misleading solutions (Everitt 1977).

Unfortunately, there has been limited work on the relationship between cluster shape and the comparative performance of alternative clustering techniques (Everitt 1977). Consequently, we can offer little guidance to help the researcher select an appropriate clustering technique with greater sensitivity to this issue. Perhaps the best, or only, solution is to compare results among several different clustering techniques. General agreement in cluster solutions usually indicates a rather pronounced structure to the data. Disagreement indicates a less clear structure to the data (e.g., nonspherical or fuzzy clusters, or no "real" clusters at all). In these situations, each procedure's bias may govern the particular cluster solution found.

3.4 The Data Set

Cluster analysis deals with problems involving a single set of variables. There is no distinction between independent variables and dependent variables. Variables can be either continuous or categorical (i.e., discrete), although usually data sets contain either all continuous or all categorical variables. There is no consensus on the valid-

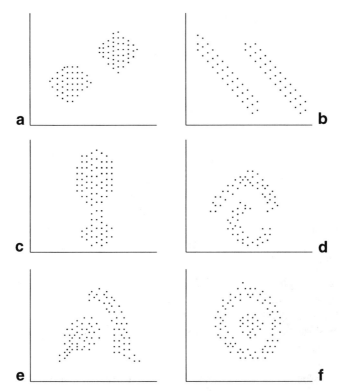

FIGURE 3.1 Hypothetical data sets showing clusters of various shapes (adopted from Everitt 1977).

ity of mixed data sets, but most experts advise against them. The only requirement of the data set is that every sampling entity be measured on the same set of variables (i.e., no missing data).

Clustering algorithms usually require a two-way data matrix as input. Many ecological studies involving CA come from the field of community ecology in which the customary input data are species abundances in a two-way samples-by-species data matrix. Applications of CA in the wildlife literature are less common, yet frequently involve the assessment of species-habitat relationships (e.g., Short and Burnham 1982; Holmes, Bonney, and Pacala 1979; Raphael and White 1984), where the objective is to group species with similar habitat requirements (e.g., to define *guilds*). The data often consist of species observations for which there is an associated mean vector of environmental measurements (i.e., each species has a mean value for each environmental variable which is the average over several samples for that species). A species-by-environmental parameters matrix consisting of N species and P environmental variables would look as follows:

	X_1	X_2	X_3	$\ldots$	X_P
1	x_{12}	x_{12}	x_{13}	$\ldots$	x_{1P}
2	x_{21}	x_{22}	x_{23}	$\ldots$	x_{2P}
3	x_{31}	x_{32}	x_{33}	$\ldots$	x_{3P}
$\vdots$	$\vdots$	$\vdots$	$\vdots$	$\ddots$	$\vdots$
N	x_{N1}	x_{N2}	x_{N3}	$\ldots$	x_{NP}

In this book, we discuss CA on a samples-by-variables matrix, where N samples represent any type of sampling unit (e.g., sites, species, specimens) for which there is an associated vector of measurements, and variables (X_i) represent any set of parameters recorded for each entity (e.g, habitat features, morphological characteristics, behavioral characteristics, species abundances). Unless otherwise noted, we refer to the samples (rows) as "entities" or "sampling entities" to reflect the generality of clustering procedures to any type of sampling unit. However, our comments are applicable to any data of suitable form.

Empirical Example. The empirical example was taken from a study on bird-habitat relationships in mature, unmanaged forest stands in the central Oregon Coast Range (McGarigal and McComb, unpubl. data). In this chapter, we use CA to group breeding bird species with similar habitat associations. Specifically, we use CA to group bird species (sample entities) on the basis of habitat similarities defined by vegetation structure characteristics (variables).

In this productive forest landscape, vegetation structure can vary dramatically from one location to another in response to physical environmental gradients and a myriad of disturbance and successional processes. These complex environments provide habitat for a number of bird species associated with mature, natural forest stands. In order to understand how vegetation structure affects the distribution and abundance of bird species, it would be helpful to know whether species where distributed independently or jointly (i.e., in groups) in relation to vegetation structure. We might hypothesize, for example, that species are affected by a few important elements of vegetation structure (e.g., dead wood, basal area, and foliage height diversity) and that these elements in combination create a set of discrete stand conditions that affect the distribution of species. We might find that many species are associated with each stand condition and that, consequently, we can succinctly describe the bird community in terms of a small number of species' groups. CA provides us with an objective means of organizing species into these groups and determining whether the grouping is natural or imposed on the data structure.

In this example, we use CA to combine 19 breeding bird species into a smaller number of groups based on 20 vegetation structure variables. For each species with greater than 4 detections, we calculated a vector of mean habitat characteristics by averaging habitat measures across 96 sample points (i.e., bird count points) weighted by number of detections at each sample point. The sample points were distributed evenly among streamside and upslope conditions in six small watersheds. The resulting data set contains 19 sample observations (rows), one for each bird spe-

OBS	SPECIES	GTOTAL	LTOTAL	TTOTAL	MTOTAL	OTOTAL	SNAGM1	SNAGM23	SNAGM45	.	.	.	FHD
1	AMRO	15.31	31.42	64.28	20.71	47.14	0.00	0.28	0.14	.	.	.	1.45
2	BHGR	5.76	24.77	73.18	22.95	61.59	0.00	0.00	1.09	.	.	.	1.28
3	BRCR	4.78	64.13	30.85	12.03	63.60	0.44	0.44	2.08	.	.	.	1.18
4	CBCH	3.08	58.52	39.69	15.47	62.19	0.31	0.28	1.52	.	.	.	1.21
5	DEJU	13.90	60.78	36.50	13.81	62.89	0.23	0.31	1.23	.	.	.	1.23
.	.	.	.	.	.	.	.	.	.				
.	.	.	.	.	.	.	.	.	.				
.	.	.	.	.	.	.	.	.	.				
19	WIWR	8.05	41.09	55.00	18.62	53.77	0.09	0.18	0.81	.	.	.	1.36

FIGURE 3.2 The example data set.

cies, with a single variable identifying bird species (SPEC) and 20 vegetation structure variables (columns) defining each species' mean habitat characteristics (Fig. 3.2). All vegetation variables are continuous and were either directly measured or mathematically derived. GTOTAL, LTOTAL, TTOTAL, MTOTAL, and OTOTAL represent percent cover within several vertical strata (less than 0.3 m, 0.3–1.3 m, 1.3–4.0 m, 4.0 m to lower canopy, and overstory canopy, respectively). SNAG variables representing densities of snags by size class (small diameter: 10–19 cm dbh; medium diameter: 20–49 cm dbh; large diameter: greater than 49 cm dbh) and decay class (1–6 represent increasing levels of decay). SNAGT is the sum of all snags across size and decay class. BAS, BAC, BAH, and BAT represent basal area (m^2/ha) of snags, conifers, hardwoods, and total basal area, respectively. FHD represents an index of foliage height diversity. Note that the variables in this data set have different measurement scales. See Appendix 3.1 for an annotated SAS script for all of the analyses presented in this chapter.

3.5 Clustering Techniques

As noted previously, there are many different clustering techniques. Several properties of clustering techniques can be used to group methods into manageable categories (Sneath and Sokal 1973; Everitt 1977; Gauch 1982). The following five properties are the most useful for succinctly describing individual techniques and for developing a classification of clustering techniques (Fig. 3.3).

1. Exclusive versus nonexclusive. Exclusive (or nonoverlapping) techniques place each entity in one, and only one, group; nonexclusive (or overlapping) techniques place each entity in one or more groups. Most techniques are exclusive. In ecological research, exclusive techniques are used almost exclusively over nonexclusive techniques.

2. Sequential versus simultaneous. Sequential techniques apply a recursive sequence of operations to the set of entities; simultaneous techniques apply a single nonrecursive operation to the entities to form clusters. Simultaneous procedures have been explored by a few researchers but have not generally been adopted. There-

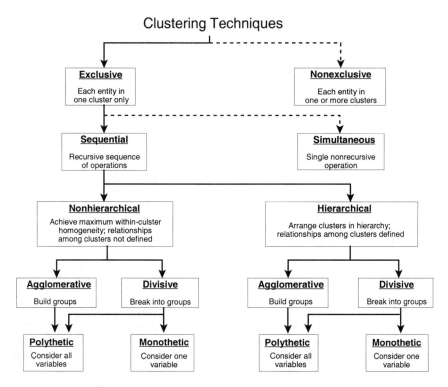

FIGURE 3.3 Classification of clustering techniques based on five properties. The techniques located on branches with solid lines are reviewed in this book.

fore, most techniques are sequential, and in ecological research, sequential techniques are used almost exclusively over simultaneous techniques.

3. Hierarchical versus nonhierarchical. Hierarchical techniques group similar entities together into groups and arrange these groups into a hierarchy that expresses the relationships among groups; nonhierarchical techniques merely assign each entity to a group, placing similar entities together to achieve within-cluster homogeneity. The nonhierarchical method does not necessarily reveal any interesting structure within clusters or definition of relationships among clusters. Both hierarchical and nonhierarchical procedures have received widespread attention, but in ecological research, most applications have used hierarchical procedures.

4. Agglomerative versus divisive. Agglomerative techniques begin with each entity in a class of its own, then fuse (agglomerate) the classes into larger classes; divisive techniques begin with all entities in a single class and divide this class into progressively smaller classes, stopping when each class contains a single member or when the predetermined limit of some "stopping rule" has been reached. Both agglomerative and divisive procedures have been used widely in ecological research, although the majority of applications have used agglomerative procedures, in part, because they are more well known than divisive procedures.

5. Polythetic versus monothetic. Polythetic techniques consider all the information for each entity when deriving cluster assignments. Monothetic techniques can only be divisive and are often used in community ecology studies involving samples-by-species data, where sets of samples are divided according to the presence or absence of a single species. Overall, however, the vast majority of ecological applications use polythetic procedures.

Currently, the most popular clustering techniques in ecological research are exclusive and sequential; that is, each entity is classified into a single cluster based on a recursive sequence of operations. Within this setting, clustering techniques may be hierarchical or nonhierarchical, agglomerative or divisive, and monothetic or polythetic. The choice of methods is entirely dependent on the intrinsic structure of the data and the objectives of the investigator. Within each of these clustering methods, there are variations in clustering algorithms arising from the emphasis placed on various classification criteria (see below). By choosing a cluster strategy (i.e., a general method and specific algorithm), the researcher has considerable control over the results. Indeed, CA is a flexible, user-directed, analytic procedure with multiple possible outcomes.

It is beyond the scope of this book to review all possible clustering methods and algorithms. There are literally hundreds of combinations. We recommend that you consult other comprehensive textbooks devoted entirely to CA in order to obtain additional understanding of the wide variety of clustering procedures used in various disciplines (e.g., Beale 1969; Tryon and Baliley 1970; Anderberg 1973; Sneath and Sokal 1973; Hartigan 1975; Everitt 1977; Gauch 1982; Digby and Kempton 1987). In this book, we discuss only exclusive and sequential techniques and focus on polythetic hierarchical methods, since these are the most popular in ecology. For comparative purposes, however, we also briefly review polythetic nonhierarchical clustering, even though nonhierarchical methods are rarely applied in wildlife studies, where we are typically interested in elucidating relationships among the clusters. Nonhierarchical methods will be discussed first because conceptually they are the simplest and often serve as precursors to hierarchical clustering.

3.6 Nonhierarchical Clustering

Ecologists sometimes collect data for a vast number of sampling entities and need to eliminate redundant information so that any underlying patterns in the data are not obscured by noise. *Nonhierarchical clustering* (NHC) techniques provide a solution to this problem. They merely assign each entity to a cluster, placing similar entities together. As a result, relationships among the clusters are not characterized. Nonhierarchical clustering is, of all clustering techniques, conceptually the simplest. Maximizing within-cluster homogeneity is the basic cluster property sought in all NHC techniques (Gauch 1982). Within-cluster homogeneity makes possible inference about an entity's properties based on its cluster membership. This one characteristic makes NHC useful for mitigating noise, summarizing redundancy, and identifying outliers. However, NHC is not effective for elucidating relationships because there is no interesting structure within clusters and no definition of relationships among clusters derived.

Because of the relatively simple requirement of within-cluster homogeneity, conceptually NHC is unlike ordination or hierarchical clustering (HC) for which numerous, quite distinctive approaches exist. Nonhierarchical clustering techniques require much less computer time than HC techniques because they involve fewer calculations and require minimal computer memory.

Nonhierarchical clustering techniques can be divided into two different approaches: polythetic agglomerative and polythetic divisive. Although the criteria of within-cluster homogeneity remains the same, agglomerative and divisive techniques have marked differences, as discussed below.

3.6.1 *Polythetic Agglomerative Nonhierarchical Clustering*

In the analysis of large multivariate data sets, it is usually desirable to address redundancy first. Since redundant entities contribute little new information, they are uninformative about broader relationships. Also, homogeneous clusters of essentially replicate entities do not contain adequate sample variation to merit analysis of within-cluster variation. Therefore, initial nonhierarchical clusters may be represented by a composite (average) sample for each cluster, and these far fewer composite samples are then more practical for subsequent hierarchical clustering to elucidate relationships among groups. Outliers, likewise, should be addressed early in the analysis, because they may overly influence the cluster solution and obscure other features of the data.

The basic role of *polythetic agglomerative nonhierarchical clustering* (PANHC) is initial clustering of large data sets. Fortunately, PANHC addresses redundancy and outliers simultaneously because it clusters redundant entities and leaves outliers solitary or in small clusters, which may then be dropped from further analyses. In addition, by clustering nearly similar entities without requiring exact similarity, PANHC is an effective means of eliminating noise. A natural sequence for analysis is first PANHC to cluster similar entities (i.e., to summarize redundancy and eliminate noise) and to identify outliers, followed by a further analysis of data relationships by hierarchical clustering and/or other multivariate techniques.

Rapid initial PANHC may be done by a technique termed *composite clustering* (Gauch 1980). There are several algorithms available for composite clustering (e.g., COMPCLUS, Gauch 1979; SAS FASTCLUS) which differ in various details. For example, they may differ in the way initial cluster seeds are selected. In all cases, the single criterion achieved is within-cluster homogeneity, and the results are, in general, similar. Composite clustering is explained most readily by visualizing entities as points in dissimilarity space; several dissimilarity measures may be used (e.g., Euclidean distance, percentage dissimilarity, complemented coefficient of community; see Section

3.6.1 for discussion of dissimilarity measures), with percentage dissimilarity generally recommended (Gauch 1982) (although SAS FASTCLUS uses Euclidean distance).

Composite clustering (as implemented in COMPCLUS) proceeds in two phases (Gauch 1979, 1980). Phase one picks an entity at random and clusters entities within a specified radius to that entity (i.e., the dissimilarity values are less than some specified value). This process is repeated until all entities are clustered, ignoring any entity previously clustered. Because center entities are selected at random, the order in which the entities of the data set are given introduces no systematic bias in the results. Phase two reassigns entities from small clusters (defined as having fewer than a specified number of members) into the nearest large cluster, provided that the entity is within a specified radius. The radius used in phase two is usually somewhat (but not greatly) larger than that used in phase one. For each resulting cluster, a composite sample is produced by averaging the entities it contains.

By controlling the radii used in phases one and two, the investigator can control the number of clusters obtained; a small radius leads to numerous clusters and a large radius to few. Unfortunately, deciding on appropriate radii or a final number of clusters is often difficult, and these factors must be determined through experimentation.

3.6.2 *Polythetic Divisive Nonhierarchical Clustering*

An alternative to PANHC is *polythetic divisive nonhierarchical clustering* (PDNHC), sometimes referred to as "K-means clustering" (Hartigan and Wong 1979). In contrast to PANHC, PDNHC operates on the original data rather than on a secondary dissimilarity matrix. This is generally considered advantageous (Gauch 1982) because all of the information contained in the original data matrix is used in the clustering process. Whereas PANHC excels at summarizing very large data sets, PDNHC performs optimally when the objective is to minimize within-cluster variation (or a ratio of among- to within-cluster variation) for a specified number of clusters (e.g., when the purpose is to reduce 200 sample entities to 50 clusters).

Polythetic divisive NHC techniques are often referred to as "optimization-partitioning techniques" because clusters are formed by optimization of a clustering criterion (Everitt 1977; Mardia, Kend, and Bibby 1979). These procedures involve splitting the entities into a specified number of groups using some criterion (for example, maximizing the ratio of among- to within-cluster variation) or simply minimizing within-cluster variation. Conceptually, the procedure based on the former criterion is like doing a one-way analysis of variance where the groups are unknown and the largest F-value is sought by reassigning members to each group. There are several different algorithms available for this procedure.

The procedure based on the minimum within-cluster variance criterion, often referred to as "minimum variance partitioning," involves several steps. First, the standardized distances between the overall centroid and each entity are computed. Hence, distances are measured in standard deviation units. The purpose of standardizing the data is to eliminate the dominating effect of variables with greater variation. Standardization is absolutely necessary if the variables have different units or scales of measurement. There is also usually an option to weight some variables more heavily than others. Second, the entity with the largest standardized distance from the overall centroid is selected to start the next cluster and this entity serves as the new cluster mean. Third, each entity is examined to see whether its distance to the new cluster mean is less than its distance to the old cluster mean. If the distance is less, the point is moved to the new cluster, thus altering the new cluster mean and minimizing the within-cluster sums-of-squares (SSW). For subsequent steps, the entity with the largest distance from its cluster mean initializes the new cluster, and all entities are examined for reclassification until the desired number of clusters is created. Note, this is a nonhierarchical process because, as the number of clusters increases, there is no constraint on cluster membership; entities shift around so that the minimum within-cluster variance criterion is achieved.

Regardless of the specific PDNHC procedure used, it is best to have a reasonable guess on how many groups to expect in the data. Adding extra clusters will eventually reveal more noise than information about the data. Percent change in the F-ratio or SSW, depending on the procedure employed, can be used to help determine the number of clusters to retain. For example, SSW usually declines asymptotically as the number of clusters increases. In a plot of SSW against the number of clusters, the point at which SSW first begins to level off signals the number of clusters to retain.

An important constraint in using PDNHC is that these procedures involve various assumptions about the form of the underlying population from which the sample is drawn (Mardia, Kent, and Bibby 1979). These assumptions often include the typical parametric multivariate assumptions, such as equal covariance matrices among clusters. Moreover, most of these techniques are strongly biased toward finding elliptical and spherical clusters. If the data contain clusters of other shapes (e.g., Figs. 3.1b–f), these methods may not find them, and in some cases, will result in a misleading solution (Everitt 1977; Mardia, Kent, and Bibby 1979). As a result of such limitations, these techniques have limited appeal for ecological data and are rarely employed.

3.7 Hierarchical Clustering

Because nonhierarchical procedures reveal nothing about the structure within clusters and the relationships among clusters, they have somewhat limited utility in most ecological studies. Hierarchical procedures, on the other hand, combine similar entities into classes or groups and arrange these groups into a hierarchy, thereby revealing interesting relationships among the entities classified. One advantage of a hierarchy is that a single analysis may be viewed at several levels of detail. For large data sets, however, hierarchies are problematic because a hierarchy with 50 or more entities is difficult to display or interpret. In addition, hierarchical techniques have a general disadvantage compared to NHC techniques since they contain no provision for reallocation of entities that may have been poorly classified at an early stage in the analysis. In other words, there is no possibility of correcting for a poor initial grouping.

Overall, *hierarchical clustering* (HC) complements NHC because HC is ideal for small data sets and NHC for large data sets, and HC helps reveal relationships in the data while NHC does not. Often, NHC can be used initially to summarize a large data set by producing far fewer composite samples, which then makes HC feasible and effective for depicting relationships. Consequently, HC often follows NHC in the analysis of large data sets.

There are many markedly different HC techniques with respect to objectives, procedures, and limitations. Hierarchical techniques can be divided into two general approaches: polythetic agglomerative and polythetic divisive. Of the many clustering approaches, polythetic agglomerative is the most frequently used. In wildlife research, this approach has been used most frequently, although divisive procedures are increasing in popularity. Therefore, we discuss agglomerative approaches in detail (illustrated with an example) and then briefly review polythetic divisive approaches.

3.7.1 *Polythetic Agglomerative Hierarchical Clustering*

Polythetic agglomerative hierarchical clustering (PAHC) techniques use the information contained in all the variables. First, each entity is assigned as an individual cluster. Subsequently, PAHC agglomerates these clusters in a hierarchy of larger and larger clusters until finally a single cluster contains all entities. This family of techniques is often referred to by the acronym SAHN: sequential agglomerative hierarchical and nonoverlapping.

Polythetic agglomerative HC involves two major steps, described in detail below. Briefly, the first step is to compute a resemblance matrix from the original data matrix. The second step is to agglomerate or fuse entities successively to build up a hierarchy of increasingly large clusters. There are numerous different resemblance measures and fusion algorithms; consequently, there exists a profusion of PAHC techniques. In the following sections, we review only the most common alternatives for each step and emphasize those with greatest potential for application in wildlife research. However, this is by no means an exhaustive review of PAHC techniques.

3.7.1.1 Assumptions

Unlike most multivariate statistical procedures, most PAHC procedures involve no explicit assumptions about the form of the underlying population; that is, there is no assumed underlying mathematical model on which the cluster solution is based (Mardia, Kent, and Bibby 1979). Hence, the purpose of PAHC is generally purely descriptive. However, there is at least one noteworthy exception for the procedures discussed in detail below. Certain resemblance measures (e.g., Euclidean distance) assume that the variables are uncorrelated within clusters, although other distance measures (e.g., Mahalanobis distance) account for these intercorrelations (Everitt 1977; Hair, Anderson, and Tatham 1987).

3.7.1.2 Standardizing Data

Before a cluster analysis is performed, it is necessary to consider scaling or transforming the variables, since variables with large variances tend to have more effect on the resulting clusters than those with small variances. With continuous variables (e.g., most environmental data), cluster results are often unsatisfactory when conducted on the raw data since they are badly affected by changing the scale of a variable. Euclidean distance is noteworthy in its susceptibility to problems of this type. In most cases, standardization of the variables to zero mean and unit variance is recommended, using the standard deviations derived from the complete set of entities, although this can have the serious effect of diluting the differences between groups on the variables which are the best discriminators (Everitt 1977). Thus, the choice of whether to standardize or not largely depends on the data set involved, the resemblance measure to be used, and whether or not each variable is to receive equal weight in the cluster analysis.

3.7.1.3 Sample Size Requirements

Unlike most multivariate statistical procedures, PAHC techniques (and cluster analysis in general) have no explicit sample size requirements. Thus, there can be many more variables than entities. Whereas most techniques require large sample sizes, PAHC actually performs poorly when the number of sampling entities exceeds about 50 because interpretation of the hierarchical relationships then becomes too complex.

3.7.1.4 Resemblance Matrix

The first step in PAHC is to compute a resemblance matrix from the original data matrix using one of many alternative resemblance measures. The resemblance matrix contains a resemblance coefficient (i.e., coefficient of similarity, or its complement, dissimilarity) for every pair of entities. The resemblance between any two entities is determined by their similarity in scores on the measured variables. The result is an entities-by-entities resemblance matrix in which resemblance coefficients are arranged in a square, symmetric matrix with diagonal elements for self-comparisons. Usually, however, the resemblance matrix is converted to a dissimilarity matrix by taking the complement of each resemblance coefficient.

Consider the simple example given in the introduction to this chapter (see Sec. 3.2). The original data matrix (primary data matrix) containing six sites (sampling entities) and four species (variables) can be converted into a dissimilarity matrix (secondary data matrix) by computing the dissimilarity in species abundances between every pair of sites using simple Euclidean distance (defined below), as follows:

Original Data Matrix (6 × 4)					Dissimilarity Matrix (6 × 6)						
	Species					Sites					
Sites	A	B	C	D	Sites	1	2	3	4	5	6
1	1	9	12	1	1	0					
2	1	8	11	1	2	1.4	0				
3	1	6	10	10	3	9.7	9.3	0			
4	10	0	9	10	4	15.9	15.2	10.9	0		
5	10	2	8	10	5	15.1	14.4	10.0	2.2	0	
6	10	0	7	2	6	13.7	12.7	13.8	8.2	8.3	0

3.7 Hierarchical Clustering

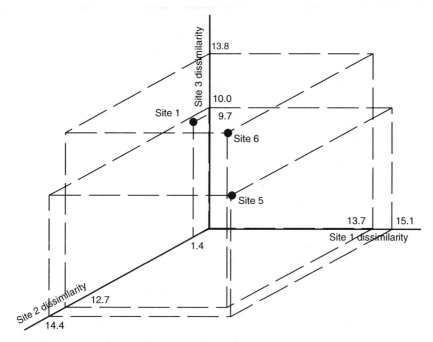

FIGURE 3.4 Graphical depiction of a three-dimensional dissimilarity space. The axes are defined by dissimilarities to sites 1, 2, and 3. The locations of three sites (shown) are given by their dissimilarity to these sites as given in the dissimilarity matrix.

The resemblance matrix may be thought of geometrically as well. Resemblance space uses either similarities (similarity space) or dissimilarities (dissimilarity space) among entities as axes, and it uses entities as points. In other words, axis 1 is the dissimilarity to sample 1, axis 2 is the dissimilarity to sample 2, and so on. Each entity has a unique location in this space based on its dissimilarity with each other entity. Consequently, resemblance space has the same number of axes as entities, in contrast to the original data space which has the same number of axes as variables.

Consider the example above. Figure 3.4 depicts a three-dimensional dissimilarity space, where the axes are defined by dissimilarities to sites 1, 2, and 3. Each site is located at the coordinates given by its dissimilarity values in the matrix above. For example, as shown, site 5 is located at coordinates 15.1 (site 1 axis), 14.4 (site 2 axis), and 10.0 (site 3 axis). The relative proximity of sites in this space indicates the degree of similarity or dissimilarity in their species abundances. Hence, on the basis of their relative proximity, it is apparent that sites 5 and 6 have greater similarity in species abundance patterns than sites 5 and 1.

Unfortunately, conversion of the original data matrix into a resemblance matrix (and the corresponding geometric conversion) is not without limitations. Recall that the original data space is the geometric equivalent of the original data matrix, containing the full information content of the variables in a samples-by-variables data matrix. Geometrically, therefore, each sampling entity occupies a location in a P-dimensional space, where each axis is defined by one of the P original variables. Hence, the relationships among sampling entities with respect to the variables is explicit; the coordinates of the sample entity give the values on each of the variables. In contrast, resemblance space contains information on relationships among sampling entities but no information on the variables. In Figure 3.4, for example, the species abundance at each site is not explicit; therefore, there is no way to determine exactly why sites 5 and 6 are more similar than sites 5 and 1. Hence, the information content of resemblance space is impoverished compared with the original data space.

There exists a wide variety of resemblance measures, most of which, however, are not commonly employed in ecological research. Those commonly used can be divided roughly into three groups: (1) distance coefficients, (2) association coefficients, and (3) correlation coefficients. The choice of resemblance measure largely depends on the type of data and the objectives of the investigator. The choice of a coefficient will frequently be guided by the type of data. When the measurement scale is such that several possible coefficients could be used, the choice is often guided by the worker's personal preference and the ease with which the resemblance measure is conceptualized. Our recommendation is to use the simplest type of coefficient for ease of interpretation. It is generally advantageous to try several different measures and weigh the results using ecological criteria (i.e., choose the measure that produces the most meaningful and interpretable results).

1. Distance coefficients. Distance coefficients measure distances between entities in various ways. Distance coefficients are measures of dissimilarity between entities and are the complement of similarity coefficients. Distance measures have inherently the greatest appeal to ecologists since they are often the easiest to visualize.

There are many different distance coefficients (Euclidean distance, Manhattan or city-block distance, Mahalanobis distance, Canberra distance, Bray–Curtis dissimilarity, and percentage dissimilarity, to name several). Many of the distance measures were developed specifically for species abundance data (i.e., meristic or count data, such as samples-by-species data) and therefore are not universally appropriate for continuous data, although most are rather generic and can be applied to either continuous, categorical, or count data. The differences among these distance measures are in the features of the data they emphasize. For example, with samples-by-species data, some measures emphasize the dominant species in the samples, while others give minor and major species the same emphasis (Gauch 1982). We briefly describe two of the most common distance coefficients here.

The most familiar and commonly used distance measure is simple *Euclidean distance* (*ED*). The *ED* between two entities (j and k) based on P variables is defined as

$$ED_{jk} = \sqrt{\sum_{i=1}^{P}(x_{ij}-x_{ik})^2}$$

where x_{ij} is the value of ith variable for jth entity and x_{ik} is the value of ith variable for the kth entity.

Another common distance measure is *Mahalanobis distance* (D^2). The Mahalanobis distance between two entities (j and k) based on P variables is defined as

$$D^2 = (x_j - x_k)' \Sigma^{-1} (x_j - x_k)$$

where Σ is the pooled within-groups variance-covariance matrix, X_j is the vector of scores for the ith entity, and X_k is the vector of scores for the jth entity.

In practice, the data are almost always standardized prior to calculating *ED* to eliminate spurious effects due to unequal variances of the variables (see below). In addition, *ED* is frequently used as a distance measure and is recommended for certain fusion strategies (average-, centroid-, and median linkage, and Ward's minimum variance; see below). Euclidean distance is rather generic and can be applied to continuous, categorical, and count data. A disadvantage of *ED* is that it assumes that variables are uncorrelated within clusters (a situation rarely encountered in wildlife research). For this reason, Mahalanobis distance has an advantage over *ED* because it adjusts for these intercorrelations (Everitt 1977; Hair, Anderson, and Tatham 1987). When the correlations among the variables are zero, D^2 is equal to *ED* measured using standardized variables. Despite this limitation, however, *ED* (or ED^2) is used most frequently in wildlife research, probably due to its ease of interpretation.

Percentage dissimilarity (*PD*) is the most commonly employed distance measure for species abundance data (i.e., meristic data) and is used frequently in community ecology studies (Gauch 1982). The calculation of *PD* begins with the computation of similarity. This similarity is then subtracted from the similarity among replicate samples to convert to a distance or dissimilarity. The value of one is almost always used as an estimate of replicate similarity (i.e., replicate samples are identical). Percentage dissimilarity between two entities (j and k) based on P variables is defined as

$$PD_{jk} = IA - \frac{2 \sum_{i=1}^{P} \min(x_{ij}, x_{ik})}{\sum_{i=1}^{P} \min(x_{ij} + x_{ik})}$$

where *IA* is the internal association (usually valued at one), x_{ij} is the value of ith variable for the jth entity, and x_{ik} is the value of ith variable for the kth entity.

2. *Association coefficients.* The greatest number of resemblance measures have been developed for categorical data (dichotomous and polytomous variables). Association coefficients are measures of the agreement between the two data rows representing the two entities, and sometimes they may be special cases of distance or correlation coefficients (Sneath and Sokal 1973; Everitt 1977). Association coefficients can also be applied to ranked, continuous, and count data. However,

this requires sacrificing information (e.g., recoding multistate count data into binary (0, 1) data).

The largest number of association coefficients have been developed for binary data (i.e., dichotomous variables) where the variates are of the presence and absence type and may be arranged in the following two-way association table in which the presence of a variable is denoted by + and its absence by −.

		Entity j		
		+	−	Total
Entity k	+	a	b	$a+b$
	−	c	d	$c+d$
	Total	$a+c$	$b+d$	n

Different association coefficients emphasize different aspects of the relation between sets of binary values and take on values in the range zero to one (Everitt 1977). There have been many proposed association coefficients, mainly because uncertainty exists over how to incorporate negative matches (d in above table) into the coefficients, and also whether or not matched pairs of variables (a and d in above table) are equally weighted or carry twice the weight of unmatched pairs (b and c in above table), or vice versa. Some coefficients exclude negative matches entirely, other coefficients may give higher weighting to matched pairs than to unmatched pairs.

Overall, these measures are infrequently used in wildlife research because we are typically interested in preserving the quantitative relationships among the entities themselves. Nevertheless, there are situations for which association coefficients can be effectively employed to address particular wildlife research questions. For example, we may be interested in clustering sites on the basis of presence/absence of a number of animal or plant species. We briefly describe three of the most commonly used coefficients here.

Complemented simple matching coefficient (*CSMC*) gives equal weight to positive and negative matches and is equal to the squared Euclidean distance based on unstandardized character states, which can take the value of zero or one. The *CSMC* between two entities (j and k) based on P variables, where a, d, n are defined in the table above, is defined as

$$CSMC_{jk} = IA - \frac{a+d}{n}$$

Complemented coefficient of Jaccard (*CCJ*) omits consideration of negative matches. The *CCJ* between two entities (j and k) based on P variables, where a, b, and c are defined in the table above, is defined as

$$CCJ_{jk} = IA - \frac{a}{a+b+c}$$

Complemented coefficient of community (CCC) is monotonic with CCJ; CCC also omits consideration of negative matches, yet gives more weight to matches than to mismatches. The CCC between two entities (j and k) based on P variables is defined as

$$CCC_{jk} = IA - \frac{2a}{2a + b + c}$$

3. Correlation coefficients. For quantitative variables, the most commonly used measure of similarity between entities is the Pearson product-moment correlation coefficient (r). The correlation between two entities (j and k) based on P variables is defined as

$$r_{jk} = \frac{\sum_{i=1}^{P}(x_{ij} - \bar{x}_j)(x_{ik} - \bar{x}_k)}{\sqrt{\sum_{i=1}^{P}(x_{ij} - \bar{x}_j)^2 \sum_{i=1}^{P}(x_{ik} - \bar{x}_k)^2}}$$

where x_{ij} is the value of ith variable for the jth entity, x_{ik} is the value of ith variable for the kth entity, $\bar{x}_j$ is the mean of all variables for the jth entity, and $\bar{x}_k$ is the mean of all variables for kth entity.

The correlation coefficient has been heavily criticized (Eades 1965; Fleiss and Zubin 1969). The measure generally has been suggested for use when the similarity of the average *profile shapes* are considered more important than the similarity of the average *profile levels*, because the correlation is unity whenever two profiles are parallel, irrespective of how far apart they are in multidimensional entity space (Everitt 1977). Unfortunately, the converse is not true, because two profiles may have a correlation of one even if they are not parallel. All that is required for perfect correlation is that one set of scores be linearly related to a second set. This troublesome feature has resulted in mixed opinions over the merits of using correlation coefficients in cluster analysis.

3.7.1.5 Agglomeration Schedule and Dendrogram

Once a resemblance matrix is computed based on the chosen metric, the next step in PAHC is to agglomerate, or fuse, entities successively to build up a hierarchy of increasingly large clusters. A key result of this fusion process is an *agglomeration schedule* (table) showing the agglomeration sequence and the corresponding dissimilarity values at which entities and clusters combine to form new clusters. In an agglomeration table, dissimilarity or distance values will vary depending on the resemblance measure and fusion strategy used.

Consider the following dissimilarity matrix computed using Euclidean distance for the simple example given in the introduction to this chapter (see Sec. 3.2) involving six sites (sampling entities) and four species (variables):

Step 1:

			Sites			
Sites	1	2	3	4	5	6
1	0					
2	1.4	0				
3	9.7	9.3	0			
4	15.9	15.2	10.9	0		
5	15.1	14.4	10.0	2.2	0	
6	13.7	12.7	13.8	8.2	8.3	0

There exist many different strategies for agglomerating entities; some of the more common strategies are discussed in detail below. All fusion strategies, however, cluster the two most similar (or least dissimilar) entities first. In this case, therefore, the first fusion is between the sites with the shortest Euclidean distance (sites 1 and 2).

Strategies differ with respect to how they fuse subsequent entities (or clusters). The differences arise because in order to determine the second fusion (and all subsequent fusions), we must determine the distance between each entity and the newly formed cluster. For the time being, let us assume that this distance is based on the distance between each entity and its nearest neighbor in the newly formed cluster. Thus, the distance between site 3 and cluster 1–2 is set at 9.3, since this is the shortest distance between site 3 and either site 1 or 2. The nearest neighbor distance to cluster 1–2 is similarly determined for the other sites and the following new dissimilarity matrix is formed:

Step 2:

			Sites			
Sites	1–2	3	4	5	6	
1–2	0					
3	9.3	0				
4	15.2	10.9	0			
5	14.4	10.0	2.2	0		
6	12.7	13.8	8.2	8.3	0	

Based on this new dissimilarity matrix, the second fusion is between sites 4 and 5. The distance between each of the remaining individual sites (3 and 6) and the two clusters is determined as before using the nearest neighbor distance. Similarly, the distance between the two clusters is based on the shortest distance between any members of the respective clusters. In this case, the distance between clusters is set to 14.4 based on the distance between sites 2 and 5. A new dissimilarity matrix is formed and the process is repeated until all entities are fused into a single cluster, as follows:

Step 3:

Sites	1–2	3	4–5	6
1–2	0			
3	9.3	0		
4–5	14.4	10.0	0	
6	12.7	13.8	8.2	0

Step 4:

Sites	1–2	3	4–5–6
1–2	0		
3	9.3	0	
4–5–6	12.7	10.0	0

Step 5:

Sites	1–2–3	4–5–6
1–2–3	0	
4–5–6	10.0	0

Step 5:

This entire agglomeration sequence can be summarized in a table, as follows:

Number of Clusters	Fusion	Minimum Distance
5	Sites 1 and 2	1.4
4	Sites 4 and 5	2.2
3	Site 6 and Cluster 4–5	8.2
2	Site 3 and Cluster 1–2	9.3
1	Cluster 1–3 and Cluster 4–6	10.0

Agglomeration tables are generally not as effective as graphical displays for elucidating relationships. Therefore, the agglomeration table is usually portrayed as a *dendrogram*. A dendrogram is a tree-like plot (vertical or horizontal) depicting the agglomeration sequence in which entities are enumerated (identified) along one axis and the dissimilarity level at which each fusion of clusters occurs on the other axis. Dendrograms are the most frequently presented result of hierarchical clustering since they display relationships effectively (although see Sneath and Sokal 1973 and Digby and Kempton 1987 for other types of graphical displays).

The dendrogram can be very effective in portraying the structure of the data and indicating the degree to which natural clusters exist in the data. The length of the branches being fused together indicates the degree of dissimilarity between members of the newly formed cluster. Consequently, tight clusters (i.e., very similar entities), or dense regions of the data cloud, will show up in the dendrogram as entities connected by relatively short branches. If the data structure contains several disjunct clusters, or dense regions of the data cloud separated by sparse regions, these will show up as separate tight clusters connected by relatively long branches. In addition, potential outliers will show up as individual entities that fuse very late in the agglomeration sequence and therefore have very long branches. Thus, the existence and number of natural clusters, compactness of the clusters, relationship among clusters, and presence of potential outliers all can be revealed quite nicely in the dendrogram.

Figure 3.5 depicts a horizontal dendrogram for the agglomeration sequence in the example above. Dissimilarity or distance is plotted on the x-axis (horizontal) and sites are depicted on the y-axis (vertical). The root of the tree (i.e., all sites in one cluster) is located on the left; branches extend to the right. The agglomeration sequence is depicted from right to left in the dendrogram; that is, each site is in a separate cluster on the far right of the dendrogram (zero dissimilarity within cluster) and as you move left, sites and clusters are sequentially fused together at increasingly greater dissimilarities. In this case, sites 1 and 2 are joined by branches that extend 1.4 units from the origin of the dissimilarity axis. Next, sites 4 and 5 are joined by branches that extend 2.2 units from the origin. Next, the branch containing site 6 extends 8.2 units from the origin and connects to the site 4–5 cluster. And so on for the remaining two fusions. Note that the dissimilarity axis will vary with different fusion strategies and dissimilarity measures.

Figure 3.5 thus depicts two distinct clusters. Sites 1 and 2 are connected by very short branches, indicating that they are very similar in species abundances. Sites 4 and 5 are similarly connected by very short branches. These two clusters, however, are connected by very long branches, indicating that they are quite different in species abundances. We can conclude that these two site clusters support very different bird communities—probably as a result of underlying environmental differences—a fact that is readily apparent in the original data matrix. In addition, site 3 is connected to the site 1–2 cluster by a relatively long branch, yet not as long as the branch that connects it to the other sites, indicating that site 3 is more like sites 1 and 2 in species abundances than sites 4, 5, and 6, yet significantly different in some respects. Indeed, upon inspection of the original data matrix, it is clear that site 3 is very like sites 1 and 2 in the abundances of species A, B, and C, yet differs dramatically in the abundance of species D.

3.7.1.6 Fusion Strategies

As noted earlier, there are many different fusion strategies that differ in a number of properties. By choosing various strategies the investigator can exert considerable control over the resulting hierarchical cluster. The choice of a particular fusion strategy will usually depend on the objectives of the investigator. In this section, we will

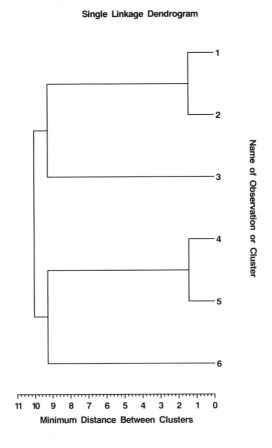

FIGURE 3.5 Horizontal dendrogram for the agglomeration sequence involving six sites (sampling entities) and four species (variables).

summarize the six most common fusion strategies roughly in order of increasing complexity. However, before doing this it is necessary to discuss one property that has a great deal of utility for comparison among strategies.

Fusion strategies may be *space-conserving* or *space-distorting* (Fig. 3.6). A space-conserving strategy preserves the multidimensional structure defined by the relationships among sampling entities in the original data matrix. Thus, the distances between entities in dissimilarity space are roughly preserved during the fusion process so that there is a high correlation between the input dissimilarities (in the original dissimilarity matrix) and the output dissimilarities defined by the lowest dissimilarity required to join any given entity pair during the fusion process. Con-

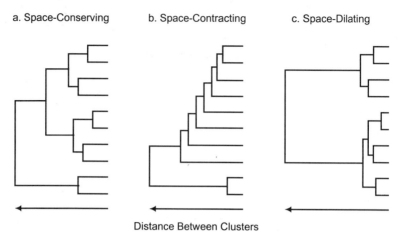

FIGURE 3.6 Hypothetical dendrograms illustrating the differences between space-conserving and space-distorting clustering strategies.

ceptually, this can be thought of as preserving the relative position of data points in the data cloud as the clusters are established. In general, space-conserving strategies are desirable when the objective is to portray the true structure of the data, as is typically the case in most ecological applications.

Alternatively, with certain fusion strategies, the model will behave as though the space in the immediate vicinity of a group has been contracted or dilated; these are described as space-distorting strategies. As a result, space-distorting strategies may not faithfully represent the spatial relationships in the input data. In a *space-contracting* system, groups appear on formation to move nearer to some or all of the remaining entities; the chance that an individual entity will add to a preexisting group rather than act as the nucleus of a new group is increased, and the system is said to "chain." This effect is usually apparent in the dendrogram as a stair-step sequence of fusions (Fig. 3.6b). Space-contracting strategies are usually undesirable, since discrete clusters are difficult to recognize. However, they can be useful in identifying major discontinuities in the data. In a *space-dilating* system, groups appear to recede on formation and growth; individual entities not yet in groups are more likely to form nuclei of new groups. Again, the effect is usually apparent in the dendrogram as tight clusters connected by relatively long branches (Fig. 3.6c). Space-dilating systems are desirable when the objective is to identify or delineate distinct clusters, perhaps for further analysis or for ease in interpretation.

1. Single linkage (nearest neighbor). Single linkage designates distance values between entities and clusters to be that of the two closest entities (Sneath and Sokal 1973), as in the example used in the previous section. An entity's dissimilarity to a cluster is defined to be equal to its dissimilarity to the closest entity in that cluster. Similarly, when two clusters agglomerate, their dissimilarity is equal to the smallest dissimilarity for any pair of entities with one entity in each cluster. Thus, fusion is based entirely on how close each entity (or cluster) is to its nearest neighbor.

Single linkage tends to produce straggly clusters, which quickly agglomerate very dissimilar entities. Consequently, clusters may contain entities with very little similarity. Because it causes entities to agglomerate rapidly, single-linkage is a space-contracting strategy and has a strong tendency to produce a chainlike dendrogram in which cluster separation is difficult (Fig. 3.6b). Single-linkage is not recommended for most wildlife research purposes, but it is mentioned here because of its relationship to complete and average linkage and because it is frequently used in taxonomic research. In addition, because it is a space-contracting strategy, only the most discrete variation among entities (i.e., disjunct clusters) will be revealed. Thus, if the purpose is to explore whether the data structure exhibits a discrete versus continuous structure, then single linkage can be useful for comparison with a space-conserving strategy.

2. Complete linkage (furthest neighbor). Complete linkage designates distance values between entities and clusters to be that of the two most widely separated entities (Sneath and Sokal 1973). An entity's dissimilarity to a cluster is defined to be equal to its dissimilarity to the furthest entity in that cluster. Similarly, when two clusters agglomerate, their dissimilarity is equal to the greatest dissimilarity for any pair of entities, with one entity in each cluster. Thus, fusion is based entirely on how close each entity (or cluster) is to its furthest neighbor. The objective of this strategy is to minimize the dissimilarity between neighbors.

Complete linkage fusion produces clusters of very similar entities that agglomerate slowly. As clusters agglomerate, groups are moved away from each other; that is, they are dilated in space. Consequently, complete linkage is a space-dilating strategy. Strategies such as this produce distinct clusters of homogeneous entities which are easily recognized (Fig. 3.6c). Hence, this strategy can be very useful when the purpose is to identify or delineate distinct clusters, perhaps for further analysis or for ease in interpretation. However, if the purpose is to reveal the true underlying structure of the data, then this strategy can produce misleading results.

3. Centroid linkage (unweighted pair-group centroid). Centroid linkage minimizes the dissimilarity between cluster centroids and is valid only when metric data are used (Sneath and Sokal 1973). An entity's dissimilarity to a cluster is defined to be equal to its dissimilarity to the cluster centroid. When two clusters agglomerate, their dissimilarity is equal to the dissimilarity between cluster centroids. Centroid linkage is a space-conserving strategy, which often aids interpretation (Fig. 3.6a). However, under certain conditions (although rare), "reversals" can occur in which a fusion takes place at a lower dissimilarity than a prior fusion (see Fig. 3.8d for an example of a reversal). Also, because the calculated centroid of two fused clusters is weighted toward the larger group, group-size distortions occur. This lack of monotonicity and group-size dependency render this strategy less useful than other space-conserving strategies.

4. Median linkage (weighted pair-group centroid). Median linkage is similar to centroid-linkage except that the centroids of newly fused groups are positioned at the median of old group centroids (Sneath and Sokal 1973). Consequently, this strategy essentially eliminates the group-size dependency problems while retaining the appeal of a space-conserving approach. Nonetheless, the potential for reversals persists.

5. *Average linkage (unweighted pair-group average).* Average linkage designates distance values between groups to be the average dissimilarity between clusters (Sneath and Sokal 1973). There are several ways to compute an average; consequently, there are several somewhat different average-linkage techniques, but the most common is the unweighted pair-group method using arithmetic means (referred to by the popular acronym UPGMA), which uses the simple, unweighted, arithmetic mean. An entity's dissimilarity to a cluster is defined to be equal to the mean of the distances between the entity and each point in the cluster. When two clusters agglomerate, their dissimilarity is equal to the mean of the distances between each entity in one cluster with each entity in the other cluster. Fusion is therefore dictated by the entities (or clusters) that have the lowest mean distance value. Average linkage is a space-conserving strategy that guarantees monotonicity and eliminates group-size dependency and reversals. This is the most frequently used technique and is recommended for PAHC when there is no specific reason for choosing some other fusion strategy. This algorithm is unique in that it appears to maximize the correlation between input dissimilarities (in the entities-by-entities dissimilarity matrix) and the output dissimilarities implied by the resulting dendrogram (using the lowest level required to join any given entity pair in the dendrogram; Gauch 1982). This is termed the *cophenetic correlation*, and its maximization implies that the patterns revealed in the cluster solution (e.g., in the dendrogram) faithfully represent the structure of the original data set (see Sec. 3.7.1.9). In other words, this strategy more or less ensures space-conservation during the fusion process.

6. *Ward's minimum-variance linkage (minimization of within-group dispersion).* Ward's minimum-variance linkage agglomerates clusters, provided that the increase in within-group dispersion (variance) is less than it would be if either of the two clusters were joined with any other cluster (Sneath and Sokal 1973). Minimum-variance fusion is similar to average-linkage fusion, except that instead of minimizing an average distance, it minimizes a squared distance weighted by cluster size. Penalty by squared distance makes minimum-variance clusters tighter than average-linkage clusters and therefore tends to be a relatively space-dilating strategy. Hence, this strategy can be very effective in displaying relationships among clusters. This strategy is conceptually very similar to K-means clustering (PDNHC). However, in contrast to K-means clustering, minimum-variance clustering operates hierarchically, and therefore does not guarantee minimum within-group variance for any given number of clusters. Furthermore, Ward's minimum-variance linkage method joins clusters to maximize the likelihood at each level of the hierarchy under the assumptions of a multivariate normal mixture, equal spherical covariance matrices, and equal sampling probabilities (Ward 1963). Thus, given the failure of most ecological data sets to meet these assumptions, this strategy has limited utility for most ecological investigations and is perhaps best used as a complement to average-linkage when the cluster structure is not readily apparent.

Empirical Example. In the example, the data were standardized ($\mu = 0$, $SD = 1$) and species were clustered based on ED (or ED^2, where appropriate) using the six different fusion methods. In each case, the agglomeration schedule provided a

3.7 Hierarchical Clustering

Average Linkage Cluster Analysis

The data have been standardized to mean 0 and variance 1.
Root-Mean-Square Total-Sample Standard Deviation = 1
Root-Mean-Square Distance Between Observations = 6.63325

NCL	Clusters	Joined	FREQ	SPRSQ	RSQ	ERSQ	CCC	PSF	PST2	Norm RMS Dist	Tie
18	WEFL	WIWR	2	0.00285	0.997	.	.	20.6	.	0.2266	
17	CBCH	DEJU	2	0.00419	0.993	.	.	17.6	.	0.2745	
16	BRCR	GCKI	2	0.00490	0.988	.	.	16.5	.	0.2971	
15	STJA	CL18	3	0.00808	0.980	.	.	14.0	2.8	0.3492	
14	CL15	SWTH	4	0.01303	0.967	.	.	11.3	2.4	0.4349	
13	RUHU	CL14	5	0.01413	0.953	.	.	10.1	1.8	0.4613	
12	CL16	CL17	4	0.01963	0.933	.	.	8.9	4.3	0.4664	
11	CL13	WIWA	6	0.01660	0.917	.	.	8.8	1.7	0.4978	
10	CL12	HAWO	5	0.02408	0.893	.	.	8.3	2.5	0.5792	
9	EVGR	HAFL	2	0.02002	0.873	.	.	8.6	.	0.6002	
8	AMRO	CL11	7	0.02858	0.844	.	.	8.5	2.6	0.6182	
7	CL10	VATH	6	0.03613	0.808	.	.	8.4	2.7	0.6966	
6	CL8	SOSP	8	0.04194	0.766	.	.	8.5	3.0	0.7338	
5	CL7	CL9	8	0.07066	0.695	.	.	8.0	3.9	0.8046	
4	CL5	GRJA	9	0.08088	0.614	.	.	8.0	3.2	1.0104	
3	CL4	HEWA	10	0.08506	0.529	0.605	-1.39	9.0	2.6	1.0541	
2	CL6	BHGR	9	0.09592	0.433	0.465	-0.44	13.0	5.4	1.0545	
1	CL2	CL3	19	0.43333	0.000	0.000	0.00	.	13.0	1.1642	

FIGURE 3.7 Agglomeration schedule resulting from average-linkage fusion using squared Euclidean distance and standardized data.

summary of the fusion process. Figure 3.7 depicts the agglomeration schedule for the average linkage fusion based on squared *ED*. The rows of the table represent the agglomeration sequence, beginning with the first fusion and continuing sequentially until all species are contained in a single cluster. The first field represents the fusion step and is equal to the number of clusters (NCL). In this case, because the data set includes 19 species, the first step (or fusion) results in 18 clusters. The second and third fields represent the species or clusters being fused. The fourth field represents the number of species in the cluster resulting from that fusion. Fields 5 through 10 will be discussed later. The last field represents the dissimilarity (in this case, normalized root mean square distance or average squared *ED*) between the two species or clusters being fused. (Note: This field varies depending on the fusion method used.)

The agglomeration table in Figure 3.7 indicates that WEFL and WIWR (first row) fused first at an average squared Euclidean distance of 0.2266 (last column entry). Next, CBCH and DEJU fused at an average distance of 0.2745. And so on until two large clusters (CL2 and CL3) fused at an average distance of 1.1642 into one overall cluster containing all 19 species (last row in agglomeration table).

For comparative purposes, the result of each fusion strategy is presented in the form of a horizontal dendrogram (Fig. 3.8a–f). The average linkage dendrogram (Fig. 3.8e) reveals two large clusters of species: (1) AMRO-BHGR, and (2) BRCR-HEWA. These two clusters fused at an average distance of 1.1642 accord-

110   3. Cluster Analysis

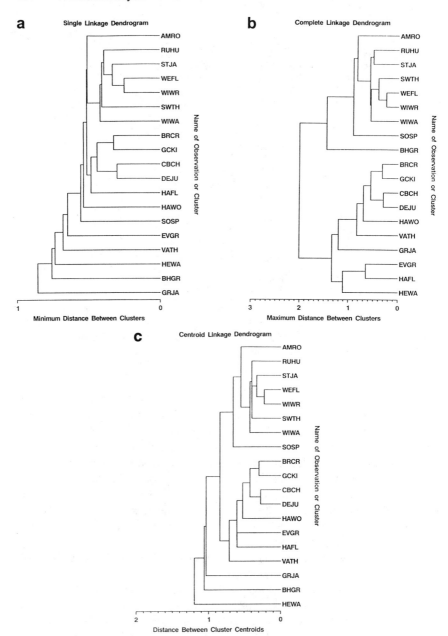

FIGURE 3.8 Dendrograms showing the different results that result from using various fusion strategies. All dendrograms were calculated using standardized data.

ing to the agglomeration schedule (last entry in last row in Figure 3.7). Within the first cluster, BHGR appears somewhat distinct from the rest of the species; it fused with the remaining species at an average distance of 1.0545. Within the second

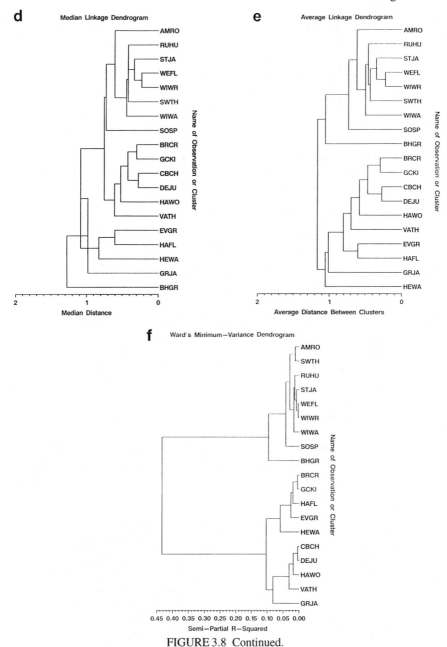

FIGURE 3.8 Continued.

cluster, HEWA and GRJA appear somewhat distinct from the rest of the species; they fused at average distances of 1.0104 and 1.0514, respectively. As expected, the results from centroid- and median linkages were consistent with those from

112   3. Cluster Analysis

average linkage, although in both cases there were minor shifts in how the three individual species (BHGR, HEWA, and GRJA) fused with the remaining species. Centroid-linkage, for example, revealed the same two large clusters, but the three individual species fused separately to a single super cluster comprising the two large subclusters. Median linkage revealed the same basic structure, albeit less clearly, although a reversal occurred with the fusion of GRJA. Complete linkage revealed the same structure; not surprisingly, the space was slightly dilated compared to average linkage. Ward's minimum-variance linkage revealed the same structure, but exacerbated the two-cluster structure by dilating the space between the two large clusters (as revealed by the very long branch connecting them). Finally, single linkage produced indistinct, unidentifiable clusters due to the chaining effect.

3.7.1.7 Cluster Membership Tables and Icicle Plots

Another product of any fusion strategy is a *cluster membership table* showing cluster membership in relation to the number of clusters. Cluster membership tables identify which cluster each entity belongs to for any specified number of clusters; they can be useful in forming composite samples or defining group membership once the number of clusters to be retained has been decided.

Cluster membership tables are generally not effective as graphical displays for elucidating relationships. Therefore, the cluster membership table is usually portrayed as an *icicle plot*. An icicle plot is a tree-like plot (vertical or horizontal) depicting cluster membership in relation to the number of clusters in which entities are enumerated (identified) along one axis and the number of clusters (cluster level) along the other axis. Icicle plots resemble dendrograms, except that instead of dissimilarity along one axis the number of clusters is plotted. Icicle plots are helpful in understanding the consequences of retaining varying numbers of clusters. Note that in contrast to dendrograms, the axes remain the same regardless of which fusion method is used.

Empirical Example. In the example, icicle plots were produced using the six different fusion methods. The icicle plot resulting from average linkage fusion is presented in Figure 3.9. Number of clusters is plotted on the x-axis (horizontal) and species on the y-axis (vertical). By drawing a vertical line through the icicle plot at any specified cluster level, the membership of each cluster can be read off the y-axis on the right. For example, a vertical line drawn through the five-cluster solution on the x-axis intersects five horizontal lines. Each horizontal line represents the stem of a separate cluster. By following each stem and all its branches to the right hand side of the tree, the species composition of each cluster can be determined. In this case, the first horizontal line intersected represents the cluster containing all the species listed between SOSP and WIWR. The second horizontal line intersected contains a single species (BHGR). And so on for the other three horizontal lines.

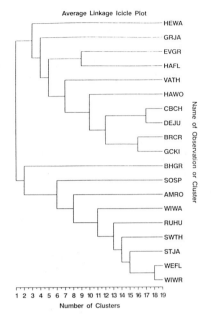

FIGURE 3.9 Icicle plot resulting from average-linkage fusion using squared Euclidean distance and standardized data.

3.7.1.8 Deciding on the Number of Significant Clusters

A problem common to all clustering techniques is the difficulty of deciding on the number of clusters present in the data. This problem can be restated in terms of deciding on whether or not each successive linkage in an agglomerative hierarchical clustering should be accepted. The null hypothesis of interest is that the two entities or clusters that are linked are sufficiently alike to be considered a single cluster (i.e., the linkage should be accepted). Under the alternative hypothesis, the two entities or clusters are distinctly different and so should be retained as separate entities or clusters (i.e., the linkage should be rejected). Once a linkage has been rejected, the linkages at greater levels of dissimilarity, which depend on the rejected linkage, are no longer defined and therefore do not need to be considered.

There are no completely satisfactory methods for determining the number of clusters, although various objective methods and stopping rules have been proposed (e.g., Beale 1969; Harper 1978; Ratliff and Pieper 1981; Strauss 1982; Bock 1985). Many of these objective procedures employ ordinary parametric significance tests, such as analysis of variance F-tests, and therefore are usually not valid for testing differences between clusters because the statistical assumptions are rarely met in ecological data sets.

A very promising resampling procedure involving *bootstrapping* has been developed to test whether or not each successive linkage in PAHC should be accepted (Nemec and Brinkhurst 1988). Briefly, in the present context, the bootstrap procedure involves drawing Q random sampling entities, with replacement, from the original data matrix (or exclusively from the subset of entities corresponding to those contained in the two clusters under consideration), where Q equals the number of entities involved in the two clusters under consideration, and randomly assigning them to the two clusters under consideration. This procedure is repeated a large number of times and the cluster analysis is applied to each bootstrap sample to calculate the fusion level (e.g., the Euclidean distance) for the two clusters under consideration. By removing any "real" group structure from the bootstrap samples, the bootstrap procedure calculates the expected distribution of the fusion level under the assumptions of the null hypothesis (i.e., that there is no real group structure to the entities and that therefore they all belong together in a single cluster). By comparing the observed fusion level to the bootstrap distribution, we can determine directly the probability (P value) of observing a fusion level that is greater than or equal to the observed fusion level. If only a small proportion of the bootstrapped dissimilarity values (e.g., not more than 5 percent for a 5 percent level of significance) is greater than or equal to the observed dissimilarity value, then the dissimilarity between the two clusters in question should be judged to be great enough to reject the null hypothesis. This entire process is repeated sequentially for each stem of the dendrogram until the null hypothesis is rejected. At this point, we conclude that further linkage of clusters is no longer warranted and that the current number of clusters represents the smallest number of clusters that differ significantly from that expected by chance if the data contained no real group structure.

Since multiple tests are performed (up to $N-1$, where N = number of entities), the overall significance level (i.e., experimentwise error rate) will, in general, be less than the significance level of an individual test (i.e., comparisonwise error rate). If there is a large number of entities being clustered and a 5 percent significance level is applied to each linkage, then there is a real possibility that even when the null hypothesis is true some of the tests (approximately 5 percent of the tests) will produce a P value less than 5 percent. One way of dealing with this "multiple comparisons" problem is to use the Bonferroni method, which adjusts the significance level of the individual tests by dividing by the total number of tests or linkages.

SAS clustering programs offer three objective criteria (cubic clustering criterion CCC; pseudo F-statistic PSF; and pseudo t^2-statistic PST2) for evaluating the number of clusters present in the data. The cubic clustering criterion is described elsewhere (Sarle 1983). The pseudo F-statistic measures the separation among all the clusters at each cluster level. The pseudo t^2-statistic measures the separation between the two clusters joined at each step. The pseudo F- and t^2-statistics may be useful indicators of the number of clusters, but they are *not* distributed as F and t^2 random variables and, because of certain restrictive assumptions, usually cannot be used to determine statistical significance.

It may be advisable to look for consensus among the three criteria, that is, local peaks (i.e. relative high values) of the cubic clustering criterion and pseudo

F-statistic and a larger pseudo t^2-statistic for the next cluster fusion. Unfortunately, the cubic clustering criterion is not determined when the number of clusters is greater than one-fifth the number of observations, limiting its usefulness for small data sets. It must be emphasized that these criteria are appropriate only for compact or slightly elongated clusters (in multidimensional space), preferably clusters that are roughly multivariate normal. Since these conditions are rarely met in most data sets, these criteria should only serve as aids and not objective decision rules.

In practice, graphical displays are often more useful than any of these objective measures. For example, a scree plot can be derived by plotting the fusion level (dissimilarity values) against the number of clusters (e.g., Fig. 3.10), and the shape of

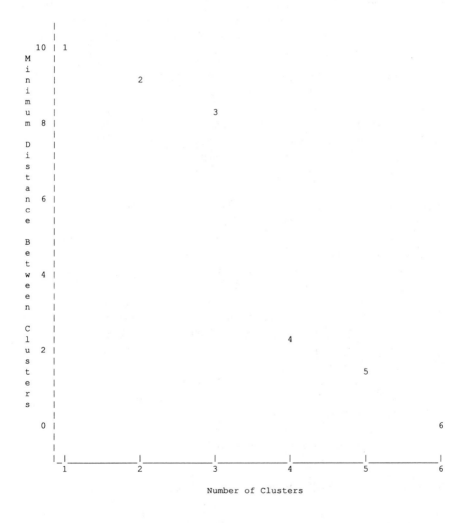

FIGURE 3.10 Scree plot of a simplified example involving six sites and four species.

the resulting curve used to evaluate the appropriate number of clusters to retain. Recall that dissimilarity values decrease sequentially as the number of clusters increases from one (when all entities are in a single large cluster) to N (when each entity is separate). Typically, the scree plot slopes steeply down initially and then gradually approaches zero when the number of clusters equals N. The point at which the curve first begins to straighten out is considered to indicate the maximum number of clusters to retain.

In ecological studies perhaps the most useful tactic for determining the number of clusters is to examine the dendrogram for large changes between fusions and to evaluate the change in interpretability of cluster compositions in relation to varying numbers of clusters. Icicle plots can be useful in this respect as well. Alternatively, ordination procedures can be used to reduce the dimensionality of the data set to a few variables (e.g., principal components), which can then be plotted and visually inspected to determine the number of natural clusters, if any, present in the data set (see Section 3.9).

Empirical Example. In the example, at least two somewhat distinct clusters (AMRO-BHGR and BRCR-HEWA) are evident in the dendrograms resulting from all fusion methods except single linkage (Fig. 3.8a–f). Ward's minimum-variance linkage, for example, clearly displays this two-cluster solution, but suggests that the latter cluster may comprise two distinct subclusters (Fig. 3.8f). The two-cluster solution is supported by the objective measures as well—a distinct peak in the pseudo F-statistic at the two-cluster level and a peak in the pseudo t^2-statistic at the one-cluster level (Fig. 3.7). Within the two larger clusters, there appears to be a core set of members (AMRO-SOSP and BRCR-HAFL, respectively) that remain together regardless of the fusion strategy. These clusters can be considered cohesive; therefore, they probably represent groups of species with real similarities in habitat with respect to vegetation structure. Cluster membership for a few species, however, appears to be less consistent. Average-, centroid-, and to a lesser extent, median linkage indicate that the two large initial clusters contain one or more single-species clusters (BHGR, HEWA, and GRJA) that differ markedly from the remaining cluster members. The fact that these three species cluster in markedly different ways depending on the fusion strategy suggests that they do not belong in any particular cluster (i.e., cluster membership is an artifact of the fusion strategy).

Based on the combined evidence, we conclude that at least two distinct clusters of species exist and that they dominate the data structure. However, the existence of additional real clusters is more problematic and subject to the interpretation of the analyst. In this case, based on the evidence presented and knowledge of each bird species, it is our opinion that five meaningful clusters are present and best describe the patterns in this data set.

3.7.1.9 Evaluating the Cluster Solution: Cophenetic Correlation

We are usually interested in knowing how well the final number of clusters portrays the original data structure. Many different ways have been suggested for evaluating the performance of the final cluster solution and comparing the performance of cluster solutions obtained by different techniques (e.g., Sneath and Sokal 1973; Fowlkes and Mallows 1983; Digby and Kempton 1987; and Nemec and Brinkhurst 1988b). The *cophenetic correlation* (CC) is perhaps the most frequently employed criteria. Cophenetic correlation is the correlation between the input dissimilarities (in the entities-by-entities dissimilarity matrix) and the output dissimilarities implied by the resulting dendrogram (using the lowest level required to join any given entity pair in the dendrogram). Cophenetic correlation is essentially a measure of how faithfully the dendrogram (i.e., the results of the cluster analysis) represents the information in the original data, and thus provides a way of objectively comparing different clustering strategies. Generally, values over 0.75 are considered good.

Consider the earlier example involving six sites and four species and the following input dissimilarity matrix computed on the basis of Euclidean distance and the corresponding output dissimilarity matrix generated from the single linkage fusion process as presented step-by-step above:

Input Dissimilarities							Output Dissimilarities					
	Sites							Sites				
Sites	1	2	3	4	5	6	1	2	3	4	5	6
1	0						0					
2	1.4	0					1.4	0				
3	9.7	9.3	0				9.3	9.3	0			
4	15.9	15.2	10.9	0			10.0	10.0	10.0	0		
5	15.1	14.4	10.0	2.2	0		10.0	10.0	10.0	2.2	0	
6	13.7	12.7	13.8	8.2	8.3	0	10.0	10.0	10.0	8.2	8.2	0

Considering each of the 15 unique pairwise combinations of sites as independent observations, the Pearson's product-moment correlation between input and output dissimilarities is equal to 0.89. Consequently, even single linkage fusion, which is a space-distorting strategy, captures the structure of the original data fairly well, indicating that the data structure is very pronounced.

Unfortunately, SAS does not compute the cophenetic correlation. Instead, SAS computes the squared multiple correlation (RSQ is the SAS acronym for R^2), which is the proportion of variation among entities in the original data accounted for by the clusters for each step in the agglomeration sequence. As such, the squared multiple correlation can be used in place of the cophenetic correlation to evaluate how well a specified number of clusters explains the variation among entities, and to compare how well alternative fusion strategies account for the variation. However, care must be exercised in using either this measure or the cophenetic correlation as the basis for choosing among alternative fusion strategies. Both measures vary with the number of clusters. For example, the squared multiple correlation varies between zero (when all entities are contained in a single cluster) and one (when each entity is in a separate cluster). Furthermore, different fusion strategies will often lead to the decision to retain and interpret different numbers of clusters. Therefore, it may be more appropriate and meaningful to compare the squared multiple correlation associated with a specified number of retained clusters for each fusion method. In this manner, R^2 can be used to compare the relative effectiveness of each fusion method.

Empirical Example. In the example, the squared multiple correlation for five clusters was fairly consistent ($R^2 = 0.695-0.715$) among complete linkage, average linkage (Fig. 3.7), and Ward's minimum-variance linkage methods; single linkage was considerably lower ($R^2 = 0.415$). Based on this measure alone, we conclude that single linkage fusion is probably an inadequate strategy for this data set; other criteria support this conclusion. Complete linkage, average linkage, and Ward's minimum-variance linkage methods perform equally with respect to the amount of variation they account for.

3.7.1.10 Describing Clusters

In most investigations involving CA, the objective is not only to establish clusters of similar entities, but also to describe, in ecological terms, how clusters differ. In other words, once the clusters have been formed, we are usually interested in describing how they differ on the basis of the original variables.

The most straightforward approach is to compare the clusters with respect to their means and variances on the various variables. A univariate analysis of variance (F-ratio test), or its nonparametric equivalent (Kruskal–Wallis ANOVA), can be employed to test for significant differences between cluster means (or medians) for each variable. From comparative profiles of the clusters on the different variables, it is usually easy to draw a thumbnail sketch of the distinguishing characteristics of the clusters.

A more complex approach is to subject the derived clusters to a discriminant analysis (see Chapter 4), either to determine which variables contributed most to the formation of the clusters, or to obtain a classification function for predicting cluster membership of a future sample of entities. Unfortunately, sample size requirements for discriminant analysis often preclude this option, since typically there will exist small clusters with only a few samples.

An alternative strategy is to cluster entities using ordination scores produced from an ordination procedure, such as principal components analysis or reciprocal averaging. In other words, to cluster entities based on their locations in ordination space rather than original entity space. This approach has a distinct advantage because it allows entities to be clustered while simultaneously providing an ecological interpretation of how clusters differ (see Section 3.7.9).

Empirical Example. In the example, we decided that the five-cluster solution based on average linkage fusion using squared ED and standardized data best

	Cluster (SOSP-WIWR)			Cluster (VATH-GCKI)				
Variable	n	Mean	CV	n	Mean	CV	F-ratio	P-Value
MGTOTAL	8	29.77	16.12	8	21.05	24.20	4.91	0.0110
MLTOTAL	8	40.06	13.27	8	60.49	19.62	7.93	0.0015
MTTOTAL	8	57.06	12.82	8	35.50	25.10	11.65	0.0002
MMTOTAL	8	19.18	12.42	8	14.01	17.27	7.31	0.0021
MOTOTAL	8	53.55	6.54	8	64.92	6.59	11.65	0.0002
MSNAGM1	8	0.08	79.03	8	0.34	63.42	3.65	0.0307
MSNAGM23	8	0.17	55.02	8	0.36	25.50	13.39	0.0001
MSNAGM45	8	0.61	49.08	8	1.87	25.84	17.97	0.0001
MSNAGL1	8	0.04	101.41	8	0.11	65.01	3.09	0.0510
MSNAGL23	8	0.36	38.41	8	0.67	57.25	42.42	0.0001
MSNAGL45	8	1.99	39.25	8	2.27	15.31	2.53	0.0873
MSNAGS6	8	1.04	29.86	8	1.78	24.41	5.82	0.0057
MSNAGM6	8	0.84	21.13	8	1.09	13.71	5.00	0.0103
MSNAGL6	8	0.81	18.51	8	1.28	29.88	11.34	0.0003
MSNAGT	8	5.94	11.38	8	9.76	14.22	25.98	0.0001
HGTCON	8	54.02	7.44	8	56.55	1.21	7.81	0.0016
HGTHARD	8	28.99	10.70	8	21.14	28.37	4.43	0.0160
MBAS	8	48.50	27.60	8	77.19	7.99	16.26	0.0001
MBAC	8	118.62	28.33	8	214.81	15.15	16.13	0.0001
MBAH	8	42.15	27.81	8	30.28	35.71	13.3	0.0001
MBAT	8	209.27	15.21	8	322.28	10.20	22.25	0.0001
MFHD	8	1.37	5.13	8	1.24	10.15	2.77	0.0691

n = number of species in sample
CV = coefficient of variation

FIGURE 3.11 Summary of the output used to describe and compare the two largest clusters in the six-cluster solution resulting from average-linkage fusion using squared Euclidean distance and standardized data.

describes the structure of the data set. We compared the two multispecies clusters with respect to their means and coefficients of variation on the vegetation structure variables (Fig. 3.11). The coefficient of variation (CV) measures the amount of variation relative to the mean vegetation characteristic among species for each cluster, and provides a useful means of determining which habitat characteristics are most useful in describing the clusters. Lower CV values indicate that species belonging to the cluster in question occupy sites that are relatively similar with respect to the vegetation characteristic being considered. We also computed F-ratios to test the null hypotheses that the mean for each habitat variable was not different between the two largest clusters (Fig. 3.11); the other three clusters were single-species clusters and could not be tested.

The two large clusters each contain eight bird species (AMRO-SOSP and BRCR-HAFL in Figure 3.8e). In comparison, species in cluster 1 occupy sites with greater cover of tall shrubs and less overstory cover, basal area of conifers, and numbers of snags. The differences in vegetation structure between clusters correspond to differences between streamside and upslope stand conditions. Therefore, we conclude that cluster 1 represents species associated with streamside areas, whereas cluster 2 represents species associated with upslope areas. Small group sample sizes preclude us from using discriminant analysis to describe cluster separation further.

3.7.2 Polythetic Divisive Hierarchical Clustering

Like PAHC, *polythetic divisive hierarchical clustering* (PDHC) techniques use the information contained in all the variables. In contrast, however, PDHC techniques begin with all sample entities together in a single cluster and successively divide the entities into a hierarchy of smaller and smaller clusters until, finally, each cluster contains only one entity or some specified number of entities. Although the theoretical merit of this approach has long been appreciated, technical challenges delayed substantial development until recently (Gauch 1982). Several different PDHC techniques have been developed; we review the two with most potential for application in wildlife research.

The simplest PDHC technique is *ordination space partitioning* (OSP) (Noy-Meir 1973; Swaine and Hall 1976; Gauch and Whittaker 1981). Any ordination technique may be used (although detrended correspondence analysis is usually recommended) to position entities in low-dimensional ordination space. Successive partitions are then drawn in the ordination to generate a divisive hierarchical clustering. The partitioning procedure can be made automatic and objective in a variety of ways, or the partitions can be placed subjectively by drawing boundaries on ordination plots by hand. According to Gauch (1982), subjective partitions can be particularly useful when: (1) divisions through sparse regions of the cloud of sample entities are desired, because none of the other clustering techniques considered here can take sparse regions into consideration; (2) field experience or previous analyses have provided a general understanding of the data that the investigator wants to incorporate into the analysis, but cannot specify precisely or supply to a computer; and (3) subjective clustering is sufficient for the purposes of a given study.

Two-way indicator species analysis (TWINSPAN) (Hill 1979; Gauch and Whittaker 1981) is the most popular PDHC technique and has been used very successfully on samples-by-species data (Gauch 1982). It is appropriate whenever a dual ordination of rows and columns in the data matrix is meaningful. For example, consider the typical samples-by-species data common in community ecology studies. Briefly, according to Gauch (1982), the data are first ordinated by reciprocal averaging (see the brief review of reciprocal averaging in Chapter 2). Then those species that characterize the reciprocal averaging axis extremes are emphasized in order to polarize the samples, and the samples are divided into two clusters by breaking the ordination axis near its middle. The sample division is refined by a reclassification using species with maximum value for indicating the poles of the ordination axis. The division process is then repeated on the two sample subsets to give four clusters, and so on, until each cluster has no more than a chosen minimum number of members. A corresponding species clustering is produced, and the sample and species hierarchical clusterings are used together to produce an arranged data matrix. The resulting sample hierarchy (and species hierarchy) may also be displayed as a dendrogram, using the sequences of divisions as integral levels or computing the levels as the average distances between samples in ordination space.

Conceptually, these two PDHC techniques are similar. They differ, however, in that: (1) OSP generally uses detrended correspondence analysis, whereas TWINSPAN uses reciprocal averaging; (2) OSP performs a single ordination having one unified coordinate system, whereas TWINSPAN uses separate ordinations for each data subset it produces at each division, and consequently, the various divisions at a given level may be focusing on different community gradients that are important for the individual data subsets; and (3) OSPs are usually imposed subjectively, whereas TWINSPAN partitions are objective and automatic (Gauch 1982).

It has been suggested that PDHC techniques are preferable over PAHC techniques for the analysis of community ecology data (Gauch and Whittaker 1981; Gauch 1982). Agglomerative techniques begin by examining small distances between similar entities. In community data, these small distances are likely to be a reflection of noise than anything else, and these initial groupings will constrain the ultimate cluster hierarchy because there are no provisions for reclassifying an entity that is poorly classified during an early stage in the clustering. Divisive techniques, however, begin by examining overall, major gradients in the data; all the available information is used to make the critical topmost divisions. In this manner, noise among entities does not seriously influence the upper level of the hierarchy. Furthermore, divisive techniques like TWINSPAN effectively hybridize ordination and cluster analysis; they combine the power of clustering for summarization with the effectiveness of ordination in revealing directions of relationship. Further advantages of TWINSPAN over agglomerative techniques are: (1) its use of the original data matrix, rather than a secondary dissimilarities matrix with information only on samples (or species); (2) integrated clustering of both samples and species (due to the use of reciprocal averaging ordination); and, consequently, (3) production of an arranged data matrix (Gauch 1982). Given these properties, it is likely that PDHC techniques such as TWINSPAN will increase in popularity over time. However, because these techniques involve a combination of ordination and clustering, they will never achieve the conceptual simplicity of most PAHC techniques.

3.8 Evaluating the Stability of the Cluster Solution

After obtaining a set of clusters, various procedures are available for evaluating the stability and usefulness of the solutions found. We briefly discuss two general procedures (Everitt 1977).

1. Data splitting. If the final cluster solution produces clusters containing several entities each (or at least some of the final clusters contain several entities), then the full set of entities can be randomly divided into two subsets and an analysis performed on each subset separately. Similar solutions should be obtained from both sets when the data are clearly structured. That is, each solution should produce the same set of clusters (call these "subclusters") produced by the full data set, but with only those entities present in each subcluster that were in the subset of data analyzed. This general procedure, referred to as *data splitting*, can be used to evaluate the stability of virtually any statistic, providing the data set is large enough to split into adequately sized subsets.

2. Variable splitting. If the data set contains a large enough number of variables, then the cluster analysis can be repeated on subsets of variables and the results compared. Variables might be divided into logical subsets based on measurement scale or some other natural division. For example, a set of vegetation variables might be broken into those pertaining to floristics (e.g., percent cover of each plant species) and those pertaining to vegetation structure (e.g., foliage height diversity, stem density). Alternatively, the variables might be randomly divided into subsets, or a single variable at a time might be removed from the set in a jackknife-like procedure. Deletion of a small number of variables from the analysis should not, in most cases, alter greatly the clusters found, if the clusters are "real" and not mere artifacts of the particular technique used.

An alternative procedure is to compare the original cluster solution with the solution using additional variables of interest which were *not* included in the original analysis. For example, these might be variables gathered during subsequent research. If differences between clusters persist with respect to these new variables, then this is some evidence that a "useful" solution has been obtained, in the sense that by stating a particular entity belongs to a particular cluster, we convey information on variables other than those used to produce the cluster.

3.9 Complementary Use of Ordination and Cluster Analysis

The clustering of ecological systems is useful because humans comfortably think and communicate in terms of classes. Note, however, that this is an observation about human thinking, not about natural systems. Whether ecological systems actually occur in discrete, discontinuous classes or types, rather than varying continuously, is a separate question. If ecological variation is discontinuous, CA is a natural framework for conceptualizing ecological systems; if ecological variation is continuous, ordination is more natural. Given that scientists rarely know with certainty whether variation is continuous or discontinuous in the ecological system under investigation, the use of ordination and CA are complementary. Moreover, most ecologists recognize the utility of CA for many practical purposes even when rather arbitrary dissections must be imposed on essentially continuous ecological variation.

Ordination and CA complement each other in a number of ways. First, ordination can be used as a precursor to CA to reduce the dimensionality of large, complex data sets and to create new, uncorrelated variables (e.g., principal components). Cluster analysis is used subsequently to group sampling entities according to their similarity in ordination space using the ordination scores (Everitt 1977). This procedure can have particular utility if the data set contains an excessively large number of variables and some form of summarization is desirable. Conceptually, this is the process employed by OSP.

Second, ordination plots depicting the relationship among entities in ordination space can be examined visually for discontinuities and natural clusters. Inspection of ordination plots can aid in determining the number of natural clusters, if any, present in the data set. This procedure can be of particular utility as a precursor to PDNHC, which requires a priori knowledge of the number of clusters to expect. There is, however, a potential danger in using ordination plots to delineate the boundaries of natural clusters by visual inspection alone (Sneath and Sokal 1973). When the points form clusters of unfamiliar shape (e.g., nonspherical), the danger arises of interpreting these according to individual whim. Some would argue that this is precisely what makes this procedure appealing, because the researcher is allowed to use his accumulated knowledge and understanding of the ecological system under investigation to aid in the analysis, whereas others would claim that subjective interpretation based on visual inspection alone is counter to our aims of objectivity and explicitness. Furthermore, ordination plots can depict at most three dimensions. Hence, natural clusters that are defined by more than three dimensions may go undetected by visual inspection of typical ordination plots. Nevertheless, inspection of ordination plots for the presence of natural discontinuities has, in practice, proved in most circumstances to be a very useful procedure.

Third, ordination and CA can be merged into a single analysis, as in OSP and TWINSPAN (see Sec. 3.7.2). In both approaches, clusters are formed based on the locations of entities in ordination space, rather than their locations in dissimilarity space calculated from the original variables. This approach has a distinct advantage over other clustering techniques because it allows entities to be clustered while simultaneously providing an ecological interpretation of how clusters differ.

3.10 Limitations of Cluster Analysis

Most of the limitations of CA have been discussed already. Nevertheless, we briefly summarize some of the most important limitations again to emphasis their importance.

- Most clustering techniques (especially hierarchical techniques) are sensitive to the presence of outliers. Thus, some form of screening for outliers is advisable (Grubbs 1950; Bliss, Cochran, and Tukey 1956; Anscombe 1960; Kruskal 1960). Fortunately, nonhierarchical clustering techniques can be used to identify outliers because they typically leave outliers solitary or in small clusters, which then may be omitted from further analyses. Perhaps the simplest way to screen for outliers is to standardize the data and then inspect the data for entities with any value more than 2.5 standard deviations from the mean around on any variable. However, caution should always be exercised in the deletion of suspect points, since careless elimination of "outliers" may result in the loss of meaningful information.

- A major problem in CA is that most clustering procedures are biased toward finding clusters of a particular shape and are usually biased specifically toward finding spherical clusters (Everitt 1977). However, there is usually no a priori reason for believing that any clusters present in the data are of one particular shape, and by using the "wrong" clustering technique (e.g., one in which the data do not meet the assumptions made by the technique), we may impose a particular structure on the data, rather than find the actual structure present (Everitt 1977). Perhaps the best or only solution is to compare results among several different clustering techniques. General agreement in cluster solutions probably indicates a rather pronounced structure to the data. Disagreement probably indicates less clear structure to the data (e.g., nonspherical or fuzzy clusters or no "real" clusters at all), and perhaps indicates that the bias of each procedure is governing the particular cluster solution found.
- Perhaps an even more serious limitation than the bias toward seeking spherical clusters is the difficulty of judging from the results of clustering the "realness" of the clusters or the number of clusters suitable for the representation of the data matrix (Everitt 1977). Many of the proposed indicators for the number of clusters, including those discussed previously, are frequently unreliable. Many authors argue that, ultimately, the validity of clusters can be judged only qualitatively by subjective evaluation and by interpretability.
- Because CA involves choosing among a vast array of different procedures and alternative measures (e.g., choice of resemblance measure and fusion strategy in PAHC), its application is more an art than a science, and it can easily be abused (misapplied) by the analyst. The choice of resemblance measures (if required) and different algorithms can and does affect the results. The analyst needs to be aware of this fact and, if possible, replicate the analysis under varying conditions and employ different clustering strategies to ensure reliable results.

Bibliography

Procedures

Anderberg, M.R. 1973. *Cluster Analysis for Applications*. New York: Academic Press.
Anscombe, F.J. 1960. Rejection of outliers. *Technometrics* 2:123–146.
Beale, E.M.L. 1969. *Cluster Analysis*. London: Scientific Control Systems.
Bliss, C.I., Cochran, W.G., and Tukey, J.W. 1956. A rejection criterion based upon the range. *Biometrika* 43:418–422.
Bock, H.H. 1985. On some significance tests in cluster analysis. *Journal of Classification* 2:77–108.
Dale, M.B. 1995. Evaluating classification strategies. *Journal of Vegetation Science* 6(3):437–440.
Digby, P.G.N., and Kempton, R.A.. 1987. *Multivariate Analysis of Ecological Communities*. New York: Chapman and Hall.
Eades, D.C. 1965. The inappropriateness of the correlation coefficient as a measure of taxonomic resemblance. *Systematic Zoology* 14:98–100.
Edwards, A.W.F., and Cavalli-Sforza, L.L.1965. A method for cluster analysis. *Biometrics* 21:362–375.

Efron, B. 1979. Computers and the theory of statistics: thinking the unthinkable. *Society for Industrial and Applied Mathematics (SIAM) Review* 21:460–480.

Efron, B. 1982. The jackknife, the bootstrap and other resampling plans. *Society for Industrial and Applied Mathematics (SIAM),* Monograph #38, CBMS-NSF. Philadelphia, PA.

Efron, B, and Gong, G. 1983. A leisurely look at the bootstrap, the jackknife, and cross-validation. *American Statistician* 37:36–48.

Everitt, B.S. 1977. *Cluster Analysis* (2nd ed.). London: Heineman Education Books, Ltd.

Fleiss, J.L., and Zubin, J. 1969. On the methods and theory of clustering. *Multivariate Behaviour Research* 4:235–250.

Gauch, H.G. 1979. *COMPCLUS—A FORTRAN Program for Rapid Initial Clustering of Large Data Sets.* Ithaca, N.Y.: Cornell University.

Gauch, H.G. 1980. Rapid initial clustering of large data sets. *Vegetatio* 42:103–111.

Gauch, H.G. 1982. *Multivariate Analysis in Community Ecology.* Cambridge: Cambridge University Press.

Gauch, H.G. and Whittaker, R.H. 1981. Hierarchical classification of community data. *Journal of Ecology* 69:537–557.

Grubbs, F.E. 1950. Sample criteria for testing outlying observations. *Annals of Mathematical Statistics*, 21:27–28.

Goodall, D.W. 1978. Numerical methods of classification. In *Classification of Plant Communities,* ed. R.H. Whittaker, pp 247–286. The Hague: W. Junk.

Gower, J.C. 1974. Maximal predictive classification. *Biometrics* 30:643–654.

Gower, J.C., and Ross, G.J.S. 1969. Minimum spanning trees and single-linkage cluster analysis. *Applied Statistics* 18:54–64.

Hair, J.F., Jr., Anderson, R.E., and Tatham, R.L. 1987. *Multivariate Data Analysis*, 2nd ed. New York: Macmillan.

Harper, C.W., Jr. 1978. Groupings by locality in community ecology and paleoecology: tests of significance. *Lethaia* 11:251–257.

Hartigan, J.A. 1975. *Clustering Algorithms.* New York: Wiley and Sons.

Hartigan, J.A., and Wong, M.A. 1979. A K-means clustering algorithm: algorithm AS 1366. *Applied Statistics* 28:126–130.

Hill, M.O. 1979. *TWINSPAN—A FORTRAN Program for Arranging Multivariate Data in an Ordered Two-Way Table by Classification of the Individuals and Attributes.* Ithaca, NY: Cornell University.

Hill, M.O., Bunce, R.G.H., and Shaw, M.W. 1975. Indicator species analysis, a divisive polythetic method of classification and its application to a survey of native pinewoods in Scotland. *Journal of Ecology* 63:597–613.

Hill R.S. 1980. A stopping rule for partitioning dendrograms. *Botanical Gazette* 141:321–324.

Jackson, D.M. 1969. Comparison of classifications. In *Numerical Taxonomy,* ed. A.J. Cole, pp 91–113. London: Academic Press.

Johnson, S.C. 1967. Hierarchical clustering schemes. *Psychometrika* 32:241–254.

Kruskal, W.H. 1960. Some remarks on wild observations. *Technometrics* 2:1–3.

Lance, G.N., and Williams, W.T. 1967. A general theory of classificatory sorting strategies. I. Hierarchical systems. *Computer Journal* 9:373–380.

Mardia, K.V., Kent, J.T. and Bibby, J.M. 1979. *Multivariate Analysis.* New York: Academic Press.

Nemec, A.F.L., and Brinkhurst, R.O. 1988a. Using the bootstrap to assess statistical significance in the cluster analysis of species abundance data. *Canadian Journal of Fisheries and Aquatic Science* 45:965–970.

Nemec, A.F.L. 1988b. The Fowlkes–Mallows statistic and the comparison of two independently determined dendrograms. *Canadian Journal of Fisheries and Aquatic Science* 45:971–975.

Noy-Meir, I. 1973. Divisive polythetic classification of vegetation data by optimized division on ordination components. *Journal of Ecology* 61:753–7660.

Ratliff, R.D., and Pieper, R.D.1981. Deciding final clusters: an approach using intra- and intercluster distances. *Vegetatio* 48:83–86.

Rohlf, F.J. 1974. Methods of comparing classifications. *Annual Review of Ecology and Systematics* 5:101–113.

Sarle, W.S. 1983. *Cubic Clustering Criterion*. SAS Technical Report A–108. Cary, N.C.: SAS Institute, Inc.

Sneath, P.H.A. 1969. Evaluation of clustering methods. In *Numerical Taxonomy*, ed. A.J. Cole, pp 257–271. London: Academic Press.

Sneath, P.H.A., and Sokal, R.R. 1973. Numerical Taxonomy. *The Principles and Practice of Numerical Classification*. San Francisco: Freeman.

Sokal, R.R. 1977. Clustering and classification: background and current directions. In *Classification and Clustering*, ed. J. Van Ryzin, pp 1–15. London: Academic Press.

Strauss, R.E. 1982. Statistical significance of species clusters in association analysis. *Ecology* 63:634–639.

Swaine, M.D., and Hall, J.B.1976. An application of ordination to the identification of forest types. *Vegetatio* 32:83–86.

Tryon, R.C., and Bailey, D.E. 1970. *Cluster Analysis*. New York: McGraw-Hill.

Ward, J.H. 1963. Hierarchical grouping to optimize an objective function. *Journal of the American Statistical Association*. 58.236–244.

Williams, W.T. 1971. Principles of clustering. *Annual Review of Ecology and Systematics* 2:303–326.

Applications

Abdelghani, M.M. 1998. Environmental correlates of species distribution in arid desert ecosystem of eastern Egypt. *Journal of Arid Environments* 38(2):297–313.

Clark, T.E. and Samways, M.J. 1996. Dragonflies (Odonata) as indicators of biotope quality in the Kruger National Park, South Africa. *Journal of Applied Ecology*. 33(5):1001–1012.

Cooper, R.J., Martinat, P.J. and Whitmore, R.C. 1990. Dietary similarity among insectivorous birds: influence of taxonomic versus ecological categorization of prey. In *Avian Foraging: Theory, Methodology, and Applications. Studies in Avian Biol. No. 13*, eds. M.L. Morrison, C.J. Ralph, J.Verner, and J.R. Jehl, pp 104–109.

Grigal, D.F., and Goldstein, R.A. 1971. An integrated ordination-classification analysis of an intensively sampled oak-hickory forest. *Journal of Ecology* 59:481–492.

Haering, R. and Fox, B.J. 1995. Habitat utilization patterns of sympatric populations of Pseudomys gracilicaudatus and Rattus lutreolus in coastal heathland: a multivariate analysis. *Australian Journal of Ecology* 20(3):427–441.

Hodder, K.H., Kenward, R.E., Walls, S.S. and Clarke, R.T. 1998. Estimating core ranges: a comparison of techniques using the common buzzard (Buteo buteo). *Journal of Raptor Research* 32(2):82–89.

Holmes, R.T., Bonney, R.E., Jr., and Pacala, S.W. 1979. Guild structure of the Hubbard Brook bird community: a multivariate approach. *Ecology* 60:512–520.

Miles, D.B. 1990. The importance and consequences of temporal variation in avian foraging behavior. In *Avian Foraging: Theory, Methodology, and Applications. Studies in Avian Biol. No. 13*, eds. M.L. Morrison, C.J. Ralph, J. Verner, and J.R. Jehl, pp 210–217.

Marsden, S.J. and Jones, M.J. 1997. The nesting requirements of the parrots and hornbill of Sumba, Indonesia. *Biological Conservation* 82(3):279–287.

Marshall, S.D. 1997. The ecological determinants of space use by a burrowing wolf spider in a xeric shrubland ecosystem. *Journal of Arid Environments* 37(2):379–393.

Miller, J.N., Brooks, R.P., and Croonquist, M.J. 1997. Effects of landscape patterns on biotic communities. *Landscape Ecology* 12(3):137–153.

Oliver, I., Beattie, A.J. and York, A. 1998. Spatial fidelity of plant, vertebrate, and invertebrate assemblages in multiple-use forest in eastern Australia. *Conservation Biology* 12(4):822–825.

Parsons, S., Thorpe, C.W. and Dawson, S.M. 1997. Echolocation calls of the long-tailed bat: a quantitative analysis of types of calls. *Journal of Mammalogy* 78(3):964–976.

Raphael, M.G., and White, M. 1984. Use of snags by cavity nesting birds in the Sierra Nevada. *Wildlife Monograph 86.*

Read, D.G. and Tweedie, T.D. 1996. Floristics of habitats of Pseudomys oralis (Rodentia, Muridae). *Wildlife Research.* 23(4):485–493.

Rogovin, K.A. and Shenbrot, G.I. 1995. Geographical ecology of Mongolian desert rodent communities. *Journal of Biogeography* 22(1)111–128.

Rumble, M.A. and Anderson, S.H. 1996. Microhabitats of Merriam's turkeys in the Black Hills, South Dakota. *Ecological Applications* 6(1):326–334.

Saetersdal, M. and Birks, H.J.B. 1993. Assessing the representativeness of nature reserves using multivariate analysis: vascular plants and breeding birds in deciduous forests, western Norway. *Biological Conservation* 65(2):121–132.

Saino, N., and Fasola, M. 1990. Habitat categorization, niche overlap measures and clustering techniques. *Avocetta* 14(1):27–36.

Short, H.L., and Burnham, K.P. 1982. Technique for structuring wildlife guilds to evaluate impacts on wildlife communities. *Special Scientific Report. Wildlife. No. 244*, USDI, U.S. Fish and Wildlife Service.

Sodhi, N.S., Choo, J.P.S., Lee, B.P.H.Y., Quek, K.C., and Kara, A.U. 1997. Ecology of a mangrove forest bird community in Singapore. *Raffles Bulletin of Zoology* 45(1):1–13.

Appendix 3.1

SAS program statements used to conduct cluster analysis on the empirical data set presented in Figure 3.2. Terms in the program statements given in lower case vary among applications with regard to characteristics of the specific data set (e.g., variable names) and the personal preferences of the user (e.g., naming of output files).

The following header information creates a library named "km" for SAS files in the specified directory, provides page formatting information, and initializes the output device for SAS graphics:

```
LIBNAME km 'd:\stats\cluster';
OPTIONS PS=60 LS=80 REPLACE OBS=MAX;
GOPTIONS DEVICE=WIN TARGET=hplj4si;
```

The following procedure computes an average-linkage fusion based on squared Euclidean distance (the default), and computes the cubic clustering criterion, pseudo-F- and t-statistics, and squared multiple correlation for each step in the agglomeration sequence. The ID variable (spec) is the variable containing the species name associated with each sampling entity. To use a different fusion method, simply replace AVERAGE with the appropriate fusion name (e.g., MEDIAN, CENTROID, COMPLETE, SINGLE, WARDS):

```
PROC CLUSTER DATA=km.cluster OUT=clustree STD M=AVERAGE CCC
    PSEUDO RSQ;
    TITLE 'Average linkage-standardized data';
    ID spec; RUN;
```

The following procedure generates a horizontal dendrogram based on the previous cluster analysis:

```
PROC TREE HORIZONTAL GRAPHICS HORDISPLAY=RIGHT;
    TITLE 'Average Linkage Dendrogram'; RUN;
```

The following procedure generates a horizontal icicle plot based on the previous cluster analysis:

```
PROC TREE HORIZONTAL GRAPHICS SORT HEIGHT=N HORDISPLAY=RIGHT;
    TITLE 'Average Linkage Icicle Plot'; RUN;
```

The following procedure generates a scree plot based on the previous cluster analysis:

```
PROC PLOT DATA=clustree;
    TITLE 'Average Linkage Scree Plot';
    PLOT _HEIGHT_*_NCL_=_NCL_; RUN;
```

The following procedure generates summary statistics for each of the original variables by cluster, based on the five-cluster solution from the previous cluster analysis:

```
PROC TREE NOPRINT OUT=clusmem NCLUSTERS=5;
    ID spec;
    COPY mgtotal-mfhd;
PROC SORT;
    BY CLUSTER;
PROC MEANS N MEAN STD CV;
    TITLE 'Average Linkage-5 Cluster Solution-Cluster Descriptions';
    VAR mgtotal-mfhd;
    BY CLUSTER; RUN;
```

The following procedure computes an F-ratio test for comparing the means of each vegetation variable between the two largest clusters from the previous cluster analysis:

```
IF CLUSTER LT 3;
PROC GLM;
    TITLE 'Average Linkage-F-Test Between Clusters 1 and 2';
    CLASS CLUSTER;
    MODEL mgtotal-mfhd=CLUSTER; RUN;
```

4
Discriminant Analysis

4.1 Objectives

By the end of this chapter, you should be able to do the following:

- Recognize the types of research questions best handled with *discriminant analysis* (DA).
- List 10 important characteristics of DA.
- Differentiate between descriptive DA using canonical functions and predictive DA using classification (or discriminant) functions.
- Given a data set, determine if sample sizes are adequate and examine whether the five assumptions of DA are met:
  — Equality of variance-covariance matrices
  — Multivariate normality
  — No singularities
  — Independent random sample (no outliers)
  — Identifiable prior probabilities
- Perform a stepwise DA and demonstrate how adjusting the F-to-enter and F-to-remove statistics can alter the list of variables selected.
- Given a set of canonical functions, assess their importance and/or significance using
  — Relative percent variance criterion
  — Canonical correlation criterion
  — Classification accuracy
  — Significance tests based on resampling procedures
  — Graphical representation of canonical scores
- Given a set of canonical functions, measure the improvement in classification accuracy over that expected by a random assignment using

- Maximum chance criterion
- Proportional chance criterion
- *Tau* statistic
- Cohen's *Kappa* statistic
• Given a set of canonical functions, interpret their meaning using
- Total structure coefficients
- Covariance-controlled partial F-ratios
- Significance tests based on resampling procedures
- Potency index
• Use the split-sample, jackknife, and randomized sample approach to validate a given set of canonical functions.
• List five limitations of DA.

4.2 Conceptual Overview

Ecologists are often interested in identifying factors that help explain the differences between two or more groups of sampling entities. The purpose might be to identify ecological or environmental factors that best explain differences in the distribution patterns of one or more organisms. For example, ecologists might be interested in determining why a species occurs in some environments and not others, or how several sympatric species are able to coexist by niche partitioning. In this instance, the emphasis is on explaining the differences among prespecified, well-defined groups of sampling entities based on a suite of discriminating variables. Alternatively, the purpose might be to predict group membership for samples of unknown membership. For example, this might involve predicting whether a species is likely to occur at a site based on a suite of ecological or environmental variables. Here, the emphasis is on prediction.

Whether the purpose is explanation or prediction, one common approach involves sampling organisms and the environment at numerous sites distributed in some random or systematic fashion throughout the study area. For example, we might sample a wide range of environments for the presence or absence of a particular species, and at each sampling location measure a suite of ecological or environmental variables that potentially influence the occurrence of the target species. Alternatively, we might sample a large number of individuals of a particular species (or a large number of locations of the same individual), and then measure a suite of variables at these sites and at random locations within the study area. In either case, the resulting data set would contain information on the species' presence or abundance and a suite of variables that characterize the environment at each sampling location. If there are consistent nonrandom patterns in the species' distribution related to one or more ecological or environmental gradients, then we should be able to identify and describe the gradients that best distinguish or discriminate between used and unused (or random) sites.

TABLE 4.1 Important characteristics of discriminant analysis.

Essentially a single technique consisting of a couple of closely related procedures (as compared with the wide variety of approaches in ordination and cluster analysis).

Operates on data sets for which prespecified, well-defined groups exist.

Assesses relationships between one set of discriminating variables and a single grouping variable; an attempt is made to define the relationship between independent and dependant variables.

Extracts dominant, underlying gradients of variation (canonical functions) among groups of sample entities (e.g., species, sites, observations) from a set of multivariate observations, such that variation among groups is maximized and variation within groups is minimized along the gradient(s).

Reduces the dimensionality of a multivariate data set by condensing a large number of original variables into a smaller set of new composite dimensions (canonical functions) with a minimum loss of information.

Summarizes data redundancy by placing similar entities in proximity in canonical space and producing a parsimonious understanding of the data in terms of a few dominant gradients of variation.

Describes maximum differences among prespecified groups of sampling entities based on a suite of discriminating characteristics (i.e., canonical analysis of discriminance).

Predicts the group membership of future samples or samples from unknown groups based on a suite of classification characteristics (i.e., classification).

Extension of multiple regression analysis if the research situation defines the group categories as dependent upon the discriminating variables, and if a single random sample (labeled N, for sample size) is drawn in which group membership is unknown prior to sampling.

Extension of multivariate analysis of variance if the values on the discriminating variables are defined as dependent upon the groups, and separate independent random samples (N_1, N_2, ...) of two or more distinct populations (i.e., groups) are drawn in which group membership is known prior to sampling.

Discriminant analysis (DA) refers to a couple of closely related procedures that share a similar goal of objectively discriminating among prespecified, well-defined groups of sampling entities based on a suite of characteristics (i.e., discriminating variables; Table 4.1). This is in contrast to ordination, which attempts to organize a single group of entities along a gradient of maximum variation, and cluster analysis (CA), which attempts to organize entities into classes or groups. Cluster analysis often serves as a precursor to DA when prespecified groups do not exist; artificial groups are created in CA, and ecological differences among the newly created groups are described using DA.

Otherwise, DA and ordination share several goals. Like ordination, DA attempts to extract dominant, underlying gradients of variation (canonical functions) from a set of multivariate observations. However, DA seeks to find gradients of variation among groups of sample entities such that variation among groups is maximized and variation within groups is minimized along these gradients. Like ordination, DA attempts to reduce the dimensionality of a multivariate data set by condensing a large number of original variables into a smaller set of new composite variables (i.e., canonical functions) with a minimum loss of information. Lastly, like ordination, DA attempts to summarize data redundancy by placing similar entities in proximity in canonical space and producing a parsimonious understanding of the data in terms of a few dominant gradients of variation.

Discriminant analysis is referred to by a wide variety of terms (see Chapter 1, Table 1.2). But in contrast to ordination and CA, which encompass a wide variety of approaches, DA is essentially a single technique; the choice of terms largely reflects personal preference rather than any difference in statistical procedure. Within the field of wildlife research, the terms "discriminant analysis" and "discriminant function analysis" are used almost exclusively. In this book, we use "discriminant analysis" exclusively to reference this technique. If the DA involves only two groups, we refer to the procedure as two-group DA; if there are more than two groups, we refer to the procedure as multiple DA. For simplicity, our discussion is fashioned after the multiple group situation. For example, we say "among" rather than "between," even though the latter is the appropriate term for the two-group situation.

As mentioned at the beginning of this chapter, there are two main objectives of DA. One objective may be to exhibit optimal separation of groups, based on certain linear transformations of the discriminating variables, and learn which variables are most related to the separation of groups; in other words, to explain the differences among groups. Such a formulation is referred to as *descriptive discriminant analysis*, and the associated linear functions are referred to as *canonical functions* (or *canonical variates*, *discriminant functions*, or *canonical discriminant functions*). In this book, we refer to the application of DA in this manner as *canonical analysis of discriminance* (CAD).[1]

The other main objective of DA is to predict group membership for an entity of unknown origin based on its measured values on the discriminating variables. Such a formulation is referred to as *predictive discriminant analysis*, and the prediction equations are called *classification functions* (or *discriminant functions*). In this book, we refer to the application of DA in this manner as *classification*. Under certain distribution assumptions, the classification approach is a logical consequence to the CAD approach. It may be shown that the canonical functions themselves may be used to develop a classification procedure equivalent to that produced by the predictive methodology (Williams 1982). In this book, "canonical" pertains to procedures and results derived from the descriptive methodology. "Classification" is reserved for procedures and results derived from the predictive methodology. The term "discriminant" denotes procedures and results that pertain to either descriptive or predictive approaches.

Both descriptive and predictive methods have been used in ecological studies, though most wildlife studies have had a descriptive orientation. In practice, the two

---

[1]. Note, however, that in wildlife literature CAD is usually referred to as *discriminant function analysis* (or, simply, *discriminant analysis*) and that the terms *canonical* and *discriminant* are used interchangeably.

methods are often used in conjunction; interpretation of the canonical function (i.e., the descriptive function) is usually followed by the formulation of classification functions, which are used to classify the entities into groups (i.e., the predictive function). The resulting misclassification rates are used to evaluate the stability and adequacy of the canonical functions. In this book, we discuss classification largely from this perspective.

4.2.1 *Overview of Canonical Analysis of Discriminance*

Canonical analysis of discriminance (CAD) seeks to describe the relationships among two or more groups of entities based on a set of two or more discriminating variables. Alternatively, we can think of CAD as a procedure for identifying boundaries among groups of entities, the boundaries being in terms of those variable characteristics that distinguish or discriminate among the entities in the respective groups.

Consider the niche separation example involving the three ground-dwelling rodents presented in Chapter 1 (see Sec. 1.2). Taking each niche variable singly, the differences among species distribution patterns along each environmental gradient are easily observed (Fig. 1.2), but with several niche variables, it is exceedingly difficult to determine the extent of multidimensional niche separation or which variables contribute most to niche separation. When three niche variables are taken together, the boundaries of each species' niche begins to take shape and it becomes possible to identify the variables that contribute most to the discrimination among niches (Fig. 1.3). However, when more niche variables are considered, it is impossible to visualize niche space; therefore, it becomes difficult to identify those variable characteristics which best discriminate among niches. Fortunately, mathematicians have developed the objective procedure of CAD for accomplishing this task.

Specifically, CAD involves deriving the linear combinations (i.e., canonical functions) of the two or more discriminating variables that will "best" discriminate among the a priori defined groups. The canonical functions are defined as weighted linear combinations of the original variables, where each variable is weighted according to its ability to discriminate among groups. Each sampling entity has a single composite canonical score derived by multiplying the sample values for each variable by a corresponding weight and adding these products together. By averaging the canonical scores for all of the entities within a particular group, we arrive at the *group mean* canonical score. This group mean is referred to as a *centroid* because it is the composite mean of several variables. When the analysis involves two groups, there are two centroids, with three groups, there are three centroids, and so forth. The centroids indicate the most typical location of an entity from a particular group, and a comparison of the group centroids shows how far apart the groups are along the dimension defined by the canonical function. The best linear combination of variables is achieved by the statistical decision rule of maximizing the among-group variance relative to the within-group variance; that is, maximizing the ratio of among-group to within-group variance in canonical scores.

For an $N \times P$ data set with G groups, where N is the number of sampling entities (rows) and P is the number variables (columns), there are Q (equal to $G - 1$ or P, whichever is smaller) possible canonical functions corresponding to the dimensions of the data set. For example, in two-group CAD, there is only one canonical function; in three-group CAD, there are two canonical functions; in eight-group CAD, there are seven canonical functions, unless the number of variables is less than seven, in which case the number of canonical functions is equal to the number of variables. The first canonical function defines the specific linear combination of variables that maximizes the ratio of among-group to within-group variance in any single dimension. In multiple CAD, the second and subsequent canonical functions define the linear combinations of variables that maximize the ratio of among-group to within-group variance in dimensions orthogonal (i.e., independent) to the preceding dimensions.

The desired outcome in two-group CAD is that the canonical function defines a gradient that significantly discriminates between groups and that a meaningful ecological interpretation can be given to the gradient based on the relative importance of the original variables in the linear function. Multiple CAD has the added desired outcome that most of the variation among groups can be explained in fewer than Q (preferably the first one or two) canonical functions. If so, group positions on several independently measured univariate gradients (which are frequently highly correlated) can be reduced to positions along a few important multidimensional gradients. The resulting reduction in dimensionality greatly simplifies elucidation of group differences and facilitates quantitative comparisons among the groups in terms of a few new highly significant variables.

4.2.2 Overview of Classification

Classification is the process by which a decision is made that a specific entity belongs to or most closely resembles one particular group. Consider the habitat relationships example involving a single bird species presented in Chapter 1 (see Sec. 1.2). Suppose that researchers characterize the habitat selected by 30 individuals and compare this to 30 random locations, as previously described. Using the CAD procedure described above, the researchers identify the combination of habitat variables that best distinguishes occupied sites from random locations. Suppose that land managers are particularly concerned over the potential impact of proposed land management activities on this species, but they find it is prohibitively costly and time-consuming to survey every proposed management unit to determine the species' presence. It would be helpful to have some way of predicting whether the species is likely to occur in a unit based on a suite of easily measured environmental variables. In this manner, managers could screen sites and determine which sites should be surveyed.

Fortunately, mathematicians have developed several ways in which an objective classification can be performed. These methods typically involve defining some notion of "distance" between the entity and each group centroid, with the entity being classified into the "closest" group. These classification procedures can use either the discriminating variables themselves or the canonical functions (Klecka 1980).

Fisher (1936) first suggested that classification should be based on a linear combination of the discriminating variables. He proposed using a linear combination that maximizes group differences while minimizing variation within the groups. An adaptation of his proposal leads us to derive a separate linear combination, called a *classification function,* for each group. For an $N \times P$ data set with G groups, these functions have the following form:

$$h_{ij} = b_{i0} + b_{i1}x_{j1} + b_{i2}x_{j2} + \cdots + b_{ip}x_{jP}$$

where h_{ij} is the score for the ith group and jth sample, b_{i0} is a constant for the ith group, b_{ik} is the classification coefficient for the ith group and the kth variable, x_{jk} is the value for the jth sample and kth variable.

A sample entity is classified into the group with the highest score (i.e., the largest h). We do not interpret the classification coefficients, because they are not standardized and there is a different function for each group. The scores also lack intrinsic value because they are merely arbitrary numbers which happen to have the property that the sample resembles most closely that group on which it has the highest score. This simple classification function is often referred to as *Fisher's linear discriminant function* and is computed using the pooled within-groups covariance matrix. Thus, we must assume equal group dispersions; otherwise, when group dispersions are not equal then quadratic classification functions can be derived, which are computed using the individual within-group covariance matrices (more on this later).

A second, and more intuitive, means of classification uses the original variables as well. It measures the distance in multidimensional space from each entity to each of the group centroids and classifies each entity into the closest group. However, when the variables are correlated and/or do not have the same measurement scales and variances, the concept of distance is not well-defined. Thus, Mahalanobis distance (Mahalanobis 1963) is used because it accounts for intercorrelations and unequal variances among the variables. After calculating the Mahalanobis distance from an entity to the centroid of each group, we classify the entity into the group to which it is closest. That group is the one in which the typical profile on the variables most closely resembles the profile of this entity. Because Mahalanobis distance has the same properties as the chi-square statistic, we can convert the distance between each entity and each group centroid into a probability that an entity belongs to each group (Klecka 1980). Moreover, Mahalanobis distance calculations can be modified to adjust for differences in group prior probabilities (i.e., the probability that an entity belongs to each group prior to sampling), perhaps to reflect differences in group sizes and to adjust for differences in the cost of misclassification among groups (e.g., if the cost of misclassifying an entity from one group is greater than another) (Klecka 1980).

Alternatively, classification can also be done with the canonical functions (computed using the CAD procedure) instead of with the original variables. In this case, we compute the canonical scores for each entity first, and then classify each entity into the group with the closest group mean canonical score (i.e., centroid). Again, procedures are available to adjust for prior probabilities and misclassification costs.

Regardless of which classification procedure is used, the final classifications will generally be identical. However, there are certain conditions under which the classifications will not necessarily be the same when using the canonical functions. One of these is when the group covariance matrices are not equal. This is because the procedure for deriving the canonical functions must use the pooled within-group covariance matrix, which is a weighted average of the individual group covariance matrices. If the group covariance matrices are not equal, the transformation is not exact, and the use of canonical distance (i.e., distance in canonical space) becomes less straightforward. Another situation in which the two procedures may yield different classifications is when one or more of the canonical functions in a multiple CAD are ignored because they are not statistically significant. In this case, some portion of the information in the original variables (albeit typically small) is not accounted for in the retained canonical functions, and the classification of entities is thus based on less information than the full set of original variables. In either of these situations, the classification results will likely be very similar with either classification method, provided that the covariance matrices are not drastically different or the retained canonical functions account for most of the variation among groups.

4.2.3 Analogy with Multiple Regression Analysis and Multivariate Analysis of Variance

Conceptually and mathematically, DA can be considered an extension of *multiple regression analysis* (MRA) and *multivariate analysis of variance* (MANOVA); (Klecka 1980; Hair et al. 1987). In some cases, it is logical to consider DA an extension of MRA (e.g., Kachigan 1982). That is, a linear combination of measurements for two or more independent variables is used to describe or predict the behavior of a single dependent variable. The key difference is that in DA the single dependent variable is categorical (i.e., it is a grouping variable) instead of continuous, as in multiple regression analysis. In this form of DA, it is logical to assume that sampling entities represent a single random sample (N) of a mixture of two or more distinct populations (i.e., groups). In other words, a single sample is drawn in which group membership (indicated by a dependent variable) is *unknown* prior to sampling.

Alternatively, in some cases it is logical to consider DA an extension of one-way MANOVA (e.g., Harris 1975; Morrison 1976). That is, the independent variable is categorical and defines group membership (typically controlled by experimental design). Populations (i.e., groups) are compared with respect to a vector of measurements for two or more dependent variables. The key difference is that in MANOVA we are interested in testing the null hypothesis that groups do not differ on any of the dependent variables (i.e., mean vectors are equal), whereas in DA we are interested in describing the linear combination of dependent variables that maximally discriminates among groups. In other words, MANOVA involves testing the null hypothesis that groups are equal, while DA involves describing the linear combination of variables that maximizes the test statistic (e.g., T^2)

used in the MANOVA procedure. In this respect, MANOVA and DA correspond nicely to the inferential and descriptive aspects of multivariate statistics. In the MANOVA form of DA, it is logical to assume that sampling units represent separate, independent random samples $(N_1, N_2, ..., N_G)$ of two or more distinct populations (i.e., groups). In other words, group membership is *known* prior to sampling, and sampling units are drawn from each population separately.

Regardless of which conceptualization is more appropriate in a given situation, the data must include samples for which there is a single categorical grouping variable and an associated vector of measurements (usually continuous variables). The measurement variables are typically referred to as *discriminating variables* to avoid reference to their dependent or independent function. If a research situation defines the group categories as dependent upon the discriminating variables, then that situation is analogous to MRA. But when the values on the discriminating variables are defined as dependent upon the groups, DA becomes an extension of MANOVA. This situation typically arises from experimental settings in which the group assignment is hypothesized to cause differences in several variables simultaneously. Whether the grouping variable is the independent variable or dependent variable, however, is unimportant to the statistical procedure. Typically, the groups are prespecified and well-defined. Often, however, natural groupings cannot be readily defined, and the categorization must be determined from the data itself using clustering procedures. In the later case, MRA is likely the more appropriate analogy.

4.3 Geometric Overview

The multivariate data set used in DA can be depicted as a multidimensional cloud of sample points (Fig. 4.1). Each dimension, or axis, of the data set is defined by one of the *P* original variables. Each sample point has a position on each axis and therefore occupies a singular location in this *P*-dimensional space. Collectively, all the sample points form a cloud in *P*-dimensional space. Each sample point also belongs to one of two or more groups. If the groups differ in their distribution with respect to the variables, we can imagine each group as being a swarm of points concentrated in a different portion of this space. For DA to be successful, it is not required of the groups that their corresponding measurements be grouped into disjunct sets; indeed, the range of measurements may be the same for all groups. It is necessary, however, for their distributions to differ. Discriminant analysis seeks to highlight among-group differences in these distributions. If indeed group distributions differ, then new canonical axes can be derived that maximize these differences. If the canonical axes represent meaningful ecological gradients, then the relative positioning of groups along the canonical axes provides insight into relationships among the groups.

Geometrically, canonical functions can be viewed as projections through the cloud of sample points at orientations that maximally separate group distributions along each axis. The first canonical axis is drawn through the grand centroid such that the ratio of among-group variation to within-group variation is maximized. Essentially, this means that the positions of group centroids on this canonical axis

138   4. Discriminant Analysis

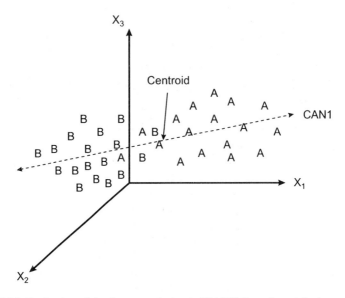

FIGURE 4.1 Derivation of the first canonical axis (CAN1) from the original axes (X_1, X_2, and X_3) in two-group discriminant analysis by translation of the origin to the grand centroid, followed by rotation of the axes around the centroid to maximize the ratio of among-group to within-group variation.

are maximally separated. The second and subsequent canonical axes are constrained by orthogonality (i.e., independence) and maximization of the remaining group differences. Geometrically, this means that the second canonical axis is perpendicular to the first axis in the direction that best separates groups. In other words, the second axis has the next most discriminating power given that it accounts for group differences not explained by the first axis. Subsequent axes follow the same constraints. Hence, in theory there is no redundancy in the canonical functions; they are completely independent and complementary (i.e., orthogonal). However, in practice they are usually slightly redundant due to nonnormal data characteristics.

Geometrically, each canonical axis can be viewed in relation to the original P axes (corresponding to the original P variables). Some of the original P axes may be oriented in the same general direction as the new canonical axes, while others may not. The degree of similarity in orientation between the original axes and the canonical axes indicates how closely related the axes are and provides a geometric means of determining which of the original P variables are most important in the new canonical functions.

4.4 The Data Set

Discriminant analysis deals with problems involving a single 2-level or higher-level categorical grouping variable, and two or more continuous, categorical, or count discriminating variables. Preferably, discriminating variables should be either all continuous or all categorical or all count variables. Discriminant analysis can be conducted on categorical data, but it is likely that certain distributional assumptions (particularly multivariate normality) are likely to be violated (see Goldstein and Dillon 1978 for a discussion of DA for categorical data). There is no consensus on the validity of "mixed" data sets (i.e., those containing both continuous and categorical variables), but most experts advise against their use. The grouping variable and discriminating variables can be considered either as dependent or independent, depending, in part, on the sampling design and particular research situation (see Sec. 4.2.3).

Several other aspects of the data set should be noted: (1) the groups of sampling entities must be mutually exclusive; that is, an entity belonging to one group cannot belong to another group; (2) every sampling entity, regardless of group membership, must be measured on the same set of variables; (3) the number of sampling entities in each group need not be the same; however, generally the greater the disparity in group sample sizes, the lower the effectiveness of the DA in distinguishing among groups; (4) there must be at least two sampling entities per group; and (5) there must be at least two more sampling entities than number of variables.

Discrimination algorithms usually require a two-way data matrix as input. The two-way data can take on many forms. Applications of DA in the wildlife literature are many and varied (see the Bibliography at the end of the chapter), but a substantial proportion concern the assessment of species-habitat associations (Williams 1981). There is a preponderance of applications with avifauna and, to a lesser degree, with small mammals. The grouping variable can vary widely from vegetation types to faunal species. The corresponding discriminating variables can range over a variety of environmental measurements, such as soil, climate, geomorphology, or vegetation characteristics, and faunal measurements, such as species abundance or behavioral and morphological characteristics. Many applications are single-species studies in which groups are determined by the presence or absence of an individual species, as described in an example earlier in this chapter. The corresponding two-way data matrix could look as follows:

	Group	X_1	X_2	...	X_P
1	A	x_{11}	x_{12}	...	x_{1P}
2	A	x_{21}	x_{22}	...	x_{2P}
⋮	⋮	⋮	⋮	⋱	⋮
n	A	x_{n1}	x_{n2}	...	x_{nP}
$n+1$	B	$x_{n+1,1}$	$x_{n+1,2}$	...	$x_{n+1,P}$
$n+2$	B	$x_{n+2,1}$	$x_{n+2,2}$	...	$x_{n+2,P}$
⋮	⋮	⋮	⋮	⋱	⋮
N	B	x_{N1}	x_{N2}	...	x_{NP}

In this book, we discuss DA on a samples-by-variables matrix, where "samples" represent any type of sampling unit (e.g., sites, species, specimens) for which there are natural groups and an associated vector of measurements, and "variables" represent any set of parameters recorded for each entity (e.g., habitat features, morphological characteristics, behavioral characteristics, species abundances). Unless otherwise noted, we refer to the samples (rows) as "entities" or "sampling entities" to reflect the generality of DA to any type of sampling unit. However, our comments are applicable to any data of suitable form.

Empirical Example. The empirical example was taken from a study on bird-habitat relationships in mature, unmanaged forest stands in the central Oregon Coast Range (McGarigal and McComb, unpublished data). In this chapter, we use DA to investigate habitat selection by Hammond's flycatchers. Specifically, we use DA to describe the vegetation structure characteristics that best distinguish occupied from unoccupied sites. The data set is similar to the one described in Chapter 3; it consists of 22 variables, including a single grouping variable (USE), 20 discriminating variables, and single variable (ID) that identifies the sampling point (Fig. 4.2). The grouping variable is dichotomous and distinguishes 48 used sites from 48 unused sites; consequently, this example represents a two-group DA with equal group sample sizes. The discriminating variables are all continuous and were either directly measured or derived for each used and unused site (see Chapter 3 for a brief description of the variables). See Appendix 4.1 for an annotated SAS script of all of the analyses presented in this chapter.

OBS	ID	USE	GTOTAL	LTOTAL	TTOTAL	MTOTAL	OTOTAL	SNAGS6	SNAGM1	SNAGM23	SNAGM45	SNAGM6	SNAGL1	SNAGL23	SNAGL45	SNAGL6	SNAGT	BAS	BAC	BAH	BAT	FHD
1	1S0	NO	21	15	75	20	30	0	0	0	0	0	0	0	1	1	0	20	40	60	1.51115
2	1S1	NO	36	15	95	15	35	0	0	0	1	0	0	1	0	2	20	20	80	120	1.35310
3	1S2	NO	30	30	70	10	55	0	0	1	2	2	1	0	1	7	140	160	0	300	1.53113
4	1S3	NO	11	50	70	20	70	0	0	0	0	1	0	0	3	1	5	60	300	0	360	1.41061
5	1S4	NO	33	40	80	15	65	0	0	1	0	0	0	0	0	1	20	160	0	180	1.47547
.	.	.	.	.	.	.	.	.	.	.	.	.	.	.	.	.	.	.	.	.	.	.
49	1U0	YES	3	15	95	20	55	3	0	0	2	1	0	1	1	2	10	80	40	80	200	1.08919
50	1U1	YES	2	15	80	30	70	5	0	0	1	3	0	0	2	0	11	80	40	180	300	1.15219
51	1U2	YES	2	65	70	15	70	0	0	0	1	0	0	0	3	0	4	60	60	120	240	1.14216
52	1U3	YES	30	55	35	25	75	0	0	0	0	3	0	0	3	2	8	20	20	80	120	1.61978
53	1U4	YES	2	20	95	10	60	2	0	0	0	1	0	0	2	2	7	20	160	40	220	0.98561
.	.	.	.	.	.	.	.	.	.	.	.	.	.	.	.	.	.	.	.	.	.	.

FIGURE 4.2 The example data set listing observation points (OBS labeled 1 through 53) and 21 variables, including a field that distinguishes used sites from unused sites (USE) and 20 vegetation variables with measured values.

4.5 Assumptions

For DA to be strictly applicable, a data set must meet several assumptions (Green 1971; Lachenbruch et al. 1973; Kranowski 1977; Wahl and Kronmal 1977; Klecka 1980; Williams 1981, 1983). Field data sets rarely, if ever, meet the assumptions precisely. Fortunately, there is evidence that certain of these assumptions can be violated moderately without large changes in correct classification results (Harris 1975; Lachenbruch 1975; Nie et al. 1975; Williams 1983). The larger the sample size, the more robust the analysis is to violations of these assumptions. Thus, particular attention should be given to meeting the assumptions when dealing with small sample sizes.

In practice, DA is often accompanied by MANOVA to test whether or not the differences among groups are significant. If inferences are to be made regarding the significance of group differences, then the model assumptions must be met rather well; for merely descriptive purposes, larger departures from ideal data structure are tolerable. Even for descriptive purposes, however, it must be remembered that DA has an underlying mathematical model and, consequently, may be applicable to one data set but not another.

In wildlife research, DA is used both as an exploratory (descriptive) and confirmatory analysis. The technique can be used to explore patterns in the data set whether the assumptions are met or not. Indeed, when the structure of the data is unknown at the time it is collected, exploratory methods are often both necessary and informative. One should recognize, however, that in the absence of statistical reliability (i.e., failure to meet assumptions) any perceived patterns are at best suggestive. Such patterns are properly regarded as preliminary and should be used to suggest hypotheses that subsequently can be tested. A common failure of researchers occurs when they bypass this important step and report the results of exploratory analysis as statistically confirmatory.

The primary assumptions of DA are listed below. The key assumption is that group dispersions (i.e., variance-covariance matrices) are homogeneous (Klecka 1980; Williams 1983). If this assumption is grossly violated, then certain desirable properties of the canonical functions are lost and some degree of distortion occurs in the canonical representations of the data. Understanding the degree to which a given data set adheres to the assumptions will allow you to determine the validity of any statistical inferences made.

4.5.1 Equality of Variance-Covariance Matrices

Discriminant analysis assumes that groups have equal dispersions (i.e., within-group variance-covariance structure is the same for all groups). Specifically, DA requires two things. First, it requires that the variances of the discriminating variables be the same in the respective populations from which the groups of entities have been drawn. Different variables can have different variances, as they most often do, but the variance of a given variable must be the same in each group. Second, DA requires that the correlation (or covariance) between any two variables be the same in the respective populations from which the different groups have been sampled. In other words, the correlation matrix of discriminating variables must be the same in each group.

The second is probably the most critical assumption of DA, and it has received widespread attention in the literature. Several important properties result from this assumption, including linearity of the canonical functions, improved efficiency of parameter estimation, and the invariance of posterior probabilities in canonical space (Williams 1981, 1983). The last property is particularly important for interpretations of canonical plots, as it enables one to display data in canonical space without distortion. With heterogeneous dispersions, the canonical transformations become distorted and the statistical relationships between distances in observation space and their canonical representations become complex and nonintuitive. Under such conditions, one cannot safely use equal frequency ellipses in canonical space for interpreting group differences. *Equal frequency ellipses* (or *confidence ellipses*) are two-dimensional confidence intervals around group means on canonical functions. The inferences drawn from the DA are not translatable back to the original entity space because distance measures are distorted by the canonical transformation. Yet, in wildlife research, equal frequency ellipses are commonly used to assess niche breadth and niche overlap among species. Certainly, before DA can be interpreted with equal frequency ellipses, distortion induced by covariance heterogeneity should be assessed.

Rarely is this assumption met in field data sets. Whether homogeneity of group dispersions is met or not, the procedure often proposed is to calculate the canonical functions and judge their ecological significance by whether they: (1) have an ecologically meaningful and consistent interpretation; (2) contribute more to group separation than any other canonical functions which fail to satisfy (1); and (3) result in significant separation of at least two groups consistent with the ecological interpretation of the functions (Green 1974). Fortunately, there is some evidence that statistical inferences from linear canonical functions are satisfactory if the covariance matrices are not too different (Cooley and Lohnes 1971; Lachenbruch 1975). In two-group CAD, the null hypothesis of no group separation is accepted more frequently when the covariance matrices are unequal (Green 1978), but there is little evidence that moderate violations significantly change classification success (Williams 1983). Thus, failure to meet this assumption exactly should not preclude the possibility of discovering important ecological relationships from the analyses.

The existence of unequal group covariance matrices can have an interesting ecological interpretation of its own. For example, if we discriminate between species based on habitat characteristics at occupied sites (e.g., nest sites), then unequal covariance matrices can indicate that one species is more specialized in its habitat associations than another species (Sakai and Noon 1991). Similarly, when comparing occupied sites to random sites, the test of the equality of covariance matrices can provide useful biological insights into the specificity of the species' habitat selection relative to random locations. In two-group CAD, the covariance test can be interpreted as a significance test for habitat selectivity, and the degree of habitat specialization within a group can be inferred from the determinant of a group's covariance matrix, which is a measure of the generalized variance (Morrison 1976) within the group (Sakai and Noon 1991). In multiple CAD, the covariance test can be interpreted similarly, although the null hypothesis tested is that all group covariances are the same. Thus, rejection implies a difference but does not tell us which groups are significantly different.

4.5.1.1 Diagnostics

(a) Univariate Homogeneity of Variance

Conduct a univariate homogeneity of variance test on each discriminating variable. Various tests are available for this. With a randomized block design, or if the assumption of normality is not met, a good test based upon the distribution of the residuals from an analysis of variance (ANOVA) is Levene's test of homogeneity of variance (Levene 1960). If the F-ratio test of the absolute value of the residuals from an ANOVA is significant, then the hypothesis of homogeneity of variance is rejected. With a more complex sampling design, visual inspection of a residual plot may be more useful. Specifically, plot the residuals against the predicted values. If the variances are homogeneous, the residuals will have a random spread about all predicted values; otherwise the distribution of the residuals should indicate the type of variance stabilizing transformation needed to correct the problem (Sabin and Stafford 1990).

It is important to note that these diagnostics are used to assess univariate homogeneity of variance; they convey nothing about multivariate variance-covariance similarities. Hence, these diagnostics are of limited use. Nevertheless, in practice these diagnostics are often used to determine whether the variables should be transformed prior to the DA. In addition, it is usually assumed that univariate homogeneity of variance is a good step toward homogeneity of variance-covariance matrices.

(b) Equality of Variance-Covariance Matrices

Conduct a Box's M-test (F-ratio test) or similar multivariate test of equal group dispersions. This is a true multivariate test of equality of variance-covariance matrices and is strongly recommended in all applications. SAS uses a chi-square test as given by Morrison (1976).

Empirical Example. In the example, we computed Levene's test of univariate homogeneity of variance for each discriminating variable. Figure 4.3 indicates, for example, that the variance of LTOTAL is significantly different between used and unused sites. Several variables exhibited unequal variances between groups. In an attempt to correct for unequal variances between used and unused sites, we transformed those variables with unequal variances through appropriate transformation functions (e.g., logit or arcsine square root transformation for the percentage variables) and conducted DA on the transformed variables. The DA results based on the original variables and the transformed variables were nearly identical. Therefore, we used the original variables in all further analyses since this finding indicates that any violation of the assumption has little effect on the results.

```
                    General Linear Models Procedure

Dependent Variable: ARES
                          Sum of              Mean
Source            DF      Squares             Square          F Value     Pr > F

Model             1       1493.3009440        1493.3009440    13.28       0.0004

Error             94      10572.5427879       112.4738594

Corrected Total   95      12065.8437319

          R-Square              CV              Root MSE            ARES Mean

          0.123763              93.23757        10.605369           11.37456597
```

FIGURE 4.3 Levene's test of homogeneity of variance of the variable LTOTAL (see text for a description of Levene's test).

```
Discriminant Analysis Test: Homogeneity Within-Group Covariance Matrices

Test Chi-Square Value = 98.493016 with 28 DF      Prob > Chi-Sq = 0.0001

Since the chi-square value is significant at the 0.1000 level,
the within-group covariance matrices will be used in the discriminant
function.

Reference: Morrison, D.F. (1976)   Multivariate Statistical Methods p 252.
```

FIGURE 4.4 Test of homogeneity of within-group covariance matrices.

In addition, we ran DA on a subset of variables identified from the stepwise procedure (see below), and tested the equality of group variance-covariance matrices with a chi-square test (the SAS method). Figure 4.4 indicates that group dispersions are not equal. This result has two direct consequences concerning the use of DA and interpretation of the results. First, since group dispersions are unequal, the individual within-group covariance matrices (W) will be used in the classification criterion (resulting in quadratic discrimination) instead of the pooled covariance matrix (resulting in linear discrimination). Second, and more important, inferences regarding the significance of group differences on canonical axes are not valid. Thus, the results of the DA will be treated solely as descriptive and exploratory. In addition, significant differences in group covariance matrices indicate that habitat variability at used sites ($\ln|W| = 34.11$) is significantly greater than at unused sites ($\ln|W| = 28.91$).

4.5.2 *Multivariate Normality*

Discriminant analysis assumes that the underlying structure of the data for each group is multivariate normal (see Chapter 2, Sec. 2.5.1 for additional discussion of multivariate normality). This permits the precise computation of tests of significance and probabilities of group membership. When this assumption is violated, the computed probabilities are not exact and will not be optimal in the sense of minimizing the number of misclassifications, even though they may still be quite useful if interpreted with caution (Lachenbruch 1975).

4.5.2.1 Diagnostics

(a) Univariate Normality of Original Variables

Conduct a univariate ANOVA on each discriminating variable with the grouping variable as the main effect, and assess the distribution of the residuals. Specifically, assess skewness, kurtosis, and normality and construct stem-and-leaf, box, and normal probability plots of the residuals. In regression and ANOVA, it is not required that the original variables be normally distributed, but rather that the residuals be normally distributed. Thus, an analysis of the residuals will generally indicate if the univariate normality assumption is adequately met, as well as reveal the transformation needed to correct any problem that exists. Because these plots are used to assess univariate normality they are of only limited use. Nevertheless, in practice these diagnostics are often used to determine whether the discriminating variables should be transformed prior to the discriminant analysis. In addition, it is usually assumed that univariate normality is a good step toward multivariate normality.

(b) Normality of Canonical Scores

Assess skewness, kurtosis, and normality and construct stem-and-leaf, box, and normal probability plots of the canonical scores within each group for each canonical function. These diagnostics are generally more appropriate for evaluating multivariate normality than the univariate diagnostics, but even normally distributed canonical scores do not guarantee a multivariate normal distribution.

Empirical Example. In the example, we subjected each discriminating variable to an ANOVA; we treated the grouping variable (USE) as the main effect and the discriminating variable as the dependent variable. We analyzed the residuals for skewness, kurtosis, and normality. Figure 4.5, for example, indicates that the residuals for LTOTAL are skewed toward one side (skewness) and have a more peaked than normal distribution (kurtosis). Moreover, the test statistic for normality (W = Shapiro–Wilk statistic) is significant ($P < 0.0001$), which indicates that the residuals are not normally distributed. The plots all reflect these findings. We conclude that LTOTAL does not adequately meet the univariate normality assumption. As expected, several of the variables are nonnormally distributed. We attempted data transformations to correct the problems, but the results of the DA remained unchanged. Therefore, we used the raw variables in all further analyses.

To assess multivariate normality, we ran DA on a subset of variables identified from the stepwise procedure (see below) and assessed the resulting canonical scores within each group for normality. Figure 4.6 indicates that the NO group (unused sites) is less normally distributed than the YES group (used sites). Skew-

```
Variable = RES                                  Variable = RES

                Moments                         Stem Leaf                              #     Boxplot
                                                  2  66888                             5        |
N             96        Sum Wgts     96           2  222222                            6        |
Mean           0        Sum           0           1  7778                              4        |
Std Dev   16.0546       Variance 257.7515         1  2222233                           7        |
Skewness  -1.1264       Kurtosis   3.2065         0  7777777788888                    13     +-----+
USS       24486.4       CSS       24486.4         0  2222233333                       10     |  +  |
CV             .        Std Mean   1.6386        -0  333322222222222222222            21     *-----*
T:Mean = 0     0        Prob >|T|  1.0000        -0  888887777777777                  15     +-----+
Sgn Rank      40        Prob >|S|  0.8846        -1  32222                             5        |
Num ^= 0      96                                 -1  8887                              4        |
W:Normal   0.9015       Prob < W   0.0001        -2  3                                 1        |
                                                 -2  8                                 1        |
             Quantiles (Def = 5)                 -3  3                                 1        0
                                                 -3
100% Max   27.7083         99%   27.7083         -4
 75% Q3     7.7083         95%   26.2292         -4
 50% Med   -2.2917         90%   22.2292         -5  3                                 1        *
 25% Q1    -7.2917         10%  -17.2917         -5  88                                2        *
  0% Min  -57.7708          5%  -27.7708            ----+----+----+----+-
                            1%  -57.7708         Multiply Stem.Leaf by 10**+1
Range     85.4791
Q3-Q1         15
Mode      -2.2916
                                          27.5+                  Normal Probability Plot
             Extremes                         |                                              +***     *
                                              |                                            *****
  Lowest   Obs      Highest   Obs             |                                            ***
 -57.7708 (10)      26.2292  (46)             |                                         +**
 -57.7708  (9)      26.2292  (96)             |                                       ****
 -52.7708 (13)      27.7083   (4)             |                                     +***
 -32.7708 (42)      27.7083  (84)             |                                   ******
 -27.7708 (14)      27.7083  (85)             |                                ******+
                                              |                             ** +++
                                              |                           ***+++
                                              |                         * +++
                                              |                       +* +
                                              |                    +++*
                                              |               +++
                                              |+
                                              |
FIGURE 4.5 Univariate summary                 |            *
statistics used to assess normality for   -57.5+*      *
the variable LTOTAL.                          +----+----+----+----+----+----+----+----+----+
                                              -2        -1        0       +1       +2
```

ness and kurtosis for the NO group indicate a slightly skewed and flatter than normal distribution. However, the YES group appears to be normally distributed. Overall, these diagnostics indicate that the multivariate normality assumption is probably not grossly violated. One important consequence of this is that the use of a quadratic classification criterion (preferable in this case because of unequal covariance matrices) will be more reliable because quadratic classification is particularly sensitive to nonnormality.

4.5.3 Singularities and Multicollinearity

Discriminant analysis requires that no discriminating variable be a linear combination of other variables in the data set being analyzed. (A variable *x* is a linear combination of one or more other variables if the sum of these variables weighted by constants gives *x*.) This prohibition against such linear combinations is the result of certain mathematical requirements that the matrix be *nonsingular*, but it also makes sense intuitively. A variable defined by a linear combination of other variables does not contain any new information beyond that which is contained in the component

4.5 Assumptions

```
------------------USE=NO------------------          ------------------USE=NO------------------

Variable = CAN1                                     Variable = CAN1

              Moments                               Stem Leaf                    #     Boxplot
                                                     -0 7                        1        |
N                48    Sum Wgts         48           -2 1                        1        |
Mean         -2.3003   Sum        -110.4120          -4 3                        1        |
Std Dev       0.9134   Variance      0.8343          -6 9                        1        |
Skewness      0.6411   Kurtosis     -0.0948          -8 50                       2        |
USS         293.1856   CSS          39.2105         -10                                   |
CV          -39.7079   Std Mean      0.1318         -12 4                        1        |
T:Mean = 0  -17.4479   Prob >|T|     0.0001         -14 21                       2        |
Sgn Rank       -588    Prob >|S|     0.0001         -16 57                       2        |
Num ^= 0         48                                 -18 61108                    5     +-----+
W:Normal      0.9415   Prob < W      0.0279         -20 800                      3     |     |
                                                    -22 7299430                  7     *--+--*
            Quantiles (Def = 5)                     -24 73                       2     |     |
                                                    -26 987                      3     |     |
   100% Max  -0.1747      99%   -0.1747             -28 00                       2     |     |
    75% Q3   -1.8906      95%   -0.4293             -30 97421886                 8     +-----+
    50% Med  -2.3038      90%   -0.8030             -32 0632                     4        |
    25% Q1   -3.0967      10%   -3.2610             -34 6                        1        |
     0% Min  -3.8828       5%   -3.5568             -36 3                        1        |
                           1%   -3.8828             -38 8                        1        |
   Range      3.7081                                    ----+----+----+----+
   Q3-Q1      1.2061                                Multiply Stem.Leaf by 10**-1
   Mode      -3.8828
                                                              Normal Probability Plot
              Extremes                         -0.1+                                      *+++
                                                   |                                    *  ++
   Lowest    Obs     Highest    Obs                |                                   *  ++
   -3.8828   (14)    -0.8030    (3)                |                                  *  ++
   -3.6328   (13)    -0.6867    (32)               |                                ** ++
   -3.5568   (10)    -0.4293    (45)               |                                ++
   -3.2978   (19)    -0.2108    (4)                |                              *+ +
   -3.2609   (26)    -0.1747    (9)                |                              +*+
                                                   |                             +*
                                                   |                            ++**
------------------USE=YES------------------        |                           ****
                                                   |                          ++**
Variable = CAN1                                    |                         ++**
                                                   |                       ++**
              Moments                              |                      ******
                                                   |                  ****+
N                48    Sum Wgts         48         |                * ++
Mean          2.3003   Sum         110.4120        |              * ++
Std Dev       1.0797   Variance      1.1657    -3.9+  *      ++
Skewness     -0.5743   Kurtosis      0.2247        +----+----+----+----+----+----+----+----+
USS         308.7646   CSS          54.7895          -2        -1        0        +1      +2
CV           46.9380   Std Mean      0.1558
T:Mean = 0   14.7603   Prob >|T|     0.0001       ------------------USE=YES------------------
Sgn Rank        583    Prob >|S|     0.0001
Num ^= 0         48                               Variable = CAN1
W:Normal      0.9724   Prob < W      0.4679
                                                  Stem Leaf                    #     Boxplot
            Quantiles (Def = 5)                    4 02                        2        |
                                                   3 555579                    6        |
   100% Max   4.2046      99%    4.2046            3 012224                    6     +-----+
    75% Q3    3.1054      95%    3.9140            2 5555778999                10    |     |
    50% Med   2.3809      90%    3.5448            2 01122233                  8     *--+--*
    25% Q1    1.5417      10%    0.7984            1 556889                    6     +-----+
     0% Min  -0.7532       5%    0.6002            1 00334                     5        |
                           1%   -0.7532            0 668                       3        |
   Range      4.9578                               0
   Q3-Q1      1.5638                              -0 0                         1        |
   Mode      -0.7532                              -0 8                         1        |
                                                     ----+----+----+----+
              Extremes
                                                              Normal Probability Plot
   Lowest    Obs     Highest    Obs            4.25+                                 +++++ *
   -0.7532   (5)     3.5448     (36)               |                                ***+* *
   -0.0354   (1)     3.6770     (35)               |                              ******
    0.6002   (6)     3.9140     (22)               |                             *****
    0.6082   (17)    3.9946     (32)               |                            ******
    0.7984   (2)     4.2046     (24)           1.75+                          ****++
                                                   |                        +****
                                                   |                      *+***
                                                   |                  +++++
                                                   |+++++       *
                                              -0.75+     *
                                                   +----+----+----+----+----+----+----+----+
                                                     -2        -1        0        +1      +2
```

FIGURE 4.6 Univariate summary statistics used to assess normality of the canonical function.

variables, so it is completely redundant. Such a redundant variable results in a singular matrix if not discarded. We refer to such a variable as a "singularity."

Singularities are often troublesome in ecological data sets since often we are interested in analyzing variables derived algebraically from field-measured variables, and such variables result in singularities if the component variables are also included in the final data set to be analyzed. A similar, although far less likely, singularity problem can occur in species presence or abundance data sets because two or more species may have identical distributions among samples. The solution, of course, is to eliminate one or more of the offending variables.

Two variables that are perfectly correlated ($r=1$), even by chance, cannot be used at the same time, since one variable can be derived explicitly from the other, resulting in a singular, or nonintuitive, matrix. This makes sense intuitively, because the two variables are redundant. In practice, rarely are two variables perfectly correlated, unless by design. However, data sets typically contain variables that are highly (but not perfectly) correlated, and it is these linear associations (called multicollinearities) that are troublesome. *Multicollinearity* refers to multiple near-linear dependencies (i.e., high correlations) among variables in the data set. The lack of multicollinearity is not a specified assumption of the discriminant model, but the influence it has on the interpretation of canonical functions is noteworthy (Williams 1981, 1983).

Traditionally, most ecological studies have involved attempts to interpret canonical functions by means of the signs and magnitudes of their standardized canonical coefficients (i.e., variable weights). Since the canonical functions are uncorrelated linear combinations of the original variables, each is assumed to represent the influence of a distinct complex of ecological factors. The meaning of these complexes is thought to be reflected in some direct way by the signs and magnitudes of the corresponding coefficients. Under certain conditions, for example, when a canonical function is dominated by a single variable (i.e., when one coefficient is much larger than all others) and the covariance structure is simple, such an interpretation is straightforward. However, when the covariance structure is complex (i.e., when there are several coefficients of significant size), interpretation of canonical functions is less direct. These difficulties arise from the fact that the original variables are correlated, some of them perhaps highly correlated. Individual coefficients, therefore, measure not only the influence of their corresponding original variables, but also the influence of other variables as reflected through the data correlation structure. The effect of a discriminating variable on the canonical coefficient is only partially given by the numerical value of the corresponding coefficient. Often, positively correlated variables have canonical coefficients with different signs. For complex structures, magnitudes and even signs of coefficients are dependent on which additional variables are included in the model. Hence, as in PCA, coefficients merely define the linear equations; they are not correlations between the original variables and the canonical functions. For this reason, significant linear dependencies are usually eliminated prior to the DA, and it is best not to base interpretations of the canonical functions on the canonical coefficients.

The most popular solution to the multicollinearity problem is to subject all variables involved in high pairwise correlations (e.g., $r = 0.7$) to a univariate, one-way ANOVA with the grouping variable as the main effect. As a general rule, for each pair of highly correlated variables with significant among-group differences (e.g., $P < 0.05$ from ANOVA), the variable retained is the one with the greatest among-group variance (or largest F value) or ease of ecological interpretation; the others are eliminated (Noon 1981). Note that since variables are often involved in several high pairwise correlations, the removal of one offending variable will often eliminate several high pairwise correlations.

Another proposed solution to the multicollinearity problem is to use principal components analysis to create new, completely independent (i.e., uncorrelated) composite variables from the original variables (McDonald 1981). These new composite variables (i.e., the principal components) can then be used in discriminant analysis. Hopefully, the canonical functions derived from the principal components will be more stable and easier to interpret than in the original canonical functions without sacrificing the discriminating ability. This approach will only be effective if the principal components represent very meaningful and easy to interpret gradients. However, even if this were the case, interpreting the canonical functions derived from composite variables is much less straightforward than interpreting functions derived from the original variables. Furthermore, because principal components represent gradients of maximum variance among all sampling entities, there is good reason to suspect that the among-group structure of particular interest in DA would be distorted or perhaps obscured completely in the principal component transformation.

A third solution to the multicollinearity problem is to use a procedure similar to ridge regression. The basic idea is to allow a certain bias in the canonical coefficients by adding a constant (k) to all the eigenvalues (DiPillo 1976; Smidt and McDonald 1976; McDonald 1981). The justification for this is that multicollinearity will generally cause some of the eigenvalues to be too small and others to be too large. In general, small characteristic roots are underestimates of the corresponding population values and large roots are overestimates. Small eigenvalues lead to the problems of instability in the coefficients. The ridge-adjusted approach works by increasing all roots by a constant amount (k). Thus, the ill-conditioning effect of underestimating small roots is countered without having much influence on the larger roots.

A final solution is to simply remove one or more of the offending variables, recompute the canonical solution, and compare the results (e.g., canonical structure coefficients; see Sec. 4.9.0). If the results remain essentially unchanged, then multicollinearity is apparently not resulting in spurious results, and the variables can be retained or eliminated without much impact on the final conclusions.

4.5.3.1 Diagnostics

Diagnostic procedures for detecting singularities are not necessary per se, since singularities either will be obvious to the researcher beforehand (e.g., in the case of algebraically derived variables) or will be made apparent when the data are analyzed using DA. However, there are procedures for detecting significant multicollinearity problems.

(a) Pairwise Correlations Among Variables

Calculate all possible pairwise correlations among the discriminating variables; high correlations (e.g., $r = 0.7$) suggest potential multicollinearity problems and indicate the need to eliminate some of the offending variables.

(b) Agreement Between Canonical Weights and Loadings

Compare the signs and relative magnitudes of the standardized canonical coefficients (weights) and structure coefficients (loadings) for disagreement. Pronounced differences, particularly in signs or rank order, indicate multicollinearity problems and highlight the need for corrective actions and caution when interpreting the canonical solutions.

Empirical Example. In the example, recall that we derived two variables (SNAGT and BAT) by linearly combining other variables in the data set. Thus, the full data set contains two singularities, and we would be unable to compute the canonical functions from the full data set as a result. However, due to multicollinearity considerations, we eliminated one or more of the variables involved in the singularities prior to the DA. Specifically, several variables were involved in high pairwise correlations. For example, Figure 4.7 indicates that LTOTAL is highly correlated with TTOTAL, BAC, BAT, and FHD. Thus, either LTOTAL or all four other variables would have to be eliminated prior to DA. We conducted a univariate ANOVA for each variable involved in a high pairwise correlation, treating the grouping variable (USE) as the main effect and each discriminating variable (e.g., LTOTAL, TTOTAL) as a dependent variable in separate ANOVAs. Fig-

CORRELATION ANALYSIS

Pearson Correlation Coefficients / Prob > |R| under Ho: Rho=0 / N = 96

LTOTAL	TTOTAL	BAC	BAT	FHD	OTOTAL	GTOTAL
1.00000	-0.80786	0.77876	0.74014	-0.69086	0.64532	-0.57822
0.0	0.0001	0.0001	0.0001	0.0001	0.0001	0.0001

SNAGT	SNAGM45	MTOTAL	BAS	SNAGS6	SNAGM1	SNAGL45
0.56892	0.52882	-0.49276	0.49103	0.37954	0.33786	0.26999
0.0001	0.0001	0.0001	0.0001	0.0001	0.0008	0.0078

SNAGL6	BAH	SNAGM6	SNAGM23	SNAGL23	SNAGL1
0.25277	-0.22070	0.21286	0.20216	0.10905	0.01296
0.0130	0.0307	0.0373	0.0482	0.2902	0.9003

FIGURE 4.7 Correlation matrix containing pairwise correlations among the original variables. Only correlations involving LTOTAL are shown.

```
                   General Linear Models Procedure

Dependent Variable: LTOTAL

                              Sum of           Mean
Source            DF          Squares          Square      F Value      Pr > F

Model              1          61155.510417     61155.510417    234.77    0.0001
Error             94          24486.395833       260.493573
Corrected Total   95          85641.906250

             R-Square               CV           Root MSE              LTOTAL Mean
             0.714084          33.95621         16.139813              47.5312500


Dependent Variable: TTOTAL

                              Sum of           Mean
Source            DF          Squares          Square      F Value      Pr > F

Model              1          59650.510417     59650.510417    143.07    0.0001
Error             94          39192.729167       416.943927
Corrected Total   95          98843.239583

             R-Square               CV           Root MSE              TTOTAL Mean
             0.603486          41.00070         20.419205              49.80208333
```

FIGURE 4.8 Univariate ANOVA results for variables LTOTAL and TTOTAL.

ure 4.8, for example, indicates that LTOTAL is a better discriminator between groups than TTOTAL because the F value is much greater for LTOTAL than TTOTAL. In fact, LTOTAL is a better discriminator than all four of the variables with which it is highly correlated. Therefore, LTOTAL will be retained in the final data set and TTOTAL, BAC, BAT, and FHD will be eliminated. Alternatively, we could eliminate LTOTAL and thereby abolish four high pairwise correlations. This latter strategy would be preferable if there were an important reason for including either TTOTAL, BAC, BAT, or FHD in the final model. As of result of this analysis, we eliminated five variables (TTOTAL, BAC, BAT, FHD, and SNAGM45) involved in high pairwise correlations from the data set, and by so doing, also avoided any singularities.

4.5.4 *Independent Random Sample and the Effects of Outliers*

Discriminant analysis assumes that random samples of observation vectors (i.e., the discriminating characteristics) have been drawn independently from respective P-dimensional multivariate normal populations and, as with most other multivariate techniques (e.g., PCA, CANCOR), DA is adversely affected by point clusters and outliers (see Chapter 2, Sec. 2.5.2 for additional discussion of these issues). Robust procedures that are less sensitive to the effects of outliers have been developed (Randles et al. 1978; Harner and Whitmore 1981), but these procedures have not been widely applied.

4.5.4.1 Diagnostics

(a) Standard Deviations

Perhaps the simplest way to screen for outliers is to standardize the data by subtracting the mean and dividing by the standard deviation of each variable (i.e., standardize to zero mean and unit variance) and then inspecting the data for entities with any value more than about 2.5 standard deviations from the mean on any variable. In this case, the standardization is done separately within each group.

(b) Univariate Plots

Construct univariate stem-and-leaf, box, and normal probability plots for each discriminating variable within each group and check for suspected outliers (i.e., points that fall distinctly apart from all others).

(c) Plots of Canonical Scores

Construct stem-and-leaf, box, and normal probability plots of the canonical scores within each group for each canonical function and check for suspected outliers. If more than two canonical functions exist, as in multiple CAD, construct scatter plots for each pair of canonical functions and check for suspect points.

Empirical Example. In the example, we computed standardized variable scores within each group and constructed univariate plots for each discriminating variable (SAS results are not shown but are similar to those in Fig. 4.5). Several points deviated noticeably from the group mean, but we did not consider any to be true outliers. Instead, we considered these points to be real and meaningful variations in patch-level vegetation characteristics. In addition, we ran DA on a subset of variables identified from the stepwise procedure (see Sec. 4.7.1) and inspected the canonical scores for potential outliers. Again, although a few entities were more than 2.5 standard units from their group mean, we did not consider any to be true outliers.

4.5.5 Prior Probabilities Are Identifiable

Discriminant analysis assumes that prior probabilities of group membership are identifiable. A prior probability represents the probability of an entity belonging to each group. It is not required that prior probabilities be equal among groups, but only that the prior probabilities of group membership be known. Prior probabilities (or "priors") may differ among groups due to unequal group population sizes, unequal sampling effort among groups, or any number of other factors. Priors influence the forms of both classification and canonical functions (Williams 1981, 1983). Since this influence is also reflected in their statistical properties, an incorrect or arbitrary specification of prior probabilities can distort or otherwise obscure any underlying structure of the data.

In ecological investigations, it is often impractical, or even impossible, to sample in such a way that reasonable estimates of priors are able to be obtained. In such cases, systematic or stratified samples commonly are collected, and priors are either guessed, determined from ancillary information (i.e., group sample sizes), or assigned arbitrarily. Unless stated otherwise, most statistical programs assume that the priors are equal for all groups. Alternatively, relative sampling intensities are often used to estimate the priors.

When priors are replaced by estimates that bear no direct relationship to them (e.g., when replaced by relative sampling intensities), an uncontrolled and largely inscrutable amount of arbitrariness is introduced into the DA. First, there is a very substantial variation in canonical coefficients, both in signs and magnitudes. Second, the variation is not always continuous or symmetric; small changes in relative sampling intensities at certain critical (and effectively unpredictable) points can result in large fluctuations in the corresponding coefficients. And third, the dominance of the top canonical variate depends on relative sampling intensities, but in a nonmonotonic fashion (i.e., not in a strictly increasing or decreasing manner).

4.5.5.1 Diagnostics

Unfortunately, we are not aware of any practical way to diagnose whether the priors have been estimated correctly.

4.5.6 Linearity

The effectiveness of DA depends on the implicit assumption that variables change linearly along underlying gradients and that there exist linear relationships among the variables such that they can be combined in a linear fashion to create canonical functions (see Chapter 2, Sec. 2.5.3, for additional discussion of the linearity assumption). Linearity is not a specified assumption of the mathematical model per se, but it determines to a large extent the appropriateness and effectiveness of DA in any particular case.

4.5.6.1 Diagnostics

Scatter Plots of Canonical Scores

In multiple CAD, construct scatter plots of each pair of canonical functions. Arched or curvilinear configurations of sample points often indicate pronounced nonlinearities.

4.6   Sample Size Requirements

There is no absolute rule for determining the appropriate sample size in DA; it largely depends on the complexity of the environment under investigation. Obviously, the more homogeneous the environment, the fewer samples needed to uncover any patterns that exist. Beyond this, there are some general rules to follow, as well as some specific rules proposed by various authors.

4.6.1 General Rules

In all eigenanalysis problems, as a minimum there must be at least as many sampling entities as variables. In DA, however, there must be at least two more entities than variables and at least two entities per group. Beyond this, enough samples of each group should be taken to ensure that means and dispersions are estimated accurately and precisely. This can be done in several ways, for example, by sampling sequentially until the mean and variance of the parameter estimates (e.g., eigenvalues, canonical coefficients) stabilize. Once the data have been gathered, the stability of the results can be assessed using a resampling procedure (see Sec. 4.8.4), and the results can be used to determine the sample size needed to estimate the parameters with a certain level of precision in future studies.

4.6.2 Specific Rules

The following specific rules have been suggested by different authors (where P = number of discriminating variables, and G = number of groups):

Rule A	$N \geq 20 + 3P$	(Johnson 1981)
Rule B	If $P \leq 4, N \geq 25G$	(Wahl and Kronmal 1977)
	If $P > 4, N \geq [25 + 12(P-4)]G$	
Rule C	Each group, $N \geq 3P$	(Williams and Titus 1988)

Rule C is the best guide for determining the appropriate sample size because it is based on extensive simulation results. In practice, however, group sample sizes are often fixed. If this is the case and the full data set does not meet sample size requirements, then one can determine roughly the number of variables to be included in the analysis given the fixed number of groups and group sample sizes. If the number of variables needs to be reduced, then variables with less ecological interpretation or management significance are eliminated, or those with less ability to discriminate among groups are taken out. Alternatively, the variables can be divided into two or more groups of related variables (e.g., those measuring vegetation floristics and those measuring vegetation structure), and separate DAs conducted on each group. Otherwise, stepwise discrimination procedures (see Sec. 4.7.1) are used to select the best set of discriminating variables from each subset of variables and combine them into one subsequent analysis. If you are forced to work with sample sizes smaller than that recommended by Rule C, the stability of the parameter estimates should be evaluated (e.g., by using a resampling procedure; see Sec. 4.8.4 and the validation procedures discussed in Sec. 4.10), and findings interpreted cautiously.

Empirical Example. In the example, the full data set contains 48 samples within each group for a total of 96 observations and $P = 20$ discriminating variables. Based on Rule A ($N \geq 20 + 3 \cdot 20 = 80$), the example data set contains an adequate number of observations, while based on Rule B ($N = [25 + (12(20-4)]2 = 434$), the example data set contains far too few observations. Based on the preferred rule, Rule C ($N \geq 3 \cdot 20 = 60$), group sample sizes are still too small. Given a fixed group sample size of 48, we can include up to 16 variables in the DA based on Rule C. Because we eliminated five variables due to multicollinearity, sample size requirements will be met.

4.7 Deriving the Canonical Functions

The first major step in CAD is to derive the canonical functions. This derivation step represents the nuts and bolts of the analysis; it includes several small steps which must be understood before we can consider how to interpret the canonical functions. These include: (1) whether to use all the variables simultaneously to discriminate among groups, or to use a stepwise procedure for selecting some "optimal" subset of variables; (2) how to compute the eigenvalues from the data matrix; and (3) how to compute the eigenvectors and canonical coefficients from the data matrix.

4.7.1 Stepwise Selection of Variables

Researchers frequently encounter situations in which they have many potential discriminating variables, but do not know which of these variables will be good discriminators. Frequently, we collect data on many suspected discriminators with the specified aim of identifying the most useful ones. In these situations, several of the variables may be poor discriminators, or they may be highly correlated with other equally powerful discriminators. Weak or redundant variables employed in the analysis only complicate procedures and may even increase the number of misclassifications (Klecka 1980).

One way to eliminate unnecessary variables is through a "stepwise selection" procedure to select the most useful subset of discriminating variables. A "forward selection" or "backward selection" procedure could be used, but stepwise procedures are more popular. Stepwise methods enter and remove variables one at a time, selecting them on the basis of some objective criterion. It is important to note that the sequence in which variables are selected does not necessarily correspond to their relative discriminatory power. Because of multicollinearity, an important discriminator may be selected late in the stepwise sequence, or not at all, because its unique contributions are not as great as those of other variables (Klecka 1980). Although stepwise procedures produce an "optimal" set of discriminating variables, they do not guarantee the best (maximal) combination. To guarantee the best combination, you would have to test every possible combination of variables. Stepwise procedures offer an efficient way of seeking the best combination even though they cannot guarantee that the final set is indeed the best. Stepwise selection of variables is optional in DA and generally is used when the objective is to describe differences among groups in the most parsimonious manner possible (i.e., with as few variables as possible).

Stepwise procedures must employ some objective measure of discrimination as the selection criterion. Several alternative criteria exist which can be used to maximize various notions of group differences. Which is "best" depends upon the research situation, although the end result will usually be the same regardless of the criterion used (Klecka 1980). The Wilks' Lambda statistic is the most commonly used criterion. This statistic considers both the differences among groups and the cohesiveness (i.e., the degree to which entities cluster near their group centroid), or homogeneity, within groups. Thus, a variable that increases cohesiveness without changing the separation among group centroids may be selected over a variable that increases separation without changing the cohesiveness (Klecka 1980). The Wilks' Lambda procedure selects the variable at each step that minimizes the overall Wilks' Lambda statistic. Wilks' Lambda is a likelihood ratio statistic for testing the hypothesis that group means are equal in the population; Lambda approaches zero if any two groups are well separated.

Most stepwise programs require a variable to pass certain minimum conditions before it is tested on the selection criterion for entry or removal. These conditions are usually a *tolerance test* and a *partial F-test*. The tolerance for a variable not yet selected equals one minus the squared multiple correlation between that variable and all variables already entered when the correlations are based on the within-groups correlation matrix (Klecka 1980). Tolerance represents the percentage of variance in a variable *not* accounted for by the variables already entered. Thus, a small tolerance indicates that most of the variation in that variable is already accounted for; that is, the variable is a near linear combination of the variables already entered. A variable with a small tolerance (e.g., less than 0.001) is likely to cause computational inaccuracies in the eigenanalysis because of the rapid accumulation of rounding errors, and may cause the matrix to be singular. Moreover, a variable with a small tolerance is highly redundant with the variables already entered, and thus has little unique information to contribute. Hence, at each step of the procedure, each potential variable-to-enter must pass some minimum tolerance level.

The partial F-test consists of both an F-to-enter and F-to-remove. The F-to-enter is a partial multivariate F statistic which tests the significance of the added discrimination introduced by the variable being considered after taking into account the discrimination achieved by the other variables already entered (Klecka 1980). If this F statistic is small and does not meet some specified significance level (e.g., if $P > 0.1$), we do not want to consider the corresponding variable for entry because it does not add enough to the overall discrimination. A variable must pass both the tolerance and F-to-enter tests before it is tested with the entry criterion (e.g., Wilks' Lambda).

The F-to-remove is also a partial multivariate F statistic, but it tests the significance of the decrease in discrimination should that variable be removed from the list of variables already selected (Klecka 1980). This test is performed at the beginning of each step to see if there are any variables that can be dropped from the model because they no longer make a sufficiently large unique contribution to discrimination. A variable may lose its unique discriminatory power because of correlations with other variables subsequently entered into the model. Often the F-to-remove is set slightly higher than the F-to-enter to make it more difficult for a variable to leave the model once it has been entered. On the final step, the F-to-remove statistic can be used to rank order the variables in terms of their unique discriminating power. The variable with the largest F-to-remove makes the largest unique contribution to the overall discrimination independent of the contributions made by all the other variables. The variable with the second largest partial F statistic is next most important, and so forth. Note that these are partial F statistics and not overall F statistics.

Empirical Example. In the example, we conducted a stepwise discriminant analysis on the data set after removing five variables involved in high pairwise correlations. We selected the SAS defaults for the F-to-enter and F-to-remove (0.15), but chose a less stringent tolerance level of 0.001. The stepwise summary in Figure 4.9 indicates that 7 of 17 variables were selected based on the Wilks' Lambda criterion. In the first step, the variable LTOTAL entered the model; it is the single variable resulting in the smallest Wilks' Lambda value. The squared canonical correlation indicates that 71 percent of the variance in LTOTAL is accounted for by group differences. In the next step, SNAGT entered the model because it resulted in the smallest overall Wilks' Lambda. It is the variable with the largest partial F-to-enter and increases the squared canonical correlation associated with the canonical function to 79 percent. After seven variables have been entered into the model, there are no other variables that contribute significantly to group separation after accounting for the separation attributed to these seven variables. In the final model, 84 percent of the variance in the canonical function is accounted for by group differences. In addition, the Wilks' Lambda statistic indicates a significant difference in group means based on either all seven variables or LTOTAL alone; recall that this test is invalid

```
                      STEPWISE discriminant analysis

         96 observations         15 variable(s) in the analysis
          2 class levels          0 variable(s) will be included

         The Method for Selecting Variables will be: STEPWISE

         Significance level to enter = 0.1500
         Significance level to stay  = 0.1500

                         Class Level Information

              USE      Frequency         Weight         Proportion

               N          48            48.0000          0.500000
               Y          48            48.0000          0.500000


Stepwise Selection:   Summary
                                                        Average
                                                        Squared
              Variable          Number    Wilks'  Prob < Canonical   Prob >
Step   Entered   Removed          In      Lambda  Lambda Correlation  ASCC
-----------------------------------------------------------------------------
  1    LTOTAL                     1     0.28591605 0.0001  0.71408395 0.0001
  2    SNAGT                      2     0.21460015 0.0001  0.78539985 0.0001
  3    BAH                        3     0.19032148 0.0001  0.80967852 0.0001
  4    GTOTAL                     4     0.17358669 0.0001  0.82641331 0.0001
  5    BAS                        5     0.16815745 0.0001  0.83184255 0.0001
  6    STUMPL                     6     0.16033173 0.0001  0.83966827 0.0001
  7    MTOTAL                     7     0.15615909 0.0001  0.84384091 0.0001
```

FIGURE 4.9 Summary of stepwise discriminant analysis on the example data set.

because group dispersions are unequal. The squared canonical correlation and Wilk's Lambda indicate that LTOTAL alone accounts for most of the discriminatory power. Based on these results, we might find it instructive to compare the results of CAD (and in particular, the correct classification rates) among models containing LTOTAL only, LTOTAL and SNAGT only, and all seven variables.

4.7.2 Eigenvalues and Associated Statistics

Computationally, DA is essentially an eigenanalysis problem, although the details of the matrix algebra are not presented here. For an $N \times P$ data set with G groups, where N is the number of sampling entities (rows) and P is the number of variables (columns), there are Q (equal to $G - 1$ or P, whichever is smaller) eigenvalues or characteristic roots. The eigenvalues are determined by solving the characteristic equation

$$|A - \lambda W| = 0$$

where A is the among-groups sums-of-squares and cross products (covariance) matrix, W is the within-groups sums-of-squares and cross products (covariance) matrix, and λ is the vector of eigenvalue solutions.

Each eigenvalue is associated with one canonical function. Consequently, for two-group DA, there exits only one canonical function, while for multiple DA there exist Q canonical functions. Eigenvalues represent the variances of the corresponding canonical functions; hence, they measure the extent of group differentiation along the dimension specified by the canonical function.

When the solution includes more than one eigenvalue (i.e., $Q > 1$), the first eigenvalue is always the largest. Therefore, the first canonical function defines the dimension or gradient with the single most discriminatory power (i.e., maximum among-group variance). The second eigenvalue measures group separation along the second canonical function; it represents the largest among-group variance in a dimension orthogonal (i.e., independent) to the first dimension. Thus, the second function provides the greatest discrimination *after* the first has done its best. The third eigenvalue corresponds to the function with the greatest among-group discrimination, given that it is orthogonal to the first and second functions; it provides the greatest additional discrimination *after* the first and second have done their best. And so on for Q eigenvalues and functions.

All of the Q eigenvalues will be positive or zero, and the larger the eigenvalue is, the more the groups will be separated on that function. Thus, the function with the largest eigenvalue (always the first function) is the most powerful discriminator, while the function with the smallest eigenvalue is the weakest. Eigenvalues near zero indicate that the corresponding function has minimal discriminatory power.

Empirical Example. In the example, we ran DA on the subset of variables identified by the stepwise procedure and computed the single eigenvalue ($\lambda_1 = 5.4037$) (Fig. 4.10). The remainder of the output in Figure 4.10 will be discussed later.

```
               Canonical Discriminant Analysis

                   Adjusted      Approx        Squared
       Canonical   Canonical    Standard      Canonical
      Correlation Correlation    Error       Correlation

1      0.918608    0.913995     0.016022      0.843841

           Eigenvalues of INV(E)*H = CanRsq/(1-CanRsq)

       Eigenvalue   Difference    Proportion    Cumulative

1        5.4037         .           1.0000        1.0000

Test of H0: The canonical correlations in the current row
             and all that follow are zero

       Likelihood
         Ratio           F        Num DF    Den DF   Pr > F

1     0.15615909      67.9326        7         88    0.0001
```

FIGURE 4.10 Eigenvalue, canonical correlation, and associated statistics for the canonical function.

4.7.3 Eigenvectors and Canonical Coefficients

The other product of eigenanalysis is an eigenvector associated with each eigenvalue. The eigenvectors are determined by solving the following system of equations:

$$[A - \lambda_i W]v_i = 0$$

where A is the among-groups sums-of-squares and cross products (covariance) matrix, W is the within-groups sums-of-squares and cross products (covariance) matrix, λ_i is eigenvalue corresponding to the ith canonical function, and v_i is the eigenvector associated with the ith eigenvalue.

The number of coefficients in an eigenvector equals the number of variables in the data matrix (P). These are the coefficients of the variables in the linear equations that define the canonical functions and are referred to as *canonical coefficients* (or *canonical weights*). Although these raw coefficients can be used for classification purposes, they are totally uninterpretable as coefficients, and the scores they produce for entities have no intrinsic meaning. In seeking to interpret the canonical functions, we would like to know which of the original P variables contribute most to each function. For this purpose, comparison of the relative magnitudes of the combining weights as given by the elements of each eigenvector is inappropriate because these are weights to be applied to the variables in raw score scales and are therefore affected by the particular unit (i.e., measurement scale) used for each variable.

A simple adjustment to these coefficients results in coefficients that give the canonical function more desirable properties. These adjusted coefficients are defined as follows:

$$u_i = v_i\sqrt{n-g}, \qquad c_i = v_i\sqrt{\frac{w_{ii}}{N-G}}$$

where u_i is the vector of unstandardized canonical coefficients (explained below) associated with the ith eigenvalue, c_i is the vector of standardized canonical coefficients associated with the ith eigenvalue, v_i is the vector of raw canonical coefficients associated with the ith eigenvalue, w_{ii} is the sums-of-squares for the ith variable, or the ith diagonal element of the within-groups sums-of-squares and cross products (covariance) matrix, N is the number of sampling entities, and G is the number of groups.

When we adjust in this manner the eigenvector coefficients that have been derived from the raw data, we refer to them as *unstandardized canonical coefficients* (or *unstandardized canonical weights*). These coefficients represent the weights that would be applied to the variables in raw form to generate standardized canonical scores. When we adjust the eigenvector coefficients that have been derived from the standardized data (i.e., variables with unit variance), we refer to them as standardized canonical coefficients (or standardized canonical weights). The matrix of standardized canonical coefficients is sometimes referred to as the *canonical pattern* (or *pattern matrix*). The standardized coefficients also can be computed directly from the unstandardized coefficients, as given in the formula above.

The adjusted coefficients represent the weights that would be applied to the variables in either raw or standardized form to generate standardized canonical scores. Note that the canonical scores are the same whether they are determined from the unstandardized coefficients and raw data or the standardized coefficients and standardized data. This means that the canonical scores over all the samples will have a mean of zero and a within-groups standard deviation of one (note that the total sample standard deviation will be greater than one). These coefficients produce canonical scores that are measured in standard deviation units. This means that each axis is stretched or shrunk such that the score for an entity represents the number of standard deviations it lies from the grand centroid (the point in space where all of the discriminating variables have their average values over all samples). In this manner, the score for an entity immediately tells how far it is from the average score.

Empirical Example. In the example, we ran DA on the subset of variables identified by the stepwise procedure and computed the unstandardized (SAS refers to these as raw) and standardized canonical coefficients (Fig. 4.11); the raw canonical coefficients (eigenvectors) as we have defined them are not printed by SAS, although they could be calculated with a little trouble.

```
                Raw Canonical Coefficients

                              CAN1

        LTOTAL         0.0548457153
        SNAGT          0.0742615501
        BAH            0.0146462641
        GTOTAL        -.0193123412
        BAS            0.0066126633
        SNAGL6         0.2651711687
        MTOTAL        -.0194043546


            Standardized Canonical Coefficients

                              CAN1

        LTOTAL         1.646736324
        SNAGT          0.397480978
        BAH            0.650438733
        GTOTAL        -0.417209741
        BAS            0.313626417
        SNAGL6         0.316969705
        MTOTAL        -0.225091687
```

FIGURE 4.11 Unstandardized and standardized canonical coefficients associated with the canonical function. Note that SAS refers to unstandardized coefficients as "raw canonical coefficients"; these are not raw coefficients, or eigenvectors, as we have defined them in the text.

4.8 Assessing the Importance of the Canonical Functions

As previously stated, the discriminatory power of each canonical function is measured by its eigenvalue. The larger the eigenvalue is, the more the groups will be separated on that function. Eigenvalues near zero indicate that the corresponding function has minimal discriminatory power. We are usually interested in knowing how small the eigenvalue has to be before we consider it to be the result of sampling or measurement error, and not a truly different dimension on which groups are separated. There are several criteria that can be used to help evaluate the importance of each canonical function and, in multiple CAD, to help determine the number of significant functions to retain and interpret.

4.8.1 Relative Percent Variance Criterion

The actual numbers representing the eigenvalues do not have any immediate interpretive value, because they reflect the scale and variance of the original variables.

However, when there is more than one canonical function, as in multiple CAD, we can compare the relative magnitudes of the eigenvalues to see how much of the total discriminatory power is accounted for by each function. To do this, we convert the eigenvalues into relative percentages. *Relative percent variance* (Φ) is defined as follows:

$$\Phi_i = \frac{\lambda_i}{\sum_{i=1}^{Q} \lambda_i}$$

where Φ_i is the relative percent variance of the ith canonical function and λ_i is the eigenvalue corresponding to the ith function.

Relative percent variance measures how much of the total discriminatory power (i.e., total among-group variance) of the variables is accounted for by each canonical function. The cumulative percent variance of all canonical functions is equal to 100 percent. Consequently, in two-group DA, relative percent variance is uninformative because there exists only one canonical function and Φ_1 is always equal to 100 percent. In multiple DA, however, it can be useful in evaluating the relative importance of each canonical function and in determining how many functions to retain. For example, if there exist four canonical functions and the first function alone accounts for 90 percent of the group differences in the data set (i.e., $\Phi_1 = 0.9$), then it might be reasonable to deem the first canonical function alone as sufficient to explain group differences.

Unfortunately, relative percent variance alone can be very misleading. If, for example, the canonical correlation (see Sec. 4.8.2) for the first canonical function is very low, then group separation along the first axis is minimal. Yet Φ_i may still be very high. This may happen because Φ does not measure the extent of group differentiation; it measures how much of the total differentiation is associated with each axis, regardless of the absolute magnitude in group differentiation. Therefore, this criterion should only be used in conjunction with other measures such as canonical correlation.

4.8.2 Canonical Correlation Criterion

Another way to evaluate the utility of a canonical function is by examining the *canonical correlation coefficient*. This coefficient is a measure of the multiple correlation between the set of discriminating variables and the corresponding canonical function (similar to multiple correlations in regression). It ranges between zero and one; a value of zero denotes no relationship between the groups and the canonical function, while large values represent increasing degrees of association. The canonical correlation for the ith function is related to the eigenvalue as follows:

$$R_{ci} = \sqrt{\frac{\lambda_i}{1 + \lambda_i}}$$

where R_{ci} is the canonical correlation of the ith canonical function and λ_i is the eigenvalue corresponding to the ith canonical function.

The *squared canonical correlation* (R_c^2) is equal to the ratio of among-group to pooled within-group variation for the corresponding canonical function. In practical terms, R_c^2 represents the proportion of total variation in the corresponding canonical function (measured by the eigenvalue) explained by differences in group means; in other words, how much of the canonical variation is due to group differences. Canonical correlation is a measure of discriminant performance and, as such, can be a valuable tool in judging the utility of the canonical functions.

Empirical Example. In the example, we ran DA on the subset of variables identified from the stepwise procedure and computed the canonical correlation of the single function (Fig. 4.10). The multiple correlation between the seven discriminating variables and the canonical function is 0.918. More important, the squared canonical correlation indicates that 84 percent of the total canonical variation is explained by group differences. In other words, the vegetation gradient defined by the canonical function is very much a function of group differences.

4.8.3 Classification Accuracy

Although researchers generally engage in classification as a means of predicting group membership of future samples or samples of unknown group membership, we can also use it as an indirect measure of the effectiveness of the canonical functions to discriminate among groups. There are several ways in which classification can be performed (see Sec. 4.2.2); the choice of classification method is not particularly important in the present context. SAS, for example, unless otherwise specified, classifies entities based on a measure of generalized squared distance between each entity and each group centroid. The classification criterion is computed using either the pooled covariance matrix (if covariance matrices are equal among groups) or the individual within-group covariance matrices (if covariance matrices are unequal among groups). The former case results in a linear classification criterion, while the latter results in a quadratic classification criterion. In either case, each sample is placed in the group from which it has the smallest generalized squared distance using the appropriate criterion.

The classification results are usually summarized in a table, referred to as a *classification matrix* or *confusion matrix*, which provides the number and percent of sample entities classified correctly or incorrectly into each group. As a direct measure of predictive accuracy, the correct classification rate (i.e., the percentage of samples classified correctly) is the most intuitive measure of discrimination. Obviously, the higher the correct classification rate, the more effective the classification criterion in predicting group membership. This percentage can also be used as an indirect measure of the amount of canonical discrimination contained in the variables. The higher the correct classification rate, the greater the degree of group discrimination achieved by the canonical functions.

We should, however, judge the magnitude of the correct classification rate in relation to the expected percentage of correct classifications if group assignments had been made randomly. A certain percentage of samples in any data set are expected to be correctly classified by chance, regardless of the classification criterion. For example, if we have two groups of equal size and prior probabilities, we can expect to get 50 percent of the classifications right by pure random assignment. The expected probability of classification into any group by chance is proportional to the group size. Therefore, as the relative size of any single group becomes predominant, the correct classification rate based on chance alone tends to increase toward unity. In extreme situations, greatly different group sizes may lead to a very high correct classification rate, but the improvement over random classification may be slight. Thus, the need for a "chance-corrected" measure of prediction (or discrimination) becomes greater with more dissimilar group sizes (or prior probabilities).

We briefly discuss four ways to assess the improvement in classification accuracy over that expected by random assignment. Regardless of which criterion is used, chance-corrected classification rates, in conjunction with canonical correlations and the four significance tests given here, can help gauge the overall significance of the canonical functions.

1. *Maximum Chance Criterion.* The simplest way to assess classification accuracy relative to chance is to compute the maximum chance criterion (C_{max}) (Morrison 1969). The premise of C_{max} is that classification based on the discriminating variables should exceed that obtained by simply assigning all samples to the group with the largest size. In other words, we should be able to do at least as well as we would do by simply assigning all individuals to the largest group. C_{max} is determined by computing the percentage of the total sample represented by the largest of the two or more groups. For example, if the group sizes are 72 and 28, the maximum chance criterion is 72 percent correct classifications. If the classification criterion based on the variables correctly classified less than 72 percent of the samples, then we could have achieved a higher correct classification rate by classifying all samples into the largest group. The maximum chance criterion should be used only when the sole objective is to maximize the overall correct classification rate. However, rarely are we interested only in maximizing the correct classification rate. We are more often interested in correctly classifying samples from all groups or, in some cases, the smallest group.

2. *Proportional Chance Criterion.* In cases where group sizes are unequal and we are interested in correctly identifying members of the two or more groups, we can use the proportional chance criterion (C_{pro}) (Morrison 1969). The premise of C_{pro} is that classification based on the discriminating variables should exceed that obtained by randomly assigning samples to groups in proportion to group sizes. With two groups, C_{pro} is computed as follows:

$$C_{pro} = p^2 + (1-p)^2$$

where p is the proportion of samples in the first group and $1-p$ is the proportion of samples in the second group.

4.8 Assessing the Importance of the Canonical Functions

For example, using the group sizes from our previous example (72 and 28), the proportional chance criterion would be 60 percent ($0.72^2 + 0.28^2 = 0.60$) correct classifications. If the classification criterion based on the variables correctly classified less than 60 percent of the samples, then we could have done better by randomly classifying 72 percent of the samples into the first group and 28 percent into the second group. A generalization of C_{pro} for G groups can be found in Mosteller and Bush (1954). The proportional chance criterion should be used when the objective is to classify correctly samples into the G groups, and it should be used in preference to C_{max} in most situations.

3. *Tau Statistic.* The previous two criteria are appropriate when prior probabilities are estimated by group sample sizes. However, when prior probabilities are known or are not assumed to be equal to sample sizes, a proportional reduction in error statistic, *Tau*, will give a standardized measure of improvement over random assignment (Klecka 1980). *Tau* is defined as follows:

$$Tau = \frac{n_o - \sum_{i=1}^{G} p_i n_i}{n - \sum_{i=1}^{G} p_i n_i}$$

where G is the number of groups, n is the total number of samples, n_o is the number of samples correctly classified, n_i is the number of samples in the ith group, and p_i is the prior probability of membership in the ith group.

The term involving the summation is the number of samples that would be correctly classified on the basis of random assignment to groups in proportion to their prior probabilities. If the groups are to be treated equally, then all the prior probabilities are set equal to one divided by the number of groups. The maximum value for *Tau* is one and it occurs when there are no errors in prediction. The minimum value of *Tau* is zero and it indicates no improvement over chance. An intermediate value of *Tau*, 0.82, for example, indicates that classification based on the discriminating variables made 82 percent fewer errors than would be expected by random assignment.

4. *Cohen's Kappa Statistic.* Cohen's *Kappa* statistic is another useful measure of chance-corrected classification equivalent to *Tau* when prior probabilities are estimated by group sample sizes (Cohen 1960; Titus et al. 1984). *Kappa* is defined as follows:

$$Kappa = \frac{p_o - \sum_{i=1}^{G} p_i q_i}{1 - \sum_{i=1}^{G} p_i q_i}$$

where G is the number of groups, p_o is the observed percentage of samples correctly classified, p_i is the percentage of samples in the ith group, and q_i is the percentage of samples classified into the ith group.

Kappa is defined here in terms of proportions in keeping with the original development of the statistic, although the terms could easily be defined using frequencies (like *Tau*). Like *Tau*, a *Kappa* of zero indicates no improvement over chance, and a *Kappa* of one indicates perfect assignment. An intermediate value of *Kappa*, 0.82, for example, indicates that classification based on the discriminating variables was 82 percent better than chance assignment. A *Kappa* that is much lower than the correct classification rate suggests that the correct classification rate, and hence group predictability, is inflated and that much of the classification power is due simply to chance.

All four of these criteria are unbiased only when computed with "holdout" samples (i.e., split-sample approach; see Sec. 4.10.1). In other words, for unbiased results, the accuracy of the classification criterion should be evaluated by comparing the classification results and chance-corrected criteria computed from a holdout or validation sample. This is because the classification functions are more accurate for the samples from which they are derived than they would be for the full population. Thus, if the samples used in calculating the classification function are the ones being classified, the result will be an upward bias in the correct classification rate. In such cases, the chance criteria would have to be adjusted upward to account for this bias. Frank et al. (1965) give methods for estimating these biases for C_{max} and C_{pro}.

Empirical Example. In the example, we ran DA on the subset of variables identified from the stepwise procedure and displayed the classification results in a classification matrix. Figure 4.12 indicates that a quadratic classification criterion was used to calculate the generalized squared distance for determining group membership. Recall that the quadratic criterion was used because group dispersions were unequal. The resubstitution results provide the list of actual and predicted group membership for each entity and the posterior probability of membership in each group. The posterior probabilities are derived from the generalized squared distance between each entity and each group centroid (see Sec. 4.2.2). Each entity is classified into the group with the highest posterior probability of membership. The resubstitution results are useful for identifying the misclassified observations.

The resubstitution summary table (classification matrix) indicates that only one observation was incorrectly classified, for a correct classification rate of 98.9 percent (95/96). Specifically, one used site was misclassified as an unused site. The maximum chance criterion and proportional chance criterion each equal 50 percent (because group sample sizes are equal) and suggest that classification based on the seven discriminating variables is substantially better than random assignment. *Tau* and *Kappa* both equal 97.9 percent and suggest that classification based on the variables is roughly 98 percent better than random assignment. However, because we classified the same entities that we used to derive the classification functions, the correct classification rates and chance-corrected classification criteria are all biased upward. Thus, the true classification accuracy is probably slightly less. Regardless, it is apparent from the high correct classification rate that groups are well separated along the canonical axis. Note, this data set was not large enough for us to reserve a holdout sample.

```
Discriminant Analysis: Classification Results for Calibration Data
    Resubstitution Results Using Quadratic Discriminant Function
           Posterior Probability of Membership in USE

        Obs    From       Classified
               USE        into USE           NO           YES

         1     NO         NO               0.9999        0.0001
         2     NO         NO               0.9999        0.0001
         3     NO         NO               0.8767        0.1233
         .     .          .                  .             .
         .     .          .                  .             .
         .     .          .                  .             .
        13     YES        NO     *         0.8263        0.1737
        14     YES        YES              0.0000        1.0000
         .     .          .                  .             .
         .     .          .                  .             .
         .     .          .                  .             .

                    * Misclassified observation

   Resubstitution Summary Using Quadratic Discriminant Function
      Number of Observations and Percent Classified into USE

    From USE              NO              YES            Total

         NO               48               0              48
                         100.00           0.00           100.00

         YES              1               47              48
                          2.08           97.92           100.00

       Total             49               47              96
       Percent          51.04            48.96           100.00

       Priors          0.5000           0.5000

                  Error Count Estimates for USE

                          NO              YES           Total

         Rate           0.0000          0.0208          0.0104

         Priors         0.5000          0.5000
```

FIGURE 4.12 Classification results obtained using the quadratic classification function. Only a portion of the classification results are shown.

4.8.4 Significance Tests

There are a number of objective tests for evaluating the statistical significance of the canonical functions when the data are from a sample as opposed to the entire population. With a sample, we can ask what is the probability that the sampling process produced entities that show the computed degree of discrimination, when in fact there are no group differences in the population. Generally, we must assume independent random samples to ensure valid probability values.

The most common approach is to use a parametric statistical procedure. When there is only one canonical function, these procedures test the null hypothesis that the canonical correlation is equal to zero in the population. When there is more than one canonical function, as in multiple CAD, they test the null hypothesis that the current canonical correlation and all smaller ones are zero in the population. Essentially, this tests whether the residual discrimination in the system *prior* to deriving that function is too small. By *residual discrimination*, we mean the ability of the variables to discriminate among groups beyond the information that has been extracted by the previously computed functions. There are different test statistics available for testing the significance of the residual discrimination. SAS, for example, computes the likelihood ratio and approximate F-statistic based on Rao's approximation to the distribution of the likelihood ratio.

Alternatively, we can use a nonparametric resampling procedure such as the jackknife, bootstrap, or randomization test procedure (see Chapter 2, Sec. 2.8.5 for a complete discussion of these procedures) to test the statistical significance of the canonical functions based on the eigenvalue, canonical correlation, correct classification rate, or any other measure of canonical performance. The resampling procedures, as given in Chapter 2, can be applied here by substituting the appropriate DA parameter in the equations.

Recall from our discussion in Chapter 2 that all three of these resampling procedures provide a measure of the statistical significance of the canonical function as measured by a particular performance criterion (e.g., canonical correlation), but they differ in the nature of the test procedure developed. Again, in this regard, we prefer the randomization test over both the jackknife and bootstrap procedures.

It is important to note that even if a canonical function is statistically significant, we may decide that it lacks substantive ecological importance for several reasons. First, it may not discriminate among the groups well enough (i.e., it may bear a small canonical correlation). Second, it may fail to correctly classify enough entities into their proper groups (i.e., it may have a poor correct classification rate). Last, it may not have a meaningful ecological interpretation as judged by the canonical coefficients. Ultimately, the utility of each canonical function must be grounded on ecological criteria. Nevertheless, objective statistical tests can be useful tools in helping to assess the importance of the canonical function(s) and, when more than one exists, in determining how many functions to retain and interpret.

Empirical Example. In the example, we ran DA on a subset of variables identified from the stepwise procedure and assessed the statistical significance of the canonical function using the likelihood ratio and approximate F-statistic based on Rao's approximation to the distribution of the likelihood ratio (Fig. 4.10). The correlation between the seven discriminating variables and the canonical function 0.918 is significantly different from zero ($P < 0.0001$). Hence, the canonical function provides significant discrimination between used and unused sites. However, this P value is not valid, because group dispersions are unequal. A more desirable test would be one based on a resampling procedure, although these must be implemented manually in SAS.

4.8.5 Canonical Scores and Associated Plots

A final way of assessing the importance of the canonical functions involves simple graphical representations of the canonical scores. Recall that each sample entity in the data set has a score on each canonical function (or a location on each canonical axis) derived by multiplying the observed values for each discriminating variable by the corresponding canonical coefficients and summing the products. If the standardized data are used, the weights are given by the standardized canonical coefficients, and the canonical scores are computed as follows:

$$z_{ij} = c_{i1}x^*_{j1} + c_{i2}x^*_{j2} + \cdots + c_{ip}x^*_{jp}$$

where z_{ij} is the standardized canonical score for the ith canonical function and jth sample, c_{ik} is the standardized canonical coefficient for the ith canonical function and kth variable, and x_{jk} is the standardized value for jth sample and kth variable.

Recall that if the original data is in raw score form, we would employ the unstandardized canonical coefficients (u_{ik}) instead of the standardized coefficients. Also, although not shown here (for simplicity), there typically is a constant added to this equation to effect a translation of the origin of the canonical axes to the grand centroid so that the mean canonical score on each function is zero.

Canonical scores are very useful for graphically displaying the sample entities in canonical space. Scatter plots (for two or more canonical axes) and histograms (for one canonical axis) graphically illustrate the relationships among entities, since entities in close proximity in canonical space are ecologically similar with respect to the environmental gradients defined by the canonical functions. Canonical plots are typically used to assess how much overlap exists in group distributions, that is, to determine how distinct the groups are. To do this, group means and confidence intervals around each group mean are often plotted in canonical space for each pair of interpreted canonical functions, resulting in *equal frequency ellipses*. However, one should be aware that unequal group dispersions (i.e., unequal variance-covariance matrices) can distort the representation of entities in canonical space. Under such conditions, one cannot safely use equal frequency ellipses for displaying differences and overlap among groups.

Empirical Example. In the example, we ran DA on the subset of variables identified from the stepwise procedure and calculated the canonical scores. Figure 4.13(a) lists a sample of the canonical scores, and Figure 4.13(b) depicts a histogram of the canonical scores by group. The histogram portrays clear separation of groups along the canonical axis.

4.9 Interpreting the Canonical Functions

Once we have derived the canonical functions and assessed their importance, we can proceed to interpreting their meaning. We find this meaning by examining the

170   4. Discriminant Analysis

```
                    Canonical Scores

            OBS      USE        CAN1

             1       NO       -2.43144
             2       NO       -2.09691
             3       NO       -0.80303
             .        .          .
             .        .          .
             .        .          .
            49      YES       -0.03543
            50      YES        0.79838
            51      YES        2.30087
             .        .          .
             .        .          .
             .        .          .

                    FREQUENCY OF CAN1
FREQUENCY

21 +                    NNNNN
   |                    NNNNN
20 +                    NNNNN
   |                    NNNNN
19 +                    NNNNN
   |                    NNNNN
18 +                    NNNNN
   |                    NNNNN
17 +           NNNNN    NNNNN
   |           NNNNN    NNNNN
16 +           NNNNN    NNNNN                           YYYYY
   |           NNNNN    NNNNN                           YYYYY
15 +           NNNNN    NNNNN                           YYYYY    YYYYY
   |           NNNNN    NNNNN                           YYYYY    YYYYY
14 +           NNNNN    NNNNN                           YYYYY    YYYYY
   |           NNNNN    NNNNN                           YYYYY    YYYYY
13 +           NNNNN    NNNNN                           YYYYY    YYYYY
   |           NNNNN    NNNNN                           YYYYY    YYYYY
12 +           NNNNN    NNNNN                           YYYYY    YYYYY
   |           NNNNN    NNNNN                           YYYYY    YYYYY
11 +           NNNNN    NNNNN                           YYYYY    YYYYY
   |           NNNNN    NNNNN                           YYYYY    YYYYY
10 +           NNNNN    NNNNN                           YYYYY    YYYYY
   |           NNNNN    NNNNN                           YYYYY    YYYYY
 9 +           NNNNN    NNNNN             YYYYY         YYYYY    YYYYY
   |           NNNNN    NNNNN             YYYYY         YYYYY    YYYYY
 8 +           NNNNN    NNNNN             YYYYY         YYYYY    YYYYY
   |           NNNNN    NNNNN             YYYYY         YYYYY    YYYYY
 7 +           NNNNN    NNNNN             YYYYY         YYYYY    YYYYY
   |           NNNNN    NNNNN             YYYYY         YYYYY    YYYYY
 6 +           NNNNN    NNNNN             YYYYY         YYYYY    YYYYY    YYYYY
   |           NNNNN    NNNNN             YYYYY         YYYYY    YYYYY    YYYYY
 5 +           NNNNN    NNNNN    YYYYY    YYYYY         YYYYY    YYYYY    YYYYY
   |           NNNNN    NNNNN    YYYYY    YYYYY         YYYYY    YYYYY    YYYYY
 4 +           NNNNN    NNNNN    NNNNN    YYYYY         YYYYY    YYYYY    YYYYY
   |           NNNNN    NNNNN    NNNNN    YYYYY         YYYYY    YYYYY    YYYYY
 3 +   NNNNN   NNNNN    NNNNN    NNNNN    NNNNN    YYYYY    YYYYY    YYYYY    YYYYY
   |   NNNNN   NNNNN    NNNNN    NNNNN    NNNNN    YYYYY    YYYYY    YYYYY    YYYYY
 2 +   NNNNN   NNNNN    NNNNN    NNNNN    NNNNN    YYYYY    YYYYY    YYYYY    YYYYY
   |   NNNNN   NNNNN    NNNNN    NNNNN    NNNNN    YYYYY    YYYYY    YYYYY    YYYYY
 1 +   NNNNN   NNNNN    NNNNN    NNNNN    NNNNN    YYYYY    YYYYY    YYYYY    YYYYY
   |   NNNNN   NNNNN    NNNNN    NNNNN    NNNNN    YYYYY    YYYYY    YYYYY    YYYYY
   ----------------------------------------------------------------------
        -4       -3       -2       -1        0        1        2        3        4

                         CAN1 MIDPOINT

            SYMBOL  USE          SYMBOL  USE

              N     NO              Y    YES
```

FIGURE 4.13 Canonical scores and histogram of the scores by group. Only a portion of the canonical scores is shown.

relationships between the individual variables and the canonical functions. Most of the methods are based in some manner on an assessment of the relative importance of the discriminating variables in the canonical functions (as we will discuss in Sec. 4.9.1).

4.9.1 Standardized Canonical Coefficients

The traditional approach to interpreting the canonical functions is to use the canonical coefficients, since these measure the contribution of each variable to the canonical scores. However, we are usually more interested in knowing the relative importance rather than the absolute importance of each discriminating variable in the canonical functions. While the unstandardized canonical coefficients do tell us the absolute contribution of a variable in determining the canonical score, this information can be misleading when the meaning of one unit change in the value of a variable is not the same from one variable to another (i.e., when the standard deviations are not the same because of differences in measurement scale or variability). To assess the relative importance of each variable, we need to look at standardized canonical coefficients. The standardized coefficients are the ones that would be obtained by adjusting the eigenvector coefficients as before to create standardized canonical scores, but using eigenvectors that have been derived from standardized data. The relative contribution of the discriminating variables to each canonical function may then be gauged by the relative magnitude of the standardized canonical coefficients; the larger the magnitude, the greater that variable's contribution. However, even the standardized canonical coefficients may distort the true relationship among variables in the canonical functions when the correlation structure of the data is complex, and thus we usually base our interpretation on the structure coefficients (see Sec. 4.9.2).

Empirical Example. In the example, we ran DA on the subset of variables identified from the stepwise procedure and computed the standardized canonical coefficients (Fig. 4.11). These coefficients could be used to define the relative importance of the discriminating variables in the canonical function. However, the relative magnitudes of the standardized coefficients do not completely agree with the total structure coefficients (see Fig. 4.14). Therefore, it is likely that the standardized coefficients are reflecting a good deal of covariance among the variables and thus do not offer a reliable means of interpreting the canonical function.

4.9.2 Total Structure Coefficients

When one canonical coefficient dominates all others in the canonical function and the correlation structure is simple, it is fairly straightforward to assess the meaning of the canonical function by the magnitude and sign of the standardized canonical coefficients. However, when the correlation structure is complex and there are several coefficients of significant size, as is usually the case in ecological data sets, interpretation is not so straightforward. The difficulty arises from the fact that the original variables in the new canonical functions are correlated, some of them perhaps highly correlated. Individual coefficients in this case reflect not only the influence of their corresponding discriminating variables, but also the influence of other variables as reflected through the correlation structure of the data. The effect of a discriminating variable on the standardized canonical coefficient is only partially given by the numerical value of the corresponding coefficient. For example, when two or more variables are correlated, the weight (reflected in the standardized canonical coefficients) may be split between them, making the associated coefficients appear relatively small. Alternatively, the coefficient for one variable may be inflated (to consider both variables), while the other variable is assigned a near zero coefficient.

The usual way of dealing with this, although it only removes severe multicollinearity, is to examine groups of highly correlated variables and retain only the variables with the greatest among-group variance (i.e., discriminating power) or ease of ecological interpretation (see Sec. 4.5.3). A better technique, or one to be used in conjunction with other multicollinearity reduction approaches, is to examine the product-moment correlations between the canonical functions and the individual variables included in the analysis. These simple bivariate correlations are not affected by relationships with other variables and therefore reflect the true relationship between each variable and the canonical function. These correlations are referred to as *structure coefficients* (or *canonical loadings*), and the matrix of structure coefficients is called the *canonical structure* (or *structure matrix*). The total structure coefficients are defined as follows:

$$s_{ij} = \sum_{k=1}^{P} r_{jk} c_{ik}$$

where s_{ij} is the total structure coefficient (correlation) for the ith canonical function and jth variable, r_{jk} is the total correlation between the jth and kth variables, and c_{ik} is the standardized canonical coefficient for the ith canonical function and kth variable.

A structure coefficient tells us how closely a variable and a function are related. When the absolute magnitude of the coefficient is large (i.e., approaching a value of one), the function is carrying nearly the same information as the variable. Conversely, when the coefficient is near zero, the function and variable have little in common. Hence, we can define a canonical function on the basis of the structure coefficients by noting those variables with the largest coefficients. Structure coefficients are intuitively appealing because most researchers are familiar with simple correlation coefficients and can visualize (in graphic terms) differences in the magnitude of the coefficient. The canonical structure provides a way to establish an ecological interpretation of each canonical function and is perhaps the single most important piece of information resulting from CAD.

Empirical Example. In the example, we ran DA on the subset of variables identified from the stepwise procedure and calculated the total canonical structure. Figure 4.14 indicates that LTOTAL is the most important variable in the canonical function,

```
        Total Canonical Structure

          Variable           CAN1

          LTOTAL           0.919908
          SNAGT            0.762435
          BAH              0.005134
          GTOTAL          -0.632135
          BAS              0.639319
          SNAGL6           0.410062
          MTOTAL          -0.452033

     Class Means on Canonical Variables

          USE                CAN1

           N             -2.300249594
           Y              2.300249594
```

FIGURE 4.14 Total canonical structure coefficients and class means on the canonical function.

which is not surprising given the stepwise selection results discussed previously. Conversely, BAH is not correlated with the canonical function at all. The structure coefficients indicate that the function represents a gradient in low shrub cover and total number of snags and, to a lesser extent, ground cover, basal area of snags, midstory cover, and large decay class 6 snags.

The class means on the canonical function indicate that used sites (YES) are positively correlated with the canonical function. This means that variables with positive structure coefficients are greater in magnitude on used sites, and variables with negative structure coefficients are lower in magnitude on used sites. Conversely, unused sites (NO) are negatively correlated with the canonical function. Hence, the canonical function defines a vegetation gradient between used and unused sites, in which used sites are characterized by greater percent cover of low shrubs and greater numbers of snags than unused sites. Note that the standardized canonical coefficients (Fig. 4.11) and the structure coefficients (Fig. 4.14) lead to two different interpretations with respect to the importance of BAH. Specifically, according to the standardized canonical coefficients, BAH is the second highest loading variable; yet, based on structure coefficients, BAH is unrelated to the canonical function. The discrepancy is indicative of the effects of multicollinearity and illustrates the difficulty of basing the interpretation on the standardized canonical coefficients, in spite of efforts to reduce multicollinearity affects.

4.9.3 *Covariance-Controlled Partial F-Ratios*

While the structure coefficients help to reveal which of the variables are, in a relative sense, most important to the overall function, they still do not allow the researcher to determine if any one variable is a significant contributor to the discriminant model, given the relationships that exist among all of the discriminating variables.

Similarly, univariate F-ratios do not meet this objective, because they ignore interrelationships among the variables, and thus are confounded by variables that contribute to the discriminant relationship in a redundant way. However, one approach to this problem is to compute a covariance-controlled partial F-ratio for each variable (Perreault et al. 1979).

The essence of this approach is to partition out the variance in the variable of interest which is already explained by the other variables. This allows the researcher to determine: (1) if significant group differences remain after the impact of other variables is considered; or (2) if a variable which appeared not to be a discriminator is in fact significant when all of the interrelationships among variables are considered. The relative importance of the variables can be determined by examining the absolute sizes of the significant partial F-ratios and ranking them. Larger partial F-ratios indicate greater discriminating power. In practice, rankings using the partial F-ratios approach are often the same as those using the structure coefficients, but the partial F-ratios indicate the associated level of significance for each variable. When a stepwise procedure for initial variable selection is employed, partial F-ratios (i.e., F-to-remove) are automatically computed for each variable at each step. Otherwise, partial F-ratios are easily computed using most standard statistical software. It is important to note that unlike the standardized canonical coefficients and structure coefficients, the partial F-ratio is an aggregative measure in that it summarizes information across the different canonical functions. Thus, if more than one canonical function exists (as in multiple CAD), the partial F-ratio measures that amount of total discrimination across functions associated with each variable.

Empirical Example. In the example, we conducted a stepwise discriminant analysis on the data set after removing five variables involved in high pairwise correlations. In the final step, after which no variables met either the F-to-enter or F-to-remove criteria, the F-to-remove was used as a partial F-ratio test for each variable in the final model. Figure 4.15 indicates that six out of the seven variables in the final model have significant partial F-ratios at the alpha level of 0.05; they account for significant group differences after removing the differences accounted for by all other variables in the

```
Stepwise Selection:   Step 8

              Statistics for Removal,   DF = 1, 88

                      Partial
       Variable        R**2            F           Prob > F

       LTOTAL         0.4997         87.886         0.0001
       MTOTAL         0.0260          2.351         0.1288
       SNAGL6         0.0506          4.690         0.0331
       BAS            0.0429          3.940         0.0503
       BAH            0.2020         22.272         0.0001
       GTOTAL         0.0866          8.342         0.0049
       SNAGT          0.0443          4.078         0.0465
```

FIGURE 4.15 Partial F-ratios (F-to-remove) for the variables in the final model.

model. Like the structure coefficients, the relative magnitudes of the F-ratios indicate that LTOTAL is a far more significant discriminator than any other variable. Otherwise, the structure coefficients and partial F-ratios lead to different interpretations. In particular, BAH has a near zero structure coefficient but has the second largest partial F-ratio. Moreover, the rank order of importance of the variables is somewhat different between the structure coefficients (Fig. 4.14) and the partial F-ratios (Fig. 4.15), especially with respect to the relative importance of BAH. Given this disagreement between the structure coefficients and partial F rankings, we prefer to base our interpretation on the structure coefficients, since the partial F-ratios involve certain parametric assumptions which likely are not met in this data set.

4.9.4 Significance Tests Based on Resampling Procedures

While partial F-ratios help to assess the relative importance and statistical significance of each discriminating variable, they do not allow us to evaluate each canonical function independently; this is because the partial F-ratio is an aggregative measure. We would like to be able to assess the statistical significance of each variable's contribution to each canonical function, as measured by the structure coefficients. This would allow us to generate an ecological interpretation of each canonical function independently, based on the significant discriminators in each function. To do this, we can employ a jackknife, bootstrap, or randomization resampling procedure (see Chapter 2, Sec. 2.8.5 for a complete discussion of these procedures). The same procedures described in Chapter 2 would be applied here, in this case substituting the structure coefficient for the relative percent variance parameter. In each case, the null hypothesis tested is that the particular structure coefficient is approximately equal to zero. Thus, rejection of the null hypothesis implies that a particular structure coefficient probably represents a real association between the corresponding variable and canonical function.

Recall from our discussion in Chapter 2 that some form of experimentwise error control, such as the Bonferroni procedure, is probably necessary because multiple tests are performed. Also, remember that statistical significance does not guarantee ecological meaningfulness. Thus, these significance tests should be used only in conjunction with other performance criteria.

4.9.5 Potency Index

When there are several canonical functions, it may be useful to develop a summary or composite index of the relative discriminating potency of each of the variables. A useful *potency index* may be computed as follows (Perreault et al. 1979):

$$PI_j = \sum_{i=1}^{M} \left[s_{ij}^2 \left(\frac{\lambda_i}{\sum_{i=1}^{M} \lambda_i} \right) \right]$$

where PI_j is the potency index for the jth variable, M is the number of significant or retained canonical functions, s_{ij} is the structure coefficient for the ith canonical function and jth variable, and λ_i is the eigenvalue corresponding to the ith canonical function.

This index is analogous to the final communality estimates used in principal components analysis and factor analysis (see Chapter 2, Sec. 2.9.4). This index has both intuitive and interpretive appeal. The squared loading (structure coefficient) of a variable on a function represents the percent of the variance in the canonical function explained by that single variable. The ratio of the eigenvalue for a function to the sum of all the significant eigenvalues represents the portion of the total explainable variance accounted for by a particular function (similar to relative percent variance discussed previously, except that only significant eigenvalues are considered here). Thus, the index measures the percent of the total discriminating power of the retained canonical functions accounted for by each variable. Like the partial F-ratios, the potency index is an aggregative measure because it summarizes information across the different canonical functions; however, it differs from the partial F-ratio in two ways. First, whereas the partial F-ratio considers all Q canonical functions, the potency index considers only those M (usually less than Q) canonical functions retained for interpretation. Second, whereas the partial F-ratio has a corresponding parametric test of significance, the potency index does not have a similar test.

A limitation of this index is that it is only a relative measure and has no meaning in an absolute sense. In other words, like relative percent variance, it only measures the amount of *existing* discriminating power attributable to each variable but does not assess how much actual discrimination is achieved. Consequently, while all of the eigenvalues could be quite small, indicating poor discrimination, the potency index for a variable could be the same as for a model with much larger eigenvalues.

4.10 Validating the Canonical Functions

Once we have derived the canonical functions, assessed their significance, and interpreted their meaning, we can test the validity of our findings to insure that they are reliable. Validating the results of CAD is an important and often overlooked step. The results of DA are reliable only if the group means and dispersions are estimated accurately and precisely, particularly when the objective is to develop classification functions for predicting group membership of future observations. The best assurance of reliable results is an intelligent sampling plan and a large sample. Sampling variability increases rapidly with increases in numbers of variables and groups, and with decreases in sample size and distances among group centroids. When the number of parameters to be estimated approaches the number of samples, there is a good likelihood that any patterns exhibited by individual canonical coefficients are fortuitous and therefore of no ecological consequence. Thus, validation becomes increasingly important as the sample size decreases relative to dimensionality (number of variables). There are several ways to validate the results of CAD, most of them involving the use of classification matrices. We briefly discuss the two most common validation approaches.

4.10.1 Split-Sample Validation

The most common validation procedure, when the sample size is large enough, is to randomly divide the total sample of entities into two groups. One subset, referred to as the *analysis, training,* or *calibration* sample, is used to derive the canonical functions, and the other, referred to as the *holdout, test,* or *validation* sample, is used to test the functions. This method of validating the functions is referred to as *split-sample validation* or *cross-validation*. Alternatively (and preferably), we could collect a fresh new sample of entities to serve as a validation data set.

Briefly, split-sample validation proceeds as follows: (1) the data set is randomly divided into two subsets; (2) the classification criterion is computed using one data set; (3) the derived criterion is used to classify samples from the second data set; and (4) the resulting correct classification rate is used to judge the reliability and robustness of the classification criterion. Poor classification rates indicate an unstable classification criterion and the fact that a larger sample may be required to obtain accurate and precise estimates of group means and dispersions. The premise of this validation approach to CAD is that unstable classification results indirectly infer that the canonical functions are also unstable.

The justification for dividing the total sample into two groups is that an upward bias will occur in the predication accuracy (i.e., correct classification rate) of the classification criterion if the samples used in deriving the classification matrix are the same as those used in deriving the classification function. That is, the classification accuracy will be too high because the function will always predict more accurately the samples it was derived from than a separate sample from the population. The implications of this upward bias are particularly important when one is concerned with the external validity of the findings; that is, when one is interested either in using the classification functions to predict group membership of future samples, or in generalizing the canonical differences among groups to entities not sampled. Because each split sample will tend to have different sampling errors, the test subset will give a better estimate of the ability to correctly predict the total population.

No guidelines have been established for determining the appropriate sizes for the analysis and holdout samples. Some recommend that they be equal, while others prefer more entities in the analysis sample. The most important consideration, however, is that the analysis sample be sufficiently large to insure stability of the coefficients. Of course, the total sample size must be sufficiently large to divide into two sets that meet the minimum sample size requirements (see Sec. 4.6). When selecting entities for the analysis and holdout samples, a proportionately stratified random sampling procedure is usually employed. For example, if a sample consisted of 25 individuals of species A and 75 individuals of species B, samples would be selected for each data set in a 1:3 ratio of individuals from species A and B.

Another strategy is to follow the split-sample procedure several times. Instead of randomly dividing the total sample into analysis and holdout samples once, the total sample is randomly divided into analysis and holdout samples several times, each time testing the validity of the functions through the development of a classification matrix and a correct classification rate. Then the several correct classification rates would be averaged to obtain a single measure. This is, in some respects, analogous to the bootstrap resampling procedure described below.

4.10.2 Validation Using Resampling Procedures

As should be apparent from the previous discussions, resampling procedures such as the jackknife, bootstrap, and randomization tests can be used in a variety of ways to assess the validity of DA results (see Chapter 2, Sec. 2.8.5 for a generalized discussion of these procedures). In the present context, we can use these resampling procedures to judge the reliability of the classification criterion, and, indirectly, as a way to validate the canonical functions.

1. *Jackknife Validation.* Although generally inferior to the bootstrap procedure, the most common validation approach based on resampling the original data, particularly when small sample sizes prohibit the use of the split-sample approach, involves the jackknife procedure. Briefly, in the present context, jackknife validation proceeds as follows: (1) a single sample is omitted from the data set; (2) the classification functions are derived; (3) the omitted sample is classified; (4) the process is repeated sequentially for each sample; and (5) the resulting jackknife correct classification rate is calculated to judge the reliability and robustness of the canonical functions. If the jackknife classification rate is much lower than the rate from the full data set, then we must suspect that the estimation of means and dispersions is not reliable, resulting in unstable canonical and classification functions, and that a larger sample size is required.

Empirical Example. In the example, we ran DA on the subset of variables identified from the stepwise procedure and used a jackknife procedure to check for arbitrariness in the classification results (Fig. 4.16). The jackknife classification summary (referred to as "cross-validation" in SAS) indicates that five samples were incorrectly classified, for a correct classification rate of 94.8 percent (91/96). This represents a classification power roughly 90 percent better than chance alone (i.e., *Tau* and *Kappa* = 89.6). Consequently, means and dispersions were probably estimated accurately and precisely, and the canonical and classification functions are probably fairly stable.

2. *Randomized Sample Validation.* An alternative approach involves the randomization test procedure. Briefly, in the present context, randomized sample validation proceeds as follows: (1) the raw data is randomized within variables (columns) to eliminate any real patterns; (2) the classification functions are derived; (3) the classification matrix is generated and the correct classification rate is computed; (4) the process is repeated many times; and (5) the resulting random permutation distribution is compared to the correct classification rate for the original data set. The mean randomized correct classification rate reflects the effects of sample bias because

```
Discriminant Analysis: Classification Results for Calibration Data

   Cross-validation Summary using Quadratic Discriminant Function

     Number of Observations and Percent Classified into USE:

       From USE            NO            YES          Total

          NO               44             4             48
                         91.67          8.33         100.00

          YES              1             47             48
                          2.08          97.92         100.00

        Total             45             51             96
       Percent           46.88          53.13         100.00

        Priors          0.5000         0.5000


               Error Count Estimates for USE:

                          NO            YES          Total

              Rate      0.0833         0.0208        0.0521

             Priors     0.5000         0.5000
```

FIGURE 4.16 Jackknife classification results obtained using the quadratic classification function. Note that SAS refers to the jackknife classification procedure as the "cross-validation" procedure; this is not the cross-validation or split-sample validation procedure as we have defined them in the text.

there are no real differences among groups in the randomized data sets. Thus, the difference between the randomized classification rate and the rate taken from the original data represents the amount of true classification achieved by the functions after eliminating that expected due to sample bias. If the difference is small, we can conclude that the power of the classification functions (and indirectly the canonical functions) is almost entirely due to the effects of sample bias. Under this circumstance, the reliability of the classification and canonical functions must be questioned. Furthermore, by comparing the original classification rate to the randomized distribution, we can determine whether the predictive power of the derived classification function is statistically greater than that achieved by random functions. Note that the randomization procedure described here provides another chance-corrected measure of prediction (or discrimination) and can be used to gauge the statistical importance of the canonical function(s).

4.11 Limitations of Discriminant Analysis

Most of the limitations of DA have been discussed already. Nevertheless, we briefly summarize some of these limitations here, other than those associated with the assumptions, to emphasize the importance of their consideration when interpreting the results of DA.

- When several canonical functions exist, as in multiple CAD, by interpreting only the first one or two canonical functions one might overlook a later axis that accounts for most of the discriminating power in some variable. Consequently, even though this variable has significant univariate discriminating power, this power is lost in the canonical transformation. Examination of all the canonical functions is necessary to assess this problem.
- Discriminant analysis is only capable of detecting gradients that are intrinsic to the data set. There may exist other more important discriminating gradients not measured by the selected variables, and these dominant, undetected gradients may distort or confuse any relationships that are intrinsic to the data.
- There are different philosophies regarding the amount of weight to give to the objective measures of discriminant performance; for example, how much weight to give significance tests when deciding on how many canonical functions to retain and interpret in multiple CAD. On the one hand, it is argued that canonical functions should be evaluated solely on the basis of whether or not they offer a meaningful ecological interpretation; little emphasis is placed on the statistics associated with the eigenvalues (i.e., significance tests, canonical correlation, relative percent variance). Others argue that canonical functions should be evaluated largely on the basis of objective performance criteria, because otherwise we can not discern whether the patterns revealed by the analysis are real or merely reflect sampling bias.
- As with a regression equation, the canonical and classification function(s) should be validated by testing their efficacy with a fresh sample of entities. The observed accuracy of prediction on the sample upon which the function was developed will always be spuriously high. The true discriminatory power of the function will be found only when it is tested with a completely separate sample. Alternatively, if the original data set is large, then data splitting may be employed to validate the functions.

Bibliography

Procedures

Carnes, B.A., and Slade, N.A. 1982. Some comments on niche analysis in canonical space. *Ecology* 63:888–893 (see also Dueser and Shugart 1979 and 1982; and Van Horne and Ford 1982).

Cohen, J. 1960. A coefficient of agreement for nominal scales. *Education Psychology Measurements* 20:37–46.

Cooley, W.W., and Lohnes, P.R. 1971. *Multivariate Data Analysis.* New York: Wiley and Sons.

Edgington, E.S. 1987. *Randomization Tests,* 2nd ed. New York: Marcel Dekker, Inc.

Efron, B. 1979. Computers and the theory of statistics: thinking the unthinkable. *Society for Industrial and Applied Mathematics (SIAM) Review* 21:460–480.

Efron, B. 1982. The jackknife, the bootstrap and other resampling plans. *Society for Industrial and Applied Mathematics (SIAM)*, Monograph #38, CBMS-NSF.

Efron, B., and Gong, G. 1983. A leisurely look at the bootstrap, the jackknife, and cross-validation. *American Statistician* 37:36–48.

Fisher, R.A. 1936. The use of multiple measurements in taxonomic problems. *Annals of Eugenics* 7:179–188.

Frank, R.E., Massy, W.F., and Morrison, D.G. 1965. Bias in multiple discriminant analysis. *Journal of Marketing Research* 2:250–258.

Goldstein, M., and Dillon, W.R. 1978. *Discrete Discriminant Analysis*. New York: Wiley and Sons.

Green, P.E. 1978. *Analyzing Multivariate Data*. Hinsdale, IL: The Dryden Press.

Green, R.H. 1971. A multivariate statistical approach to the Hutchinsonian niche: bivalve mollusca of central Canada. *Ecology* 52:543–556.

Green, R.H. 1974. Multivariate niche analysis with temporally varying environmental factors. *Ecology* 55:73–83.

Habbema, J.D.F., and Hermans, J. 1977. Selection of variables in discriminant analysis by F-statistic and error rate. *Technometrics* 19:487–494.

Hair, J.F., Jr., Anderson, R.E., and Tatham, R.L. 1987. *Multivariate Data Analysis*, 2nd ed. New York: Macmillan.

Harner, E.J., and Whitmore, R.C. 1977. Multivariate measures of niche overlap using discriminant analysis. *Theoretical Population. Biology.* 12:21–36.

Harris, R.J. 1975. *A Primer of Multivariate Statistics*. New York: Academic Press.

Horton, I.F., Russell, J.S. and Moore, A.W. 1968. Multivariate-covariance and canonical analysis: a method for selecting the most effective discriminators in a multivariate situation. *Biometrics* 24:845–858.

Hudlet, R., and Johnson, R. 1977. Linear discrimination and some further results on best lower dimensional representations. In *Classification and Clustering*, ed. J.V. Ryzin. pp 371–394.

Johnson, D.H. 1981. How to measure—a statistical perspective. In *The Use of Multivariate Statistics in Studies on Wildlife Habitat*, ed. D.E. Capen, pp 53–58. U.S. Forest Service General Technical Report RM–87.

Kachigan, S.K. 1982. *Multivariate Statistical Analysis: A Conceptual Introduction*. New York: Radius Press.

Klecka, W.R. 1980. Discriminant analysis. Sage University Paper series. *Quantitative Applications in the Social Sciences*, Series No. 07–019. Beverly Hills and London: Sage Publications.

Krzanowski, W.J. 1977. The performance of Fisher's linear discriminant function under nonoptimal conditions. *Technometrics* 19:191–200.

Lachenbruch, P.A. 1975. *Discriminant Analysis*. New York: Hafner Press.

Lachenbruch, P.A., and Goldstein, M. 1979. Discriminant analysis. *Biometrics* 35:69–85.

Lachenbruch, P.A., Sneeringer, C., and Revo, L.T. 1973. Robustness of the linear and quadratic discriminant functions to certain types of non-normality. *Communications in Statistics* 1:39–56.

Levene, H. 1960. Robust test for equality of variances. In *Probability and Statistics*, ed. I. Olkin, pp 278–292. Palo Alto: Stanford University Press.

Mahalanobis, P.C. 1963. On the generalized distance in statistics. *Proceedings National Institute of Science, India* 12:49–55.

Marcus, M., and Minc, H. 1968. *Elementary Linear Algebra*. New York: The Macmillan Company.

McDonald, L.L. 1981. A discussion of robust procedures in multivariate analysis. In *The Use of Multivariate Statistics in Studies on Wildlife Habitat*, ed. D.E. Capen, pp 242–244. U.S. Forest Service General Technical Report RM–87.

Morrison, D.G. 1969. On the interpretation of discriminant analysis. *Journal of Marketing Research* 6:156–163.

Morrison, D.F. 1976. *Multivariate Statistical Methods*, 3rd ed. New York: McGraw-Hill.

Morrison, M.L. 1984. Influence of sample size on discriminant function analysis of habitat use by birds. *Journal Field Ornithology* 55:330–335.

Mosteller, F., and Bush, R.R. 1954. Selective quantitative techniques. In *Handbook of Social Psychology, Vol. 1*, ed. G. Lindzey. Reading: Addison-Wesley.

Perreault, W.D., Jr., Behrman, D.N., and Armstrong, G.M. 1979. Alternative approaches for interpretation of multiple discriminant analysis in marketing research. *Journal of Business* 7:151–173.

Randles, R.H. 1978. Generalized linear and quadratic discriminant functions using robust estimates. *Journal of the American Statistical Association*. 73:564–568.

Tatsuoka, M.M. 1970. Discriminant analysis: the study of group differences. *Selected Topics in Advanced Statistics—An Elementary Approach. Number 6*. Champaign: Institute for Personality and Ability Testing.

Titus, K., Mosher, J.A., and Williams, B.K. 1984. Chance-corrected classification for use in discriminant analysis: ecological applications. *The American Midlands Naturalist* 111:1–7.

Wahl, P.W., and Kronmal, R.A. 1977. Discriminant functions when covariances are unequal and sample sizes are moderate. *Biometrics* 33:479–484.

Williams, B.K. 1981. Discriminant analysis in wildlife research: theory and applications. In *The Use of Multivariate Statistics in Studies on Wildlife Habitat*, ed. D.E. Capen, pp 59–71. U.S. Forest Service General Technical Report RM–87.

Williams, B.K. 1982. A simple demonstration of the relationship between classification and canonical variates analysis. *American Statistician* 36:363–365.

Williams, B.K. 1983. Some observations on the use of discriminant analysis in ecology. *Ecology* 64:1283–1291.

Williams, B.K., and Titus, K. 1988. Assessment of sampling stability in ecological applications of discriminant analysis. *Ecology* 69:1275–1285.

Van Horne, B., and Ford, R.G. 1982. Niche breadth calculation based on discriminant analysis. *Ecology* 63:1172-1174.

Van Ness, J.W., and Simpson, C. 1976. On the effects of dimension in discriminant analysis. *Technometrics* 18:175–187.

Weiner, J.M., and Dunn, O.J. 1966. Elimination of variates in linear discrimination problems. *Biometrics* 22:268–275.

Applications

Able, K.P., and Noon, B.R. 1976. Avian community structure along elevational gradients in the northeastern United States. *Ecologia* 26:275–294 (use with Noon 1981).

Anderson, S.H., and Shugart, H.H. Jr. 1974. Habitat selection of breeding birds in an east Tennessee deciduous forest. *Ecology* 55:828–837.

Bertin, R.I. 1977. Breeding habitats of the wood thrush and veery. *Condor* 79:303–311.

Block, W.M., Morrison, M.L., and Scott, P.E. 1998. Development and evaluation of habitat models for herpetofauna and small mammals. *Forest Science* 44(3):430–437.

Buechner, H.K., and Roth, H.D. 1974. The lek system in Uganda kob antelope. *American Zoologist* 14:145–162.

Buzas, M.A. 1967. An application of canonical analysis as a method for comparing faunal areas. *Journal of Animal Ecology* 36:563–577.

Cody, M.L. 1978. Habitat selection and interspecific territoriality among the sylviid warblers of England and Sweden. *Ecological Monographs* 48:351–396.

Conley, W. 1976. Competition between Microtus: a behavioral hypothesis. *Ecology* 57:224-237.

Conner, R.N., and Adkisson, C.S. 1976. Discriminant function analysis: a possible aid in determining the impact of forest management on woodpecker nesting habitat. *Forest Science* 22:122–127.

Conroy, M.J., Gysel, L.W., and Dudderer, E.K. 1979. Habitat components of clear-cut areas for snowshoe hares in Michigan. *Journal of Wildlife Management* 43:680–690.

Converse, K.A., and Morzuch, B.J. 1981. A descriptive model of snowshoe hare habitat. In *The Use of Multivariate Statistics in Studies on Wildlife Habitat*, ed. D.E. Capen, pp 232–241. U.S. Forest Service General Technical Report RM–87.

Crawford, H.S., Hooper, R.G. and Titterington, R.W. 1981. Songbird population response to silvicultural practices in central Appalachian hardwoods. *Journal of Wildlife Management* 45:680–692.

Dueser, R.D., and Shugart, H.H., Jr. 1978. Microhabitats in a forest-floor small mammal fauna. *Ecology* 59:89–98 (use with Dueser and Shugart 1979).

Dueser, R.D., and Shugart, H.H., Jr. 1979. Niche pattern in a forest-floor small mammal fauna. *Ecology* 60:108–118 (use with Dueser and Shugart 1978; see also Van Horne and Ford 1982; Carnes and Slade 1982; and Dueser and Shugart 1982).

Dueser, R.D., and Shugart, H.H., Jr. 1982. Reply to comments by Van Horne and Ford and by Carnes and Slade. *Ecology* 63:1174–1175 (see also Dueser and Shugart 1979; Van Horne and Ford 1982; and Carnes and Slade 1982).

Fielding, A.H., and Haworth, P.F. 1995. Testing the generality of bird-habitat models. *Conservation Biology* 9(6):1466–1481.

Goldingay, R.L., and Price, M.V. 1997. Influence of season and a sympatric congener on habitat use by Stephen's kangaroo rat. *Conservation Biology* 11(3):708–717.

Goldstein, R.A., and Grigal, D.F. 1972. Definition of vegetation structure by canonical analysis. *Journal of Ecology*. 60:277–284.

Grigal, D.F., and Goldstein, R.A. 1971. An integrated ordination-classification analysis of an intensively sampled oak-hickory forest. *Journal of Ecology* 59:481–492.

Grue, C.E., Reid, R.R. and Silvy, N.J. 1981. A windshield and multivariate approach to the classification, inventory and evaluation of wildlife habitat: an exploratory study. In *The Use of Multivariate Statistics in Studies on Wildlife Habitat*, ed. D.E. Capen, pp 124–151. U.S. Forest Service General Technical Report RM–87.

Hale, P.E. 1983. Use of discriminant function to characterize ruffed grouse drumming sites in Georgia: a reply. *Journal of Wildlife Management* 47:1152.

Hale, P.E., Johnson, A.S., and Landers, J.L. 1982. Characteristics of ruffed grouse drumming sites in Georgia. *Journal of Wildlife Management*. 46:115–123 (see also Magnusson 1983 and Hale 1983).

Harner, E.J., and Whitmore, R.C. 1981. Robust principal component and discriminant analysis of two grassland bird species' habitat. In *The Use of Multivariate Statistics in Studies on Wildlife Habitat*, ed. D.E. Capen, pp 209–221. U.S. Forest Service General Technical Report RM–87.

James, F.C. 1971. Ordinations of habitat relationships among breeding birds. *Wilson Bulletin*. 83:215–236.

Jorgensen, E.E., Demarais, S., Sell, S.M., and Lerich, S.P. 1998. Modeling habitat suitability for small mammals in Chihuahuan desert foothills of New Mexico. *Journal of Wildlife Management* 62(3):989–996.

Kowal, R.R., Lechowicz, M.J. and Adams, M.G. 1972. The use of canonical response curves in physiological ecology. *Flora* 165:29–46.

Ladine, T.A., and Ladine, A. 1998. A multiscale approach to capture patterns and habitat correlations of Peromyscus leucopus (Rodentia, Muridae). *Brimleyana* 25:99–109.

Langbein, J., Strech, J., and Scheibe, K.M. 1998. Characteristic activity patterns of female mouflons (Ovis orientalis musimon) in the lambing period. *Applied Animal Behavior Science* 58(3–4):281–292.

Maehr, D.S. and Cox, J.A. 1995. Landscape features and panthers in Florida. *Conservation Biology* 9(5):1008–1019.

Magnusson, W.E. 1983. Use of discriminant function to characterize ruffed grouse drumming sites in Georgia: a critique. *Journal of Wildlife Management*. 47:1151–1152.

Marsden, S.J., and Jones, M.J. 1997. The nesting requirements of the parrots and hornbill of Sumba, Indonesia. *Biological Conservation* 82(3):279–287.

Martinka, R.P. 1972. Structural characteristics of blue grouse territories in southwestern Montana. *Journal of Wildlife Management* 36:498–510.

Massolo, A. and Meriggi, A. 1998. Factors affecting habitat occupancy by wolves in northern Apennines (northern Italy): a model of habitat suitability. *Ecography* 21(2):97–107.

Montanucci, R.R. 1978. Discriminant analysis of hybridization between leopard lizards, Gambelia (Reptilia, Lacertilia, Iguanidae). *Journal of Herpetology* 12:299–307.

Noon, B.R. 1981. The distribution of an avian guild along a temperate elevational gradient: the importance and expression of competition. *Ecological Monographs* 51:105–124 (use with Able 1976).

Norris, J.M., and Barkham, J.P. 1970. A comparison of some Cotswold beechwoods using multiple-discriminant analysis. *Journal of Ecology*. 58:603–619.

Raphael, M.G. 1981. Interspecific differences in nesting habitat of sympatric woodpeckers and nuthatches. In *The Use of Multivariate Statistics in Studies on Wildlife Habitat*, ed. D.E. Capen, pp 142–151. U.S. Forest Service General Technical Report RM–87.

Reese, D.A., and Welsh, H.H. 1998. Habitat use by western pond turtles in the Trinity River, California. *Journal of Wildlife Management* 62(3):842–853.

Rice, J.C. 1978a. Behavioral interactions of interspecifically territorial vireos. I. Song discrimination and natural interactions. *Animal Behaviour* 26:527–549.

Rice, J.C. 1978b. Behavioral interactions of interspecifically territorial vireos. II. Seasonal variation in response intensity. *Animal Behaviour* 26:550–561.

Rice, J.C., Ohmart, R.D., and Anderson, B. 1981. Bird community use of riparian habitats: the importance of temporal scale in interpreting discriminant analysis. In *The Use of Multivariate Statistics in Studies on Wildlife Habitat*, ed. D.E. Capen, pp 186–196. U.S. Forest Service General Technical Report RM–87.

Ricklefs, R.E. 1977. A discriminant function analysis of assemblages of fruit-eating birds in Central America. *Condor* 79:228–231.

Rogovin, K.A., and Shenbrot, G.I. 1995. Geographical ecology of Mongolian desert rodent communities. *Journal of Biogeography* 22(1)111–128.

Sakai, H.F., and Noon, B.R. 1991. Nest-site characteristics of Hammond's and Pacific-slope flycatchers in northwestern California. *Condor* 93:563–574.

Smith, K.G. 1977. Distribution of summer birds along a forest moisture gradient in an Ozark watershed. *Ecology* 58:810–819.

Squibb, R.C., and Hunt, G.L., Jr. 1983. A comparison of nesting-ledges used by seabirds on St. George Island. *Ecology* 64:727–734.

Tolley, K.A. 1998. Assessing the population structure of the harbor porpoise (Phocoena phocoena) by discriminant analysis. *Marine Mammal Science* 14(3):646–649.

Tomblin, D.C., and Adler, G.H. 1998. Differences in habitat use between two morphologically similar tropical forest rodents. *Journal of Mammalogy* 79(3):953–961.

Van Horne, B., and Ford, R.G. 1982. Niche breadth calculation based on discriminant analysis. *Ecology* 63:1172–1174 (see also Dueser and Shugart 1979 and 1982; and Carnes and Slade 1982).

Welsh, H.H., and Lind, A.J. 1995. Habitat correlates of the Del Norte salamander, Plethodon elongatus (Caudata, Plethodontidae), in northwestern California. *Journal of Herpetology* 29(2):198–210.

Appendix 4.1

SAS program statements used to conduct discriminant analysis on the empirical data set presented in Figure 4.2. Anything given in lower case varies among applications with regard to characteristics of the specific data set (e.g., variable names) and the personal preferences of the user (e.g., naming output files).

The following header information creates a library named "km" for SAS files in the specified directory and provides page formatting information:

```
LIBNAME km 'd:\stats\da';
OPTIONS PS=60 LS=80 REPLACE OBS=MAX;
```

The following macro is used to conduct Levene's test of homogeneity of variance and compute descriptive univariate statistics and plots for each of the original variables in order to assess univariate homogeneity of variance and normality assumptions:

```
%MACRO RES(Y);
   DATA A; SET km.da;
      TITLE 'Univariate ANOVA on the Original Variables';
      PROC GLM;
         CLASS use;
         MODEL &Y=use;
         OUTPUT OUT=resid R=res; RUN;
   DATA resid2; SET resid;
      ares=ABS(res); RUN;
      TITLE 'Levene?s Test of Homogeneity of Variance';
      PROC GLM;
         CLASS use;
         MODEL ares=use; RUN;
     TITLE 'Assessing Univariate Normality of the Original Variables';
     PROC UNIVARIATE NORMAL PLOT;
        VAR &Y; RUN;
%MEND;
```

The following line should be repeated for each variable in the data set, substituting the appropriate variable names:

```
%RES(ltotal);
```

The following procedure is used to compute pairwise correlations among all variables in order to evaluate multicollinearity:

```
DATA A; SET km.da;
   TITLE 'Pairwise Correlations Among Original Variables';
```

```
PROC CORR RANK;
    VAR ltotal--fhd; RUN;
```

The following procedure is used to conduct a stepwise DA on the full variable set, after eliminating variables involved in high pairwise correlations:

```
DATA A; SET km.da;
    DROP ttotal bac bat fhd snagm45;
    TITLE 'Stepwise DA on Vegetation Structure Variables';
    PROC STEPDISC SINGULAR=.001;
        VAR ltotal--snagt;
        CLASS use; RUN;
```

The following procedure is used to conduct DA on the subset of variables identified from the stepwise procedure, and to test for homogeneity of variance-covariance matrices and compute descriptive univariate statistics and plots for the canonical function in order to assess multivariate homogeneity of variance-covariance and normality assumptions:

```
DATA B; SET km.da;
    TITLE 'Assessing Homogeneity of Group Dispersions';
    PROC DISCRIM POOL=TEST CAN OUT=scores SHORT;
        VAR ltotal snagt bah gtotal bas snag16 mtotal;
        CLASS use; RUN;
DATA B; SET scores;
    TITLE 'Assessing Normality of the Canonical Function';
    PROC SORT;
        BY use;
    PROC UNIVARIATE NORMAL PLOT;
        BY use;
        VAR CAN1; RUN;
```

The following procedure is used to inspect the standardized data for possible outliers, including only the subset of variables identified from the stepwise procedure:

```
DATA A; SET km.da;
    PROC STANDARD OUT=standard MEAN=0 STD=1;
        VAR ltotal snagt bah gtotal bas snag16 mtotal;
        BY use; RUN;
DATA B; SET standard;
    ARRAY XX ltotal--mtotal;
        DO OVER XX;
            IF XX GT 2.5 THEN OUTLIER=1;
            ELSE XX='.';
        END; RUN;
DATA C; SET B;
    IF OUTLIER=1;
    TITLE 'List of Potential Outlier Observations';
    PROC PRINT;
        VAR Id ltotal--mtotal; RUN;
```

The following procedure is used to conduct DA on the subset of variables identified from the stepwise procedure, including computing the canonical function, classification matrix, and jackknife classification matrix, and generating a list and histogram of the canonical scores by group:

Appendix 4.1

```
DATA A; SET km.da;
   TITLE 'Canonical Analysis of Discriminance on the Vegetation
Structure Variables';
   PROC DISCRIM POOL=TEST CAN LIST CROSSVALIDATE OUT=scores;
      VAR ltotal snagt bah gtotal bas snag16 mtotal;
      CLASS use; RUN;
DATA B; SET scores;
   PROC SORT;
      BY use;
   TITLE 'List of Canonical Scores by Group';
   PROC PRINT;
      VAR use CAN1; RUN;
   TITLE 'HISTOGRAM OF CANONICAL SCORES BY GROUP';
   PROC CHART;
      VBAR CAN1 / MIDPOINTS=-4 TO 4 BY 1 SUBGROUP=use; RUN;
```

5
Canonical Correlation Analysis

5.1 Objectives

By the end of this chapter, you should be able to do the following:

- Recognize the types of research questions best handled with *canonical correlation analysis* (CANCOR).
- List nine important characteristics of CANCOR.
- Draw a diagram depicting the canonical correlation between the two canonical variates for a hypothetical situation involving two variables to demonstrate a geometric understanding of CANCOR.
- Given a data set, determine if sample sizes are adequate and examine whether the four assumptions of CANCOR are met:
  — No singularities
  — Multivariate normality
  — Independent random sample
  — Linearity
- Explain the difference between using the covariance and correlation matrix in the eigenanalysis.
- Given a set of canonical variate pairs, assess their importance and/or significance using
  — Canonical correlation criterion
  — Canonical redundancy criterion
  — Significance tests based on resampling procedures
  — Graphical representation of canonical scores
- Given a set of canonical variate pairs, interpret their meaning using
  — Structure coefficients
  — Canonical cross-loadings
  — Significance tests based on resampling procedures

- Use the split-sample and jackknife procedures to validate a given set of canonical variates.
- List five limitations of CANCOR.

5.2 Conceptual Overview

Ecologists have long realized that a number of environmental variables (e.g., slope, aspect, soil organic matter, soil texture) simultaneously contribute to a smaller number of major factors (such as moisture or soil fertility), and likewise that the distributions of numerous plant and animal species may be determined by these few dominant factors. In other words, there often exists a relatively simple set of major factors (or gradients) which effectively summarizes the totality of environmental factors and largely determines species distributions. Hence, ecologists are often interested in directly relating patterns in the distribution of organisms with patterns in the environment, or, more generally, directly relating the patterns in one set of variables to patterns in a second set. For example, they might want to associate the major patterns in the composition of the plant community directly to physical environmental factors. Or they might want to associate the abundance of several animal species (i.e., community structure) directly with the composition and structure of the plant community. Or they might wish to associate the composition of the animal community directly with the composition and configuration (i.e., structure) of a landscape.

As described in Chapter 2, these and many other similar questions could be approached in a two-step process using unconstrained ordination methods. First, the researchers would use unconstrained ordination methods, such as principal components analysis (PCA) or detrended correspondence analysis (DCA), to extract the dominant patterns in the data. Then they would attempt to associate these patterns with a second set of explanatory variables in a subsequent, secondary analysis. Unfortunately, the association of ordination axes with patterns in a second set of variables often produces ambiguous results. One solution, described in Chapter 2, is to directly ordinate the first set of variables on axes that are combinations of the second set of variables; that is, to constrain the ordination of the first set of variables by their relationships to a second set, particularly when the first set of variables consists of species scores (e.g., abundances) and the second set consists of environmental variables (e.g., habitat characteristics). In this manner, the gradients in the species data set can be described directly in terms of the environmental variables. Here, the emphasis is on extracting the major gradients in the first data set (i.e., the dependent variables, usually species abundances) and explaining these gradients in terms of the measured explanatory variables in the second data set (i.e., the independent variables, usually environmental factors). This process is called constrained ordination, and canonical correspondence analysis (CCA)[1] is the best constrained ordination technique developed to date.

---

[1]. Note the similarity in names between canonical correspondence analysis (CCA) and canonical correlation analysis (CANCOR).

5.2 Conceptual Overview

An alternative solution is to simultaneously ordinate both sets of variables, subject to the constraint that the derived axes in one data set are maximally correlated with corresponding axes from the second data set. Canonical correlation analysis (CANCOR) can be viewed as a formal mathematical approach to describing the relationship between the major factors in each data set (Gauch and Wentworth 1976). Here the emphasis is on finding the relationship between two (or more) sets of variables, and although one variable set can often be treated as dependent on the second variable set, as in the constrained ordination case above, it is not necessary to do so. Thus, this approach has somewhat broader application than the constrained ordination approach.

Consider the hypothetical forest system described in the introduction to Chapter 1. Suppose that researchers wish to determine whether or not the arthropod community is structured in relation to environmental factors; that is, whether the arthropod community changes in composition or structure along important environmental gradients. Suppose they sample 30 locations distributed randomly or systematically throughout the study area, and at each location they measure the abundance of each of the 10 arthropod species and a suite of 10 environmental variables.

As before, these data can be tabulated into a two-way data matrix, in this case with 30 rows and 20 columns, where each row represents a site (or sampling location) and the columns represent variables, partitioned into two groups (species abundances and environmental factors). In this case, it makes sense to consider these variable groups as two separate data spaces. Each site has a value on each variable and simultaneously occupies a unique location in both 10-dimensional spaces. The shape, clumping, and dispersion of each data cloud individually and in relation to each other describes the ecological patterns of this study system.

In this case, species abundances could be considered dependent variables whose response can be explained by the set of independent environmental variables. Therefore, both the constrained ordination and CANCOR approaches are applicable. The choice of techniques depends on subtle differences in the emphasis sought by the researchers. If the emphasis is on finding dominant gradients in the arthropod community and then explaining those gradients in terms of the environmental variables, then constrained ordination is the best approach. If, on the other hand, the emphasis is on determining whether dominant gradients in the arthropod community are strongly related to dominant gradients in the environment, that is, whether there exists joint structure to the two variable sets, then CANCOR is the best approach. Conceptually, the differences between these approaches is subtle; therefore, it may be advantageous to use both approaches in a complementary analysis.

The remainder of this chapter is devoted to CANCOR. Specifically, CANCOR is a technique for analyzing the relationship between two or more sets of variables. This is in contrast to unconstrained ordination and cluster analysis (CA), which assess relationships within a single set of interdependent variables, and discriminant analysis (DA), which assesses relationships among groups of entities based on a single set of discriminating or predictive variables. Although CANCOR can be used to analyze the relationships among any number of sets of variables, we are typically only interested in relating two sets. Thus, we will limit our discussion to applications involving two sets of variables, but our comments are easily generalized to applications involving multiple sets. Table 5.1 summarizes some of the important characteristics of CANCOR.

Conceptually, CANCOR is similar to DA in some respects and ordination in others. Like DA, and in contrast to ordination, linear dependencies or high correlations among the variables (i.e., multicollinearity) are troublesome in CANCOR. Steps are generally taken to eliminate severe multicollinearity problems. Like both ordination and DA, CANCOR attempts to extract dominant, underlying gradients of variation from a set of multivariate observations. However, CANCOR seeks to find gradients of variation within one set of variables that are maximally correlated with gradients in a second set of variables. In this respect, CANCOR is similar to doing a PCA on two sets of variables, with the added constraint that the components in one set be derived such that they are maximally correlated with the components in the other set. Thus, instead of deriving a linear combination of variables that explains the maximum variance *within* a single set of variables (PCA), we derive two linear functions that explain the maximum correlation *between* two sets of variables. Like both

TABLE 5.1 Important characteristics of canonical correlation analysis (CANCOR)

A single technique, as compared to the variety of approaches in ordination and cluster analysis.

Organizes sampling entities (e.g., species, sites, observations) along pairs (or larger groupings) of continuous ecological gradients.

Assesses relationships between two (or more) sets of variables; often an attempt is made to define the relationship between a set of independent variables and a set of dependent variables.

Extracts pairs (or larger groupings) of dominant, underlying gradients of variation (canonical variates) among sampling units from two (or more) sets of multivariate observations.

Reduces the dimensionality of a multivariate data set by condensing a large number of original variables into a smaller set of new composite dimensions (canonical variates) with a minimum loss of information.

Summarizes data redundancy by placing similar entities in proximity in canonical space and producing a parsimonious understanding of the data in terms of a few pairs (or larger groupings) of maximally correlated gradients of variation.

Defines new composite dimensions within a set of variables as weighted, linear combinations of the original variables (canonical variates) that are maximally correlated with linear combinations from the opposing set of variables.

Describes the joint or common structure of two (or more) sets of variables; that is, describes the degree to which sample entities occupy a multidimensional space jointly or simultaneously defined by two (or more) sets of variables.

A generalized case of multiple regression analysis where there is more than one dependent variable; or a generalized case of multivariate analysis of variance, where the latter is the special case of CANCOR with one of the two sets of variables consisting entirely of categorical variables (typically representing experimental treatments).

ordination and DA, CANCOR attempts to reduce the dimensionality of a multivariate data set by condensing a large number of original variables into a smaller set of new composite dimensions (canonical variates) with a minimum loss of information. Similarly, like both ordination and DA, CANCOR attempts to reduce data redundancy by placing similar entities in proximity in canonical space and producing a parsimonious description of the data in terms of a few dominant gradients, only in this case, CANCOR condenses the information into pairs of correlated gradients.

Canonical correlation analysis can be considered the generalized case of multiple regression analysis (MRA), in which one attempts to examine the interrelationships between two sets of variables simultaneously. Whereas MRA examines the relationship between a *set* of continuous independent variables and *one* continuous dependent variable, CANCOR examines the relationship between *multiple* independent variables and *multiple* dependent variables. Similarly, N-way multivariate analysis of variance (MANOVA) can be considered a special case of CANCOR in which one of the two sets of variables consists entirely of categorical, group-membership variables. Unlike MRA and MANOVA, however, in CANCOR it is not necessary to make a distinction between independent and dependent variables, since both sets of variables can serve either function. In ecological studies, however, it is often logical to treat one set as independent and the other as dependent. For example, we are often interested in investigating organism responses (e.g., animal abundances, animal morphology) to a suite of environmental conditions (e.g., microclimate, habitat characteristics). And in this case, it would be logical to treat the set of animal data as the dependent variables and the environmental parameters as the independent variables.

The main purpose of CANCOR is to reveal the common structure of two multivariate data sets. To accomplish this, CANCOR constructs linear combinations of variables from each data set, called *canonical variates*, such that the correlation between the two canonical variates is maximized. The canonical variates are defined as weighted linear combinations of the original variables, where each variable is weighted according to its contribution to the variance of the corresponding canonical variate. Each sample entity has a single composite canonical score for each canonical variate, derived by multiplying the sample values for each variable by the corresponding weight and adding these products together. Thus, each sample entity has a score on each canonical variate from the first data set and a score on the corresponding variate from the second data set. Canonical correlation analysis finds the linear combination of variables in each data set that maximizes the relationship between the corresponding sample scores.

For an $N \times (P + M)$ data set (where N equals number of samples, P equals number of variables in the first variable set, and M equals number of variables in the second variable set), there exist P or M (whichever is smaller) pairs of canonical variates (note that this can be generalized to any number of data sets). For example, in a data set containing 8 environmental variables and 15 bird abundance variables, there exist 8 pairs of canonical variates. The first pair of variates defines the linear combination of variables in the two data sets that result in the maximum correlation between resulting canonical scores. The second and subsequent pairs of variates define the linear combinations of variables that maximize the correlation between sample scores given that they are independent (i.e., uncorrelated) to the preceding canonical variates.

The correlation between each pair of canonical variates is termed *canonical correlation*. The major interpretation problem with canonical correlation analysis is that canonical correlations do not represent correlations between sets of original variables; they represent correlations between pairs of canonical variates calculated from the original variables. Thus, in CANCOR, canonical correlations must be interpreted differently than they are in DA, since they do not directly involve the original variables. A significant correlation may be obtained between two canonical variates even though these variates do not extract large or meaningful proportions of variance from the respective sets of original variables. To help overcome this interpretational problem, a measure called *redundancy* has been developed, whereby one can calculate the amount of variance in one set of original variables that is explained by the canonical variate of the other data set. By calculating redundancy for all variates of a data set and summing the results, the proportion of variance of one set that is accounted for by the other set can be calculated.

The desired outcome in CANCOR is that most of the variation in one data set (often considered the dependent data set) can be explained by the other data set (often considered the independent data set), or vice versa when no independent-dependent distinction is made. Moreover, it is hoped that most of the explained variation can be accounted for by the first few pairs of canonical variates and that these variates can be meaningfully interpreted based on the relative importance of the original variables in the linear functions. If so, then the joint structure of two complex, multivariate data sets can be effectively summarized in a few important multidimensional gradients. The resulting reduction in dimensionality greatly simplifies elucidation of sample relationships.

Canonical correlation analysis has not been applied widely in ecological research. In part, this is because it is poorly understood. But it is also because of the availability of more familiar techniques, the difficulty of interpreting the results, and the tendency of the results to be situation-specific and not generalizable (Thorndike and Weiss 1973). Canonical correspondence analysis (CCA) is widely used as an alternative method to quantify the multivariate relationships between sets of independent and dependent ecological variables. The advantages of CCA are that linearity is not assumed (probably the greatest limitation of CANCOR), multicollinearity is not an impediment (another major limitation of CANCOR) and the ordination is easily interpretable. However, CANCOR is still a viable technique which ecologists should consider. When its assumptions are met, it can elucidate aspects of the relationship between data sets that cannot be obtained with other methods. Successful applications of CANCOR have included relating animal morphology to environmental conditions (Calhoon and Jameson 1970; Vogt and Jameson 1970; Jameson et al. 1973), relating animal behavior to environmental conditions (James and Porter 1979), and relating marine invertebrate communities to aspects of the marine environment (McIntire 1978; Poore and Mobly 1980). Few studies on terrestrial community ecology have used CANCOR, and most that have, have suffered from small sample-to-variable ratios (e.g., Herrera 1978). Canonical correlation analysis has shown potential to be a very useful tool in wildlife habitat studies (Boyce 1981; Folse 1981). A common goal in many wildlife habitat studies is to relate features of the habitat to species abundances. The traditional approach has been to assess each species' habitat association separately using MRA, DA, PCA, or similar procedures, or to assess community summary attributes such as diversity or richness. Canonical correlation analysis, like CCA, offers an alternative in which habitat associations among all species can be assessed simultaneously.

5.3 Geometric Overview

Perhaps one reason CANCOR is poorly understood and rarely applied in wildlife research is because, geometrically, it is difficult to conceptualize the canonical correlations and canonical variates. In CANCOR, the data set can be depicted as a multidimensional cloud of sample points occupying a space defined by not one but two (or more) sets of axes, corresponding to the two sets of variables (Fig. 5.1). Thus, each set of variables defines a multidimensional space, where each dimension (i.e., axis) is defined by one of the P or M original variables. Each sample has a position on each axis and therefore simultaneously occupies a position in the P- and M-dimensional spaces. Geometrically, we can think of CANCOR as a procedure for measuring the extent to which entities occupy the same relative position in the space defined by each of the variable sets. For example, we might be interested in the extent to which sites occupy the same relative position in vegetation structure space as in bird community space.

Figure 5.1 portrays a simple two-dimensional geometric representation of CANCOR. As described by Smith (1981), variable set one (X set) comprises a set of original variables X_1 and X_2, defining plane one. Variable set two (Y set) is composed of variables Y_1 and Y_2 and defines plane two. This example of two sets, each with two original variables, is the simplest canonical correlation situation. Canonical correlation analysis will reveal a composite of X_1 and X_2 (an axis called canonical variate 1_X) that is correlated with a composite of Y_1 and Y_2 (an axis called canonical variate 1_Y). The objective is to create 1_X and 1_Y such that they will be maximally correlated. This will be the first canonical correlation (R_1). The first canonical correlation is equal to the cosine of the angle between 1_X and 1_Y (dotted line in Fig. 5.1). As the angle approaches zero degrees, R_1 approaches 1.0; that is, the more similar the orientation of 1_X and 1_Y, the greater the canonical correlation.

When each set of variables contains two or more variables, CANCOR will define a second pair of canonical variates or axes (2_X and 2_Y), which will produce the maximum possible correlation between the two variates (R_2), subject to the constraint that these two new variates be independent of the first two canonical variates (1_X and 1_Y). The process of constructing canonical variates continues until the number of pairs of canonical variates equals the number of variables (P or M) in the smaller data set. Each successive pair of variates has a smaller canonical correlation, each are uncorrelated with any of the preceding variates, and each account for successively less of the variation shared by the two sets of variables.

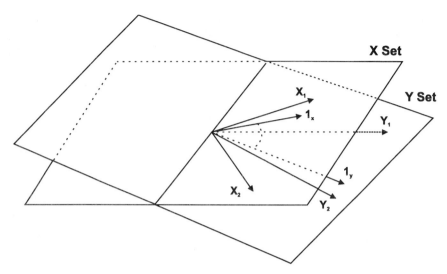

FIGURE 5.1 Geometric representation of canonical correlation analysis; derivation of the first canonical variate in the X-variable set (1_X) from original variables X_1 and X_2, which define plane one, and the first canonical variate in the Y-variable set (1_Y) from original variables Y_1 and Y_2, which define plane two. See text for a full explanation. The cosine of the angle between 1_X and 1_Y (dotted line) is equal to the canonical correlation (adopted from Thorndike 1978).

Geometrically, each canonical variate can be viewed as a product of their respective X- or Y-axes. For example, with respect to the canonical variates in the X variable set ($1_X, 2_X, \ldots, P_X$), the degree of similarity in orientation between the original X-axes and the canonical axes varies. Some may be oriented in the same general direction as the canonical variates, while others may not. The degree of similarity in orientation between the original axes and the canonical axes indicates how closely related the axes are and provides a geometric means of determining which of the original X variables are most important in each canonical variate.

5.4 The Data Set

Canonical correlation analysis deals with problems involving two or more sets of continuous, categorical, or count variables. In general, variables within each data set should either be all continuous, all categorical, or all count variables. As in DA, the validity of mixed data sets is not clear. One set of variables is often considered the independent set and the other is considered the dependent set, although it is not required that any distinction be made.

Several other aspects of the data sets should be noted: (1) each sample entity must be measured on both complete sets of variables; (2) there must be more samples than variables in both data sets combined (i.e., $N > P + M$); (3) the number of variables in each data set need not be the same (i.e., $P \neq M$); and (4) only variables that clearly relate to the same phenomenon should be included in the respective data sets.

Canonical correlation algorithms usually require a two-way data matrix as input in the form of a samples-by-variables matrix, where the variable list includes both sets of variables. Canonical correlation analysis is generally used to correlate two "related" sets of variables, for example, where a presumed cause-and-effect relationship exists between two sets of variables. In wildlife research, for example, CANCOR might be used to examine animal-habitat associations at the community level; specifically, it might be used to relate measured characteristics of the habitat with estimates of abundances for several species. The corresponding two-way data matrix could look as follows:

	X_1	X_2	$\cdots$	X_P	Y_1	Y_2	$\cdots$	Y_M
1	x_{11}	x_{12}	$\cdots$	x_{1P}	y_{11}	y_{12}	$\cdots$	y_{1M}
2	x_{21}	x_{22}	$\cdots$	x_{2P}	y_{21}	y_{22}	$\cdots$	y_{2M}
$\vdots$	$\vdots$	$\vdots$	$\ddots$	$\vdots$	$\vdots$	$\vdots$	$\ddots$	$\vdots$
N	x_{N1}	x_{N2}	$\cdots$	x_{NP}	y_{N1}	y_{N2}	$\cdots$	y_{NM}

In this book, we discuss CANCOR on a samples-by-variables matrix, where samples represent any type of sampling unit (e.g., sites, species, specimens) for which there are associated two vectors of measurements corresponding to the two sets of variables, and variables represent any set of parameters recorded for each sampling unit (e.g., habitat features, morphological characteristics, behavioral characteristics, species abundances). Unless otherwise noted, we refer to the samples (rows) as "entities" or "sampling entities" to reflect the generality of CANCOR to any type of sampling unit. However, our comments are applicable to any data of suitable form.

Empirical Example. The empirical example was taken from a study on bird-habitat relationships in mature, unmanaged forest stands in the central Oregon Coast Range (McGarigal and McComb, unpublished data). In this chapter, we use CANCOR to examine relationships between vegetation structure characteristics and bird community structure. Specifically, we use CANCOR to identify and describe vegetation gradients that are maximally correlated with avian species abundance patterns. The data set is similar to the one described in Chapter 3. Here it consists of 40 variables, including 20 vegetation structure variables (see Chapter 3, Sec. 3.4, for a brief description of these variables), 19 variables corresponding to each bird species' abundance, and a single variable (ID) that identifies the sampling point (Fig. 5.2). Only bird species with four or more detections are included. The vegetation variables are all continuous and were either directly measured or derived for each site ($N = 96$). Note that the variables have different measurement scales and that the variables SNAGT and BAT are linear combinations of other variables. (See Appendix 5.1 for an annotated SAS script for all of the analyses presented in this chapter.)

OBS	ID	GTOTAL	LTOTAL	TTOTAL	MTOTAL	OTOTAL	SNAGS6	SNAGM1	SNAGM23	SNAGM45	SNAGM6	SNAGL1	SNAGL23	SNAGL45	SNAGL6	SNAGT	AFHD	BMRO	BHGR	EEJU	DBCH	CRCR	GVGR	GCKI	HAFL	HRJA	HAWO	REWA	SUHU	SOSP	STJA	VWTH	WATH	WEFI	WIWA	WIWR
1	1S0	21	15	75	20	30	0	0	0	0	0	0	0	0	1	1	1.5	1	0	0	2	1	0	0	0	0	0	0	0	2	0	11	0	0	8	2
2	1S1	36	15	95	15	35	0	0	0	0	1	0	0	1	0	2	1.4	0	0	2	0	0	1	0	0	0	0	0	0	0	0	5	0	1	5	3
3	1S2	30	30	70	10	55	0	0	0	1	2	2	1	0	1	7	1.5	0	0	2	0	0	0	0	0	1	0	0	0	1	0	0	0	2	4	
4	1S3	11	50	70	20	70	0	0	0	0	0	1	0	0	3	15	1.4	0	0	5	0	0	0	0	0	0	0	0	4	0	0	1	0	1	2	3
5	1S4	3	40	80	15	65	0	0	1	0	0	0	0	0	0	1	1.5	0	0	2	2	0	2	1	0	0	0	0	1	0	0	0	0	2	1	7
.	.	.	.	.	.	.	.	.	.	.	.	.	.	.	.	.	.	.	.	.	.	.	.	.	.	.	.	.	.	.	.	.	.	.	.	.
92	6U3	7	80	10	5	75	0	0	0	1	0	0	0	0	1	2	1.0	0	0	3	2	0	0	0	0	3	0	1	1	0	0	3	0	0	3	1
93	6U4	20	75	25	10	55	0	0	0	1	0	0	0	2	2	5	1.4	1	0	3	0	0	0	0	0	0	0	0	1	0	0	4	0	1	1	4
94	6U5	25	65	2	1	70	6	0	1	2	1	0	0	1	4	9	1.0	0	0	8	9	0	0	2	0	2	0	0	0	0	0	1	0	0	4	4
95	6U6	2	85	20	2	75	3	0	0	0	1	0	0	1	3	8	0.8	0	0	5	4	0	0	2	0	2	1	0	0	0	0	1	0	0	9	0
96	6U7	2	99	1	2	80	2	0	0	1	1	0	0	3	1	8	0.4	0	0	6	3	0	0	4	0	6	0	1	1	0	0	1	0	0	3	1

FIGURE 5.2 The example data set listing observation points (OBS labeled 1 through 96) and 39 variables, including 20 vegetation and 19 bird variables with measured values.

5.5 Assumptions

For canonical correlation analysis to be strictly applicable, a data set must meet several assumptions, including the usual multivariate statistical assumptions (see Chapters 2 and 4 for additional discussion). Understanding the degree to which a given data set adheres to these assumptions allows the validity of any statistical inferences made to be determined. Field data sets rarely, if ever, meet the requirements precisely. If inferences are to be made regarding the significance of the canonical correlations, then the model assumptions must be met rather well; for merely descriptive purposes, larger departures from ideal data structure are tolerable. Even for descriptive purposes, however, it must be remembered that CANCOR has an underlying mathematical model and, consequently, may be applicable to one data set but not another.

In wildlife research, CANCOR is generally used as an exploratory method to aid in the generation of testable hypotheses. If the structure of the data is unknown at the time it is collected, exploratory methods often are both necessary and informative. Patterns obtained from descriptive analyses are properly regarded as preliminary and should be used to suggest hypotheses that subsequently can be tested. A common failure in the wildlife literature is to bypass this important step and report the results of exploratory analysis as statistically confirmatory.

5.5.1 Multivariate Normality

Canonical correlation analysis does not require an underlying multivariate normal distribution per se (see Chapter 2, Sec. 2.5.1 for additional discussion of multivariate normality), but the closer the data approach normality, the more realistic and interpretable the relationships between variables will be (Morrison 1976). If significance tests of the canonical correlations are to be considered valid, then at least one of the two sets of variables should have an approximate multivariate normal distribution.

5.5.1.1 Diagnostics

(a) Univariate Normality of Original Variables

Assess skewness, kurtosis, and normality and construct stem-and-leaf, box, and normal probability plots of each original variable. It is usually assumed that univariate normality is a good step toward multivariate normality.

(b) Normality of Canonical Scores

Assess skewness, kurtosis, and normality and construct stem-and-leaf, box, and normal probability plots of the canonical scores for each canonical variate. These diagnostics are generally more appropriate for evaluating multivariate normality than the univariate diagnostics, but even normally distributed canonical scores do not guarantee a multivariate normal distribution.

Empirical Example. In the example, we assessed each variable for univariate normality. Although the output is not shown here, it is similar to that shown in Figure 2.3. Not surprisingly, few variables are normally distributed. Meristic data, such as bird counts, often require a logarithmic transformation to normalize the distribution. Logarithmic transformations of the bird count data are only partially successful in normalizing the distributions. Nevertheless, we used the log-transformed bird variables in the CANCOR. (Note: The results of CANCOR were similar using the untransformed data.) We attempted several transformations of the vegetation variables. None of the transformations significantly improved the distributions or altered the results of the CANCOR. Consequently, we used the untransformed vegetation variables in the CANCOR.

To assess multivariate normality, we conducted CANCOR on the log-transformed bird variables and original vegetation variables and assessed the resulting canonical scores for normality. The normality diagnostics indicate that most of the canonical variates are approximately normally distributed. However, the first habitat variate (HAB1) has a bimodal distribution (Fig. 5.3), reflecting a strong discontinuity in the habitat gradient defined by the first habitat canonical variate. From this, we conclude that the vegetation structure gradient defined by the first canonical variate may be better characterized as two relatively distinct vegetation types rather than one continuous gradient. Indeed, this canonical variate describes a gradient between conifer-dominated upslope patches and hardwood-dominated streamside areas, two relatively distinct vegetation types within mature, unmanaged forest stands. Overall, however, these diagnostics indicate that the multivariate normality assumption is probably not grossly violated.

5.5.2 Singularities and Multicollinearity

Canonical correlation analysis, like DA, requires that the data matrix be nonsingular (i.e., no variable can be a linear combination of other variables in the data set). It is also subject to multicollinearity problems caused by near linear associations (see Chapter 4, Sec. 4.5.3 for additional discussion of these issues). The solutions to the multicollinearity problem are the same as presented in Chapter 4, with a slight modification to the first solution offered. Specifically, for variables involved in high (e.g., $r = 0.7$) within-set pairwise correlations (i.e., correlations between variables from the same

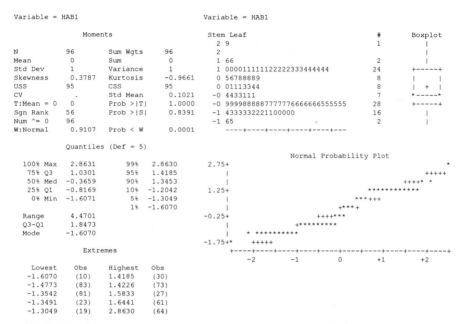

FIGURE 5.3 Univariate summary statistics used to assess normality of the canonical variates. Only output for the first habitat variate is presented.

set), the technique is as follows: Instead of retaining the offending variable with the greatest F-ratio derived from an ANOVA between groups, as in DA, we compute the multiple correlation coefficient (from MRA) between each offending variable (treated as the dependent variable) and the full set of opposing variables (treated as the independent variables). As a general rule, for each pair of highly correlated variables (based on within-set correlations), we retain either the variable with the greatest multiple correlation coefficient with variables from other variable sets, or with the greatest ease of ecological interpretation, and we eliminate the others. When a distinction is made between a dependent and independent variable set, we are more interested in multicollinearity within the independent set, as in multiple regression analysis. Note that since variables are often involved in several high correlations, the removal of one offending variable will often eliminate several high correlations.

5.5.2.1 Diagnostics

Diagnostic procedures for detecting singularities are not necessary per se, since singularities either will be obvious to the researcher beforehand or will be made apparent when the data is analyzed using CANCOR. However, there are procedures for detecting significant multicollinearity problems.

(a) Pairwise Correlations Among Variables

Calculate all possible pairwise correlations among the variables within each variable set; high correlations (e.g., $r = 0.7$) suggest potential multicollinearity problems and indicate the need to eliminate some of the offending variables.

(b) Agreement Between Canonical Weights and Loadings

Compare the signs and relative magnitudes of the standardized canonical coefficients (weights) and structure coefficients (loadings) for disagreement. Difference in the sign and/or rank order indicate multicollinearity problems.

Empirical Example. In the example, recall that we derived two vegetation structure variables (SNAGT and BAT) by linearly combining other variables in the data set. We would be unable to compute the canonical variates from the complete data set because of these two singularities. In this case, we chose to eliminate these two variables prior to CANCOR. Alternatively, we could have eliminated one of the field-measured variables involved in each linear dependency.

We computed the bivariate product-moment correlations between all habitat variables and discovered only three high ($r=0.7$) correlations (Fig. 5.4). Note that because

```
               Correlations Among the Original Variables

                 Correlations Among the VAR Variables

                AMRO         BHGR          BRCR         CBCH          DEJU

AMRO          1.0000       -0.0794       -0.0152      -0.0244       -0.1021
BHGR         -0.0794        1.0000       -0.0656       0.0844        0.1256
BRCR         -0.0152       -0.0656        1.0000       0.2360        0.1732
CBCH         -0.0244        0.0844        0.2360       1.0000        0.2934
DEJU         -0.1021        0.1256        0.1732       0.2934        1.0000

                 Correlations Among the WITH Variables

               LTOTAL       TTOTAL        MTOTAL       OTOTAL        SNAGM1

LTOTAL        1.0000       -0.8079       -0.4928       0.6453        0.3379
TTOTAL       -0.8079        1.0000        0.3966      -0.6162       -0.2833
MTOTAL       -0.4928        0.3966        1.0000      -0.1881       -0.2922
OTOTAL        0.6453       -0.6162       -0.1881       1.0000        0.2356
SNAGM1        0.3379       -0.2833       -0.2922       0.2356        1.0000

         Univariate Multiple Regression Statistics for Predicting
                the WITH Variables from the VAR Variables

              Squared Multiple Correlations and F-Tests

                 19 numerator DF      76 denominator DF

                           Adjusted   Approx 95% CI for RSQ       F
              R-Squared    R-Squared  Lower CL   Upper CL   Statistic  Pr > F

BAC           0.757838     0.697298     0.568      0.792     12.5179   0.0001
LTOTAL        0.748773     0.685966     0.553      0.784     11.9219   0.0001
TTOTAL        0.661658     0.577073     0.418      0.703      7.8224   0.0001
FHD           0.617689     0.522111     0.353      0.661      6.4627   0.0001
```

FIGURE 5.4 Correlation matrices containing pairwise correlations among variables in the respective variable sets and the squared multiple correlation coefficients for predicting the habitat variates from the bird variates. Only a portion of the correlation matrices and squared multiple correlation coefficients for variables involved in high pairwise correlations are shown.

we are making an independent-dependent distinction between variables, we are more interested in multicollinearity within the independent variable set (habitat variables). LTOTAL has a strong positive correlation with BAC and negative correlation with TTOTAL and FHD. To determine whether it would be preferable to eliminate LTOTAL or the other three variables, we computed the squared multiple correlations for predicting each of these habitat variables from the full set of bird variables. Figure 5.4 indicates that BAC has the highest R-square, followed closely by LTOTAL. Based on these findings, we would probably do better to eliminate LTOTAL from the data set; this might yield slightly more stable canonical coefficients. However, we conducted the full canonical analysis with and without LTOTAL, and the results were invariant. Consequently, we retained LTOTAL in the data set because it represents an important attribute of habitat structure. We also noted pronounced differences in both signs and relative magnitudes of the standardized canonical coefficients and structure coefficients. This indicates that multicollinearity is probably adversely affecting the results and that we should interpret the results cautiously.

5.5.3 Independent Random Sample and the Effects of Outliers

Canonical correlation analysis assumes that random samples of observation vectors have been drawn independently from P- and M-dimensional multivariate, normal populations, respectively, and, as with most other multivariate techniques (e.g., PCA, DA), is adversely affected by point clusters and outliers (see Chapter 2, Sec. 2.5.2 for additional discussion of these issues).

5.5.3.1 Diagnostics

(a) Standard Deviations

A simple way to screen for outliers is to standardize the data by subtracting the mean and dividing by the standard deviation for each variable (i.e., to standardize to zero mean and unit variance) and then inspect the data for entities with any value more than about 2.5 standard deviations from the mean on any variable.

(b) Univariate Plots

Construct univariate stem-and-leaf, box, and normal probability plots for each variable and check for suspected outliers (i.e., points that fall distinctly apart from all other points).

(c) Plots of Canonical Scores

Construct stem-and-leaf, box, and normal probability plots of the canonical scores for each canonical variate and check for suspected outliers. In addition, construct scatter plots of each pair of canonical variates and check for suspect points.

Empirical Example. In the example, we constructed stem-and-leaf, box, and normal probability plots for each of the original variables and inspected them for potential outliers (SAS results are not shown, but are similar to those in Figure 5.3). A few points

5.5 Assumptions

deviated noticeably from the norm, be we did not consider any to be true outliers. Instead, we considered these points to be real and meaningful variations in vegetation conditions that potentially influence bird species distributions. We also conducted CANCOR on the log-transformed bird variables and original vegetation variables and constructed plots of the canonical scores for each canonical variate (e.g., Fig. 5.3 and 5.5). We identified a few outlying points, but did not consider any to be true outliers.

5.5.4 Linearity

The effectiveness of CANCOR depends on the implicit assumption that variables change linearly along underlying gradients and that there exist linear relationships among the variables such that they can be combined in a linear fashion to create canonical variates (Levine 1977; Johnson 1981; see Chapter 2, Sec. 2.5.3 for additional discussion of this assumption). Linearity is not a specified assumption of the mathematical model, per se, but it determines to a large extent the appropriateness and effectiveness of CANCOR in any particular case.

5.5.4.1 Diagnostics

(a) Scatter Plots of Canonical Scores

Construct scatter plots of each pair of canonical variates. Arched or curvilinear configurations of sample points often indicate pronounced nonlinearities.

Empirical Example. In the example, we inspected scatter plots of the first three pairs of canonical variates for nonlinearities and detected none (scatter plot for first pair is given in Fig. 5.5).

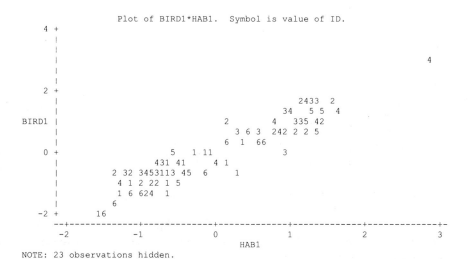

FIGURE 5.5 Scatter plot of the first pair of canonical variates.

5.6 Sample Size Requirements

There is no absolute rule for determining appropriate sample size in CANCOR. The more homogeneous the environment, the fewer samples needed to uncover any patterns that exist. Beyond this, there are some general rules to follow as well as some specific rules proposed by various authors.

5.6.1 General Rules

As in all eigenanalysis problems, there must be at least as many samples as variables. Thus, in CANCOR there must be more samples than variables in both data sets combined ($P + M$). Beyond this, enough samples should be taken to ensure that the canonical coefficients and loadings are estimated accurately and precisely. This can be done in several ways, for example, by sampling sequentially until the mean and variance of the parameter estimates (e.g., eigenvalues, canonical coefficients) stabilize. Once the data have been gathered, the stability of the results can be assessed using a resampling procedure (see Sec. 5.10.2), and the results can be used to determine the sample size needed to estimate the parameters with a certain level of precision in future studies. If the goal of CANCOR is the generalization of results to other situations, then the canonical variates should reflect the true underlying relationships, and stability in the estimates of canonical coefficients should be expected when sample sizes vary.

5.6.2 Specific Rules

The following specific rules have been suggested by different authors (where P = number of variables in the first set and M = number of variables in the second set):

Rule A	$N > 100 + (P + M)^2$	Thorndike (1978)
Rule B	$N > 10(P + M) + 100$	Thorndike (1978)
Rule C	$N > 3(P + M)$	Smith (1981)

Few ecological studies have approached sample sizes determined using Rule A or B. Rule C is probably adequate in most ecological studies. The inherent problem in using smaller sample sizes is that as sample size approaches number of variables, the canonical correlation approaches one, and statistical significance of the first canonical correlation is assured. Biological significance, however, should be questioned in such situations, and results based on CANCOR with equal or near equal numbers of sample and variables should be viewed with skepticism.

In practice, sample sizes are often either fixed or far too small given the large number of variables typically measured. Under these conditions it may be necessary to eliminate variables from one or both data sets. In wildlife habitat studies, for example, data sets often contain highly redundant variables, some of which will need to be eliminated because of multicollinearity considerations. Even after eliminating some of the offending variables, the reduced data set still may contain too many variables. One approach to solving this problem is to conduct CANCOR on a full or reduced variable set, remove variables with low weights (or loadings), and reanalyze the reduced data set. In general, if one is forced to deal with small sample sizes, the coefficient stability should be evaluated using one or more of the resampling procedures mentioned above, and the findings interpreted cautiously.

Empirical Example. In the example, the full data set contains 96 samples and a total of 37 (the value of $P+M$) variables, including 19 bird variables (P), 18 vegetation variables (M) (recall that we eliminated 2 vegetation variables due to singularities), and a single variable (ID) that identifies the sampling point. Based on the most liberal rule, Rule C, ($N = 3 \cdot 37 = 111$), the example data set still contains about 15 too few samples. There are several options: First, we can eliminate several variables from one or both data sets (perhaps redundant vegetation variables). Second, we can obtain additional samples (although in many applications this is unrealistic). Third, we can ignore the strict sample size requirements, consider the results as exploratory, and not attempt to generalize any relationships detected to other areas without first validating. For the purposes of this book, we choose the last strategy and proceed on.

5.7 Deriving the Canonical Variates

The first major step in CANCOR is to derive the canonical variates. This derivation step represents the nuts and bolts of the analysis; it includes several small steps that must be understood before we can consider how to interpret the canonical functions. These include: (1) whether to use the raw covariance or correlation matrix in the eigenanalysis; (2) how to compute the eigenvalues and canonical correlations from the covariance or correlation matrix; and (3) how to compute the canonical coefficients from the covariance or correlation matrix.

5.7.1 The Use of Covariance and Correlation Matrices

Like other eigenanalysis procedures (e.g., PCA and DA), the CANCOR solution is not derived directly from the original data matrix; that is, the eigenvectors and eigenvalues that define and describe the relationship between the canonical variates are not computed from the original two-way data matrix. Rather, the eigensolution is derived from either the raw covariance (or variance-covariance) matrix or correlation matrix, which have themselves been derived from the original data matrix. Recall that the raw covariance matrix is computed from the raw, unstandardized scores on the original variables; whereas, the correlation matrix is equivalent to the covariance matrix computed from standardized data.

In contrast to PCA, in CANCOR the choice between the raw covariance and correlation matrices is relatively trivial. The canonical solutions are largely the same regardless of which matrix is used. Specifically, the eigenvalues and associated canonical correlations, the structure coefficients (loadings), and the standardized canonical scores are invariant. The most notable difference is in the raw canonical coefficients. When the raw covariance matrix is used, the raw canonical coefficients will be in units proportional to those of the respective responses in each set, and the dimensionality of the respective canonical variates will have meaning (Dillon and Goldstein 1984). In contrast, canonical variates based on the correlation matrix are dimensionless. Because we rarely interpret the raw canonical coefficients in ecological studies, the choice between raw or standardized data is generally not important. Nevertheless, when the scale or unit of measurement differs among variables, it is customary to standardize the data or use the correlation-based approach. Statistical algorithms (e.g., SAS) generally use the raw covariance matrix in the eigenanalysis, unless otherwise specified.

5.7.2 Eigenvalues and Associated Statistics

Computationally, CANCOR is essentially an eigenanalysis problem similar to DA and PCA. For an $N \times (P + M)$ data set, where N equals the total number of sampling entities (rows of the data matrix), P equals the number of variables in one data set, and M equals the number of variables in the other set, there exists P or M (whichever is smaller) eigenvalues, or characteristic roots. (Note: This can be generalized to any number of variable sets.) The eigenvalues are determined by solving the following characteristic equations:

$$\left| S_{xx}^{-1} S_{xy} S_{yy}^{-1} S_{xy}' - \lambda I \right| = 0$$

$$\left| S_{yy}^{-1} S_{xy}' S_{xx}^{-1} S_{xy} - \lambda I \right| = 0$$

where S_{xy} is the between-sets covariance matrix, S_{xx} is the within-set covariance matrix for the X variable set, S_{yy} is the within-set covariance matrix for the Y variable set, λ is the vector of eigenvalue solutions, and I is the identity matrix.

Note that the correlation matrices can be used instead of the raw covariance matrices in the characteristic equations. Each eigenvalue is associated with one pair of canonical variates and is equal to the squared canonical correlation (R_c^2) between the canonical variates in the corresponding pair. The squared canonical correlation equals the proportion of variance in the canonical variate of one set that is accounted for by the variate of the other set. Note that the relationship being described is between the variates, not between the original variables. Hence, the eigenvalues measure how well the canonical variates from one data set "relate" to the canonical variates from the second data set, and not how well the original variables relate to each other.

The first eigenvalue is always the largest and thus corresponds to the pair of canonical variates that have the maximum possible correlation. Therefore, the first eigenvalue is a measure of the best possible relationship between linear combinations of variables from the two data sets. The second eigenvalue measures the proportion of shared variance in the pair of linear combinations that have the second highest correlation. We constrain the second set of combinations to be independent of the first pair. The third eigenvalue corresponds to the pair of variates with the next highest correlation, given that it is uncorrelated to the first and second pairs; it is a measure of the best possible relationship between variable sets *after* the first and second have done their best. And so on, for the rest of the eigenvalues. The rationale here is that we want to identify statistically independent patterns of linkage between the two variable sets.

All of the eigenvalues will be positive or zero, and the larger the eigenvalue, the greater the relationship between the corresponding variates. Thus, the pair of canonical variates with the largest eigenvalue will have the maximum correlation, while the pair with the smallest eigenvalue will have the lowest correlation. Eigenvalues near zero indicate that the corresponding pair of canonical variates is almost totally unrelated.

Empirical Example. In the example, we conducted CANCOR on the log-transformed bird variables and original vegetation variables and computed the 18 eigenvalues corresponding to the number of variables in the smaller set (Fig. 5.6). In this

	Canonical Correlation	Adjusted Canonical Correlation	Approx Standard Error	Squared Canonical Correlation
1	0.937884	0.912802	0.012350	0.879627
2	0.833372	0.750352	0.031343	0.694509
3	0.776835	0.639305	0.040683	0.603473
4	0.758709	.	0.043538	0.575640
5	0.696174	0.556717	0.052873	0.484659
6	0.666571	.	0.057012	0.444316
7	0.595947	0.452593	0.066160	0.355153
8	0.542454	0.389693	0.072408	0.294256
9	0.499017	0.389148	0.077049	0.249018
10	0.441945	0.322294	0.082559	0.195315
11	0.363315	.	0.089055	0.131998
12	0.347894	.	0.090180	0.121030
13	0.272132	.	0.095000	0.074056
14	0.233917	.	0.096984	0.054717
15	0.203505	.	0.098349	0.041414
16	0.121558	.	0.101082	0.014776
17	0.066764	.	0.102141	0.004457
18	0.030466	.	0.102503	0.000928

Eigenvalues of INV(E)*H = CanRsq/(1-CanRsq)

	Eigenvalue	Difference	Proportion	Cumulative
1	7.3075	5.0341	0.4502	0.4502
2	2.2734	0.7515	0.1401	0.5902
3	1.5219	0.1654	0.0938	0.6840
4	1.3565	0.4160	0.0836	0.7675
5	0.9405	0.1409	0.0579	0.8255
6	0.7996	0.2488	0.0493	0.8747
7	0.5508	0.1338	0.0339	0.9087
8	0.4169	0.0854	0.0257	0.9344
9	0.3316	0.0889	0.0204	0.9548
10	0.2427	0.0907	0.0150	0.9697
11	0.1521	0.0144	0.0094	0.9791
12	0.1377	0.0577	0.0085	0.9876
13	0.0800	0.0221	0.0049	0.9925
14	0.0579	0.0147	0.0036	0.9961
15	0.0432	0.0282	0.0027	0.9987
16	0.0150	0.0105	0.0009	0.9997
17	0.0045	0.0035	0.0003	0.9999
18	0.0009	.	0.0001	1.0000

FIGURE 5.6 Canonical correlations, eigenvalues, and associated statistics. Note that SAS defines the eigenvalues differently than we have defined them in the text.

book, we have defined the eigenvalue to be equal to the squared canonical correlation (R_c^2). SAS, however, computes the eigenvalues from $W^{-1}A$ (where W = pooled within-groups sums-of-squares and cross-products matrix, and A = among-groups sums-of-squares and cross-products matrix), as is done in DA, instead of computing the eigenvalues from $S_{xx}^{-1}S_{xy}S_{yy}^{-1}S_{xy}'$ as we have discussed for CANCOR. SAS makes use of the mathematical relationship between DA and CANCOR to derive the eigenvalues for $S_{xx}^{-1}S_{xy}S_{yy}^{-1}S_{yx}'$ as follows:

$$\lambda_i^{DA} = \frac{\lambda_i^{CANCOR}}{1 - \lambda_i^{CANCOR}}$$

where λ_i^{DA} are the eigenvalues of $W^{-1}A$ and λ_i^{CANCOR} are the eigenvalues of $S_{xx}^{-1}S_{xy}S_{yy}^{-1}S_{xy}'$. Using our definition, the eigenvalue associated with the first pair of canonical variates is 0.88; that is, 88 percent of the variance in the first bird variate (BIRD1) is accounted for, or shared by, the first habitat variate (HAB1), and vice versa. Similarly, 69 percent of the variance in the second bird variate (BIRD2) is shared by the second habitat variate (HAB1); and so forth. The remainder of the output in Figure 5.6 will be discussed later.

5.7.3 Eigenvectors and Canonical Coefficients

The other product of eigenanalysis is two eigenvectors associated with each eigenvalue. The eigenvectors correspond to the two sets of variables and are determined by solving the following system of equations:

$$\left| S_{xx}^{-1}S_{xy}S_{yy}^{-1}S_{xy}' - \lambda_i I \right| a_i = 0$$

$$\left| S_{yy}^{-1}S_{xy}'S_{xx}^{-1}S_{xy} - \lambda_i I \right| b_i = 0$$

where λ_i is the eigenvalue associated with ith pair of canonical variates, a_i is the eigenvector associated with the ith eigenvalue in the X variable set, b_i is the eigenvector associated with the ith eigenvalue in the Y variable set, and the other parameters are defined as before.

Again, note that the correlation matrices can be used instead of the covariance matrices. The number of coefficients in each eigenvector equals the number of variables in the respective variable set. These are the coefficients of the variables in the linear equations that define the canonical variates and are referred to as *canonical coefficients* (or *canonical weights*). These coefficients define the contributions of the original variables to the variances of the raw canonical variates. Because these are weights to be applied to the variables in raw score scales, and are therefore affected by the particular unit (i.e., measurement scale) used for each variable, they have no interpretive value.

The canonical coefficients are typically adjusted to generate standardized canonical scores (i.e., scores with zero mean and unit variance) for each sample entity. For each canonical variate, the canonical scores over all the samples will have a mean of zero and a standard deviation of one. In other words, each axis is stretched or shrunk such that the score for an entity represents the number of standard deviations it lies from the overall centroid. In this manner, the score for an entity immediately tells how far it is from the average score.

Recall from Section 5.7.1 that the canonical coefficients will differ depending on whether the raw covariance matrix or correlation matrix is used in the eigenanalysis. If the raw covariance matrix is used, then the adjusted coefficients represent the weights to be applied to the raw scores of the original variables and are referred to as *unstandardized* canonical coefficients (also sometimes referred to as "raw" canonical coefficients, as in SAS). If the correlation matrix is used, then the coefficients represent the weights to be applied to standardized scores of the original variables and are referred to as *standardized* canonical coefficients. Most statistical programs compute and print the standardized canonical coefficients, regardless of whether the raw covariance matrix or correlation matrix is used.

Empirical Example. In the example, we conducted CANCOR on the log-transformed bird variables and original vegetation variables and computed the canonical coefficients for the 18 pairs of canonical variates (Fig. 5.7). The unstandardized coefficients (referred to as "raw" coefficients in the SAS output) define the absolute contribution of each variable to the variance of the canonical variate; in this case, they have no real interpretive value because the data were not standardized prior the analysis. The standardized coefficients represent the relative contributions of the variables to the variance of the canonical variate and can be multiplied by the standardized data values to generate the standardized canonical scores. Because the interpretation of the canonical variates is typically based on the canonical structure, rather than the canonical coefficients, these coefficients have little utility other than for comparison with the canonical structure to provide evidence of multicollinearity effects.

5.8 Assessing the Importance of the Canonical Variates

Once the canonical variates have been derived, the first important task is to assess the importance of each pair of canonical variates and determine the number of significant variates to retain and interpret. Specifically, we must assess how well the canonical variates from one variable set relate to the variates from the other set; that is, how well the common structure of the data set has been captured by the reduced set of canonical variates. We review four criteria that can be useful in this assessment.

5.8.1 Canonical Correlation Criterion

Recall that each eigenvalue is associated with a single pair of canonical variates and measures how closely related the variate from one set of variables is to the corresponding variate from the other set of variables. The larger the eigenvalue is, the more closely related the two canonical variates are. Conversely, eigenvalues near

Raw Canonical Coefficients for the VAR Variables				Raw Canonical Coefficients for the WITH Variables			
	BIRD1	BIRD2	BIRD3		HAB1	HAB2	HAB3
AMRO	1.108	2.166	1.215	GTOTAL	0.003	-0.026	-0.003
BHGR	-0.038	1.066	-0.088	LTOTAL	0.007	0.006	-0.014
BRCR	0.598	0.315	1.092	TTOTAL	-0.005	0.019	-0.011
CBCH	0.627	-0.612	0.170	MTOTAL	-0.007	-0.023	-0.002
DEJU	0.737	0.028	0.598	OTOTAL	0.004	0.026	-0.000
EVGR	0.128	2.649	-1.231	SNAGM1	-0.084	0.287	0.057
GCKI	0.591	1.021	0.376	SNAGM23	0.076	0.436	0.251
GRJA	0.202	5.169	3.036	SNAGM45	0.128	0.001	0.120
HAFL	0.576	0.032	-2.022	SNAGL1	-0.114	-0.510	-0.090
HAWO	1.265	0.385	0.057	SNAGL23	0.139	0.447	0.181
HEWA	1.674	-1.495	2.725	SNAGL45	0.069	-0.131	0.035
RUHU	-0.004	-0.829	-0.346	SNAGS6	-0.035	0.204	-0.130
SOSP	0.417	-2.527	2.857	SNAGM6	-0.075	0.052	-0.374
STJA	-0.519	-0.952	1.837	SNAGL6	0.039	0.089	0.002
SWTH	-1.351	1.730	0.073	BAS	0.003	-0.001	0.011
VATH	-0.280	-1.389	2.503	BAC	0.002	-0.007	-0.002
WEFL	-0.698	-0.931	1.118	BAH	-0.004	0.004	0.005
WIWA	-0.216	0.421	-1.009	FHD	-0.225	-0.059	1.386
WIWR	-0.305	-0.091	0.950				

Standardized Canonical Coefficients for the VAR Variables				Standardized Canonical Coefficients for the WITH Variables			
	BIRD1	BIRD2	BIRD3		HAB1	HAB2	HAB3
AMRO	0.087	0.170	0.095	GTOTAL	0.070	-0.552	-0.055
BHGR	-0.006	0.167	-0.013	LTOTAL	0.204	0.188	-0.432
BRCR	0.193	0.101	0.353	TTOTAL	-0.163	0.613	-0.346
CBCH	0.194	-0.189	0.052	MTOTAL	-0.076	-0.267	-0.023
DEJU	0.126	0.004	0.102	OTOTAL	0.081	0.487	-0.008
EVGR	0.010	0.212	-0.098	SNAGM1	-0.058	0.199	0.039
GCKI	0.199	0.344	0.126	SNAGM23	0.049	0.281	0.162
GRJA	0.023	0.587	0.345	SNAGM45	0.222	0.002	0.208
HAFL	0.176	0.009	-0.620	SNAGL1	-0.038	-0.169	-0.030
HAWO	0.173	0.052	0.007	SNAGL23	0.210	0.677	0.274
HEWA	0.201	-0.179	0.327	SNAGL45	0.123	-0.235	0.063
RUHU	-0.000	-0.115	-0.048	SNAGS6	-0.055	0.328	-0.209
SOSP	0.044	-0.271	0.306	SNAGM6	-0.070	0.049	-0.349
STJA	0.009	-0.164	0.316	SNAGL6	0.046	0.106	0.002
SWTH	-0.421	0.539	0.022	BAS	0.134	-0.050	0.521
VATH	-0.035	-0.174	0.314	BAC	0.218	-0.775	-0.179
WEFL	-0.172	-0.230	0.277	BAH	-0.198	0.187	0.201
WIWA	-0.064	0.124	-0.299	FHD	-0.068	-0.017	0.419
WIWR	-0.078	-0.023	0.244				

FIGURE 5.7 Unstandardized and standardized canonical coefficients for the first three pairs of canonical variates. Note that SAS refers to unstandardized coefficients as "raw canonical coefficients"; these are not raw coefficients, or eigenvectors, as we have defined them in the text.

zero indicate that the corresponding pair of variates have little in common and do not serve well to define the common structure of the two data sets. We are usually interested in knowing how small the eigenvalue has to be before we consider it to be the result of sampling or measurement error rather than a truly important dimension of the data structure.

The square root of the eigenvalue is equal to the product-moment correlation between the two canonical variates. We refer to this as the *canonical correlation* (R_c). The canonical correlation measures, in familiar terms, how correlated the canonical variate derived from one set of variables is to the corresponding variate derived from the other set of variables. Geometrically, the canonical correlation is equal to the cosine of the angle between vectors formed by the canonical variates

(Fig. 5.1). If the angle between vectors is small, then the canonical correlation is high and the two gradients defined by the respective variates are closely related. It is often more useful to think of the relationship between variates in terms of the percent of variance in one variate accounted for by the other, as measured by the squared canonical correlation (R_c^2), or the eigenvalue (λ) as we have defined it. Note that the squared canonical correlation (or the canonical correlation itself) is a symmetric index in the sense that the proportion of variance in one variate accounted for by the opposing variate is independent of which variate is considered first.

No generally accepted guidelines have been established regarding acceptable sizes for canonical correlations. Rather, the decision is usually made based on the contribution of the findings to a better understanding of the research problem being studied. When attempting to use the canonical correlation coefficient or the squared canonical correlation to assess canonical performance, it is important to note two things: First, the relationship being described by either measure is the one between the canonical variates, not the original variables. This makes these coefficients difficult to interpret because the correlation between two canonical variates may be very high when in fact the variates do not explain the original data very well. In other words, a relatively strong canonical correlation between two linear combinations (canonical variates) may be obtained even though these linear combinations may not extract significant portions of variance from their respective sets of variables. Consequently, there may be a temptation to assume that the canonical analysis has uncovered substantial relationships of conceptual and practical importance. Before such conclusions are warranted, however, further analysis involving measures other than canonical correlation must be undertaken (e.g., Sec. 5.8.2). The second important point to remember is that the canonical correlation is affected by the sample-to-variable ratio. Specifically, as the sample-to-variable ratio approaches one, the canonical correlation approaches one and statistical significance is assured (e.g., Sec. 5.8.2). Nevertheless, even with these limitations, R_c and R_c^2 can be helpful in evaluating the importance of each pair of canonical variates when used carefully in conjunction with other performance criteria.

Empirical Example. In the example, we conducted CANCOR on the log-transformed bird variables and original vegetation variables and computed the canonical correlations between each of the 18 pairs of canonical variates (Fig. 5.6). The first pair of canonical variates (BIRD1 and HAB1) have a product-moment correlation of 0.94; thus, these canonical variates are very closely related. In fact, 88 percent of the variance in BIRD1 is accounted for by the variance in HAB1, and vice versa. This relationship is obvious in the scatter plot of Figure 5.5 as well. In practical terms, this means that samples with high scores on HAB1 also have high scores on BIRD1, and, conversely, samples with low scores on HAB1 also have low scores on BIRD1. Thus, samples are positioned similarly along these corresponding canonical axes. Geometrically, this means that the gradients defined by HAB1 and BIRD1 are closely aligned, that is, oriented in very similar directions. The second pair of canonical variates are also closely related ($R_c = 0.83$).

SAS also prints the adjusted canonical correlation for each canonical correlation coefficient (less biased) and approximate standard error of the canonical correlation. Based on the canonical correlation information alone, we would conclude that the first two to four pairs of canonical variates do an excellent job of relating the habitat variables and bird variables. However, remember that canonical correlations do not represent the relationship between original variables. The canonical correlations can be very high even when the original variables do not relate very well (see Sec. 5.8.2).

5.8.2 *Canonical Redundancy Criterion*

To help overcome the inherent bias and uncertainty in interpreting R_c and R_c^2, a measure called *redundancy* has been developed (Stewart and Love 1968) whereby we can calculate the amount of variance in one set of original variables that is explained by a canonical variate from the other data set. Note that the relationship is between original variables in one set and a canonical variate from the other set. The redundancy measure is the equivalent of computing the squared correlation coefficient between one set of variables (represented by a canonical variate) and each individual variable in the other set, and then averaging these squared coefficients to arrive at an average squared correlation coefficient. It provides a summary measure of the ability of one set of variables (taken as a set) to explain the variation in the other variables (taken one at a time). As such, the redundancy measure is perfectly analogous to the R^2 statistic in MRA, and its value as a statistic is similar.

Specifically, the redundancy coefficient is a measure of the amount of variance in one set of variables that is "redundant" or shared with the variance in the other set of variables. A redundancy coefficient can be computed for the X-set given the Y-set, denoted by $R^2_{x|y}$, as well as for the Y-set given the X-set, denoted by $R^2_{y|x}$. Hence, redundancy is not a symmetric index like R_c^2. When an independent-dependent distinction is made between variable sets (e.g., habitat characteristics and species abundances), the redundancy in the dependent data set is of primary interest; that is, how much of the variation in the dependent variables (e.g., species abundances) can be accounted for by the canonical variates from the independent set (e.g., habitat variables). The reverse redundancy, although it could be calculated, is less meaningful.

The redundancy coefficient is computed from two parameters: (1) the squared canonical correlation (R_c^2), and (2) the proportion of a set's variance accounted for by its own canonical variate. The squared canonical correlation is derived by solving the characteristic equations, as discussed in Section 5.7.2. The second parameter, however, must be derived from the canonical structure coefficients, as defined in Section 5.9.2 below. The structure coefficients represent the product-moment correlations between the original variables and the respective canonical variates. Thus, a structure coefficient reflects the degree to which a variable is represented by a canonical variate. Each variable has a structure coefficient for each canonical variate in its respective set of variables. The squared structure coefficient represents the percent of the variable's variance accounted for by the respective canonical variate. By averaging the squared structure coefficients from all P (or M) variables, we can

5.8 Assessing the Importance of the Canonical Variates

determine the proportion of variance in a set of variables accounted for by a particular canonical variate (the second parameter needed to calculate the redundancy coefficient). Note that this is analogous to the relative percent variance measure described for PCA. The proportion of explained variance in the Y-set variables that is accounted for by a particular canonical variate is given by

$$R^2_{i(y)} = \frac{\sum_{j=1}^{M} s^2_{ij(y)}}{M}$$

where $R^2_{i(y)}$ is the proportion of variance in the Y-set variables accounted for by the ith canonical variate from the same set, $s_{ij(y)}$ is the structure coefficient (correlation) for the ith canonical variate and jth variable in the Y-set, and M is the number of variables in the Y-set.

Similarly, the proportion of variance in the X-set variables accounted by a particular canonical variate is given by

$$R^2_{i(x)} = \frac{\sum_{j=1}^{P} s^2_{ij(x)}}{P}$$

where $R^2_{i(x)}$ is the proportion of variance in the X-set variables accounted for by the ith canonical variate from the same set, $s_{ij(x)}$ is the structure coefficient (correlation) for the ith canonical variate and jth variable in the X-set, and P is the number of variables in the X-set.

The redundancy coefficients associated with the ith canonical variate in the Y-set and X-set, respectively, are given by

$$R^2_{i(y|x)} = \frac{\lambda_i \sum_{j=1}^{M} s^2_{ij(y)}}{M} = \lambda_i R^2_{i(y)}$$

$$R^2_{i(x|y)} = \frac{\lambda_i \sum_{j=1}^{P} s^2_{ij(x)}}{P} = \lambda_i R^2_{i(x)}$$

What is the minimum acceptable redundancy coefficient needed to justify the interpretation of a canonical variate? Just as with canonical correlations, no generally accepted guidelines have been established. Again, the decision is usually made on the basis of how much the canonical pair contribute to the ecological understanding of the phenomenon under investigation. When using the redundancy coefficient to assess individual canonical pairs or overall canonical performance, it is important to remember three points: First, in order to have a high redundancy coefficient, you must have not only a high canonical correlation but also a high degree of variance explained by the respective canonical variate. A high canonical correlation alone does not ensure a high redundancy. Second, remember that the amount of variance in one set of variables explained by the canonical variates in another set does not necessarily equal the amount of variance of the second set explained by the first (i.e., the redundancy measure is not symmetrical). Therefore, when there is no independent-dependent distinction between variable sets, the redundancy coefficients must be calculated for both sets of variables. Third, the first canonical correlation need not and often does not have the highest redundancy. Remember, the canonical variates are derived to maximize the canonical correlation coefficient, not the redundancy coefficient, although the latter would seem more desirable. Indeed, an alternative type of canonical analysis based on maximizing the redundancy coefficient (redundancy analysis) has been developed. Redundancy analysis successively derives orthogonal variates of the independent variables (X-set) that maximize the variance explained among the dependent variables (Y-set), or vice versa (Wollenberg 1977). Overall, the redundancy coefficients are probably the most important piece of information resulting from CANCOR.

Thus far, we have used redundancy to evaluate the adequacy of each pair of canonical variates. However, once we have calculated the redundancy coefficient for each canonical variate, we can sum the coefficients and determine the proportion of variance in one set of variables accounted for by *all* of the canonical variates in the other set. We can use this "total" redundancy measure as a way to assess the overall performance of the full set of canonical variates. Note that the overall performance of each set of canonical variates must be determined separately. Typically, when there is an independent-dependent distinction between variable sets, we are only interested in how well the canonical variates from the independent set account for the variation in the dependent variables.

In addition to measuring the performance of the *full* set of canonical variates, we can also use redundancy to help determine *how many* pairs of variates to retain for interpretation. To do this we can employ a measure similar to the relative percent variance criterion discussed in Chapters 2 and 4. Here we simply calculate the relative percent redundancy for each canonical variate; that is, the percent of the total redundancy accounted for by each canonical variate. The cumulative percent redundancy of all canonical variates always equals 100 percent. Thus, by examining the change in cumulative percent redundancy as we move from the first to the last canonical variate, we can determine how well the total redundancy is captured by any specified number of canonical variates. Once we have decided on how many pairs of canonical variates to retain for interpretation, we can sum the redundancy coefficients for those canonical variates retained. This measures how well the *retained* canonical variates capture the variation in the other set and provides us with a more realistic measure of how well the canonical transformation has summarized the data structure.

Empirical Example. In the example, we conducted CANCOR on the log-transformed bird variables and original vegetation variables and computed the canonical redundancy (Fig. 5.8). Because we are considering the bird variables as dependent upon the habitat variables, only the redundancy in the bird data is of interest;

5.8 Assessing the Importance of the Canonical Variates

Canonical Redundancy Analysis

Raw Variance of the VAR Variables Explained by

	Their Own Canonical Variables			The Opposite Canonical Variables	
	Proportion	Cumulative Proportion	Canonical R-Squared	Proportion	Cumulative Proportion
1	0.2727	0.2727	0.8796	0.2399	0.2399
2	0.0632	0.3359	0.6945	0.0439	0.2838
3	0.0571	0.3930	0.6035	0.0344	0.3182
4	0.0301	0.4232	0.5756	0.0174	0.3356
5	0.0336	0.4567	0.4847	0.0163	0.3519
6	0.0561	0.5128	0.4443	0.0249	0.3768
7	0.0402	0.5530	0.3552	0.0143	0.3911
8	0.0458	0.5987	0.2943	0.0135	0.4045
9	0.0432	0.6420	0.2490	0.0108	0.4153
10	0.0222	0.6641	0.1953	0.0043	0.4196
11	0.0373	0.7014	0.1320	0.0049	0.4245
12	0.0239	0.7254	0.1210	0.0029	0.4274
13	0.0396	0.7649	0.0741	0.0029	0.4304
14	0.0469	0.8118	0.0547	0.0026	0.4329
15	0.0274	0.8393	0.0414	0.0011	0.4341
16	0.0631	0.9024	0.0148	0.0009	0.4350
17	0.0357	0.9381	0.0045	0.0002	0.4351
18	0.0341	0.9722	0.0009	0.0000	0.4352

Standardized Variance of the VAR Variables Explained by

	Their Own Canonical Variables			The Opposite Canonical Variables	
	Proportion	Cumulative Proportion	Canonical R-Squared	Proportion	Cumulative Proportion
1	0.1577	0.1577	0.8796	0.1387	0.1387
2	0.0671	0.2248	0.6945	0.0466	0.1853
3	0.0561	0.2809	0.6035	0.0339	0.2192
4	0.0389	0.3198	0.5756	0.0224	0.2416
5	0.0362	0.3560	0.4847	0.0175	0.2591
6	0.0381	0.3941	0.4443	0.0169	0.2760
7	0.0452	0.4393	0.3552	0.0161	0.2921
8	0.0460	0.4853	0.2943	0.0135	0.3056
9	0.0562	0.5415	0.2490	0.0140	0.3196
10	0.0368	0.5783	0.1953	0.0072	0.3268
11	0.0415	0.6199	0.1320	0.0055	0.3323
12	0.0441	0.6639	0.1210	0.0053	0.3376
13	0.0428	0.7067	0.0741	0.0032	0.3408
14	0.0556	0.7623	0.0547	0.0030	0.3438
15	0.0592	0.8215	0.0414	0.0025	0.3463
16	0.0510	0.8724	0.0148	0.0008	0.3470
17	0.0478	0.9202	0.0045	0.0002	0.3473
18	0.0398	0.9601	0.0009	0.0000	0.3473

FIGURE 5.8 Canonical redundancy analysis results. Only the redundancy coefficients for the bird data set (VAR variables) is shown.

the reverse redundancy (i.e., how much of the habitat variation can be explained by the bird variates) is not particularly meaningful. SAS prints out both parameters used to calculate redundancy. The first column in Figure 5.8 identifies the canonical variate. The second column is the proportion of variance in the bird (VAR) variables accounted for by the bird canonical variates ($R^2_{i(y)}$). The third column is the cumulative percent variance accounted for by the first M canonical variates. The fourth column is the eigenvalue (λ_i) or squared canonical correlation coefficient (R_{ci}^2). The fifth column is the redundancy coefficient for each canonical variate ($R_{i(y|x)}^2$); recall that it is computed by multiplying columns 2 and 4. The last column is the cumulative redundancy. The cumulative redundancy for all 18 variates is equal to the total redundancy for the full set of variates. To calculate the relative percent redundancy for each canonical variate, divide the ith redundancy coefficient by the total redundancy of all 18 variates. For example, the relative percent redundancy of the first canonical variate is 55 percent ((0.2399 / 0.4352) · 100). In other words, the first variate accounts for slightly more than half of the total redundancy in the bird data.

The top table in Figure 5.8 indicates that, collectively, all 18 habitat variates explain 44 percent of the raw variance in the bird data (i.e., 44 percent of the unstandardized, original bird abundance data); the first habitat variate alone accounts for 24 percent of the variation in bird abundances. The redundancy was much less for the standardized bird abundance data, seen in the bottom table of Figure 5.8 (total redundancy = 35 percent). The reason redundancy is so low, despite such high canonical correlations, is that the bird variates are not very effective in explaining the variation in bird abundances; the first bird variate explains only 27 percent of the raw variance in bird abundances and 16 percent of the standardized variance. Therefore, although the habitat and bird canonical variates are closely related, the habitat variables do not explain a greater portion of the variation in bird abundances because the bird variates do a poor job of capturing the variation in bird abundances. These findings highlight the need to interpret canonical correlations cautiously and furthermore demonstrate the importance of reporting redundancy estimates in all cases.

5.8.3 Significance Tests

There are a number of objective tests for evaluating the statistical significance of the canonical correlations when the data are from a sample as opposed to the entire population. With a sample, we can ask, "What is the probability that the sampling process produced entities that show the computed degree of canonical correlation between sets of variables, when in fact there are no real relationships between variable sets in the population?" Generally we must assume independent random samples to ensure valid probability values.

The most common approach is to use a parametric statistical procedure. Such procedures essentially test the null hypothesis that the current canonical correlation and all smaller ones are zero in the population. The first test is a general test of whether there is a significant link between any of the canonical variates; that is, whether there is any relationship between the Y-set variables and X-set variables. If this test is significant (i.e., the null hypothesis is rejected), then the contribution of the first canonical pair can be removed and the significance of the remaining canonical variate pairs assessed. This is continued sequentially until all P or M (whichever is smaller) canonical variate pairs have been removed or until the null hypothesis has been accepted. There are several different parametric test statistics available for testing the significance of the canonical correlations. The most widely used test, and the one normally printed out in computer packages, is the approximate F-statistic based on Rao's approximation to the distribution of the likelihood ratio. There is a similar parametric test for the significance of the redundancy coefficient, although it has not been widely utilized (Alpert et al. 1975).

It is important to remember three points concerning the use of these parametric tests for assessing the canonical performance: First, these tests measure the significance of the shared *canonical* variation, not shared variation in the original variables. Thus, two variates may have a significant correlation, but they may lack any substantive explanation of the original variance. Second, these significance tests are only valid when Y-set and X-set variables are jointly distributed according to a multivariate normal distribution (Bartlett 1941; Lawley 1959). Third, the tests for significant canonical correlations, other than the first general test, are extremely conservative. Therefore, unless the canonical correlations removed are close to one, the test is likely to indicate that statistically significant canonical variate pairs remain (Dillon and Goldstein 1984).

Alternatively, we can use a nonparametric resampling procedure, such as the jackknife, bootstrap, or randomization test procedure (see Chapter 2, Sec. 2.8.5 for a general discussion of these procedures) to test the statistical significance of the canonical correlations based on the eigenvalue, canonical correlation, redundancy coefficient, or any other measure of canonical performance. The same procedures described previously can be applied here by substituting the appropriate parameter. For example, we could use these procedures to conduct a nonparametric test of significance for each canonical correlation coefficient and redundancy coefficient. In each case, the null hypothesis tested is that the particular coefficient is approximately equal to what would be expected from the same data but without any real (i.e., nonrandom) structure. Thus, rejection of the null hypothesis implies that a particular coefficient probably represents a real solution and is not a reflection of sample or measurement bias.

Recall from our previous discussions in Chapter 2 that all three of these resampling procedures will provide a measure of the statistical significance of the canonical solution as measured by a particular performance criterion (e.g., canonical correlation), but they differ in the nature of the test procedure developed. Again, we prefer the randomization test over both the jackknife and bootstrap procedures.

It is important to note that even if a canonical correlation or any other performance measure is statistically significant, we may decide that it lacks substantive ecological importance for two reasons: first, the solution may not explain enough of the sample variance-covariance structure to warrant interpretation (e.g., low redundancy coefficients), yet be statistically significant; second, the canonical solution may not have a meaningful ecological interpretation as judged, for example, by the structure coefficients. Ultimately, the utility of each canonical solution must be grounded on ecological criteria. Nevertheless, objective statistical tests can be useful tools in helping to assess the importance of each canonical correlation and in determining how many functions to retain and interpret, as long as they are treated in conjunction with other performance criteria.

	Likelihood Ratio	Approx F	Num DF	Den DF	Pr > F
1	0.00030569	2.0632	342	826.7307	0.0001
2	0.00253951	1.5993	306	791.1352	0.0001
3	0.00831290	1.4028	272	754.5572	0.0002
4	0.02096428	1.2577	240	717	0.0128
5	0.04940210	1.0962	210	678.4679	0.1982
6	0.09586288	0.9723	182	638.9664	0.5846
7	0.17251340	0.8387	156	598.5028	0.9088
8	0.26752607	0.7377	132	557.0868	0.9831
9	0.37906960	0.6476	110	514.7319	0.9970
10	0.50476553	0.5556	90	471.4571	0.9996
11	0.62728342	0.4729	72	427.29	0.9999
12	0.72267479	0.4244	56	382.2725	0.9999
13	0.82218390	0.3415	42	336.4716	1.0000
14	0.88794087	0.2915	30	290	0.9999
15	0.93933881	0.2315	20	243.0635	0.9998
16	0.97992176	0.1257	12	196.0771	0.9998
17	0.99461850	0.0675	6	150	0.9988
18	0.99907181	0.0353	2	76	0.9653

Multivariate Statistics and F Approximations

S = 18   M = 0   N = 28.5

Statistic	Value	F	Num DF	Den DF	Pr > F
Wilks' Lambda	0.00030569	2.0632	34	826.7307	0.0001
Pillai's Trace	5.21934353	1.6335	34	1368	0.0001
Hotelling-Lawley Trace	16.23258886	2.7107	2	1028	0.0001
Roy's Greatest Root	7.30748437	29.2299	19	76	0.0001

FIGURE 5.9 Test statistics for the null hypothesis that the ith canonical correlation and all smaller correlations are zero.

Empirical Example. In the example, we conducted CANCOR on the log-transformed bird variables and original vegetation variables and tested the significance of the canonical correlations using an approximate F-statistic based on Rao's approximation to the distribution of the likelihood ratio (Fig. 5.9). This tests the null hypothesis that the canonical correlations associated with the ith pair of canonical variates and all subsequent pairs are zero. Since the assumption of multivariate normality does not appear to be grossly violated, we place higher confidence in the validity of these test results. Figure 5.9 indicates that the data set contains canonical correlations significantly different from zero until after the fourth pair of canonical variates have been extracted. Based on these findings alone, we would choose to retain the first four pairs of canonical variates for subsequent ecological interpretation. SAS also prints out four different multivariate statistics (e.g., Wilks' Lambda) for testing the null hypothesis that all canonical correlations are zero in the population. The Wilks' Lambda test is equivalent to Rao's likelihood ratio test for the first canonical correlation.

5.8.4 Canonical Scores and Associated Plots

A final way of assessing the performance of each pair of canonical variates involves simple graphical representations of the canonical scores. Recall that each sample entity in the data set has a score on each canonical variate derived by multiplying the standardized values of the original variables by the standardized canonical coefficients and adding the products. Scatter plots of each pair of canonical variates graphically illustrate the relationship between canonical variates. These plots are basically graphical representations of the canonical correlations and are similar to the familiar scatter plot used to portray the results of a simple linear regression, only here we are regressing one canonical variate (often representing independent variables) against another canonical variate (often representing dependent variables). These plots illustrate relationships among sampling entities, since entities in close proximity occupy the same relative position in the joint space defined by the two sets of variables. Remember that these plots portray the correlation between corresponding canonical variates. Thus, a high canonical correlation will appear in a scatter plot as a strongly linear distribution of sample points, even though the respective gradients defined by the two variates may not account for a meaningful amount of variation in the original variables. Consequently, plots of canonical scores do not display intuitive relationships among samples and are therefore somewhat difficult to interpret.

Empirical Example. In the example, we conducted CANCOR on the log-transformed bird variables and original vegetation variables and computed the canonical scores for each canonical variate (Fig. 5.10). The scores given in Figure 5.10 were calculated by multiplying the standardized canonical coefficients by the original data in standardized form. These scores are used to assess the linearity and multivariate normality assumptions, as discussed previously, and can serve as new variables in subsequent analyses. We also created scatter plots of the first three pairs of canonical variates; a plot of the first pair (HAB1 versus BIRD1) is displayed in Figure 5.5. The strong linear relationship between HAB1 scores and BIRD1 scores reflects the high canonical correlation ($R_c = 0.94$) between these canonical variates. Otherwise, these plots are not particularly informative.

OBS	BIRD1	BIRD2	BIRD3	HAB1	HAB2	HAB3
1	-0.83376	0.67365	0.20840	-1.17447	0.20530	0.36710
2	-1.22792	1.08154	-0.53629	-1.30395	0.43624	-0.03584
3	0.11558	-0.59783	-0.07723	-0.12392	-0.85339	0.73816
4	-0.51896	-0.66292	-0.29370	0.25190	-0.78811	-0.43002
5	0.07084	0.06539	-0.23961	-0.26551	0.41036	-0.05687
.	.	.	.	.	.	.
.	.	.	.	.	.	.
.	.	.	.	.	.	.
92	0.39492	-0.22970	-1.39080	0.66987	-0.51579	-0.95514
93	-0.81836	0.86924	0.31625	-0.11851	0.55585	1.08146
94	0.90530	0.22792	-0.94375	1.21569	-0.21700	-0.95754
95	1.10233	0.73703	-2.31503	0.78271	0.12099	-2.10068
96	1.57843	-0.02889	-1.35493	1.17074	-0.11398	-2.27221

FIGURE 5.10 Canonical scores for the first three pairs of canonical variates. Only a sample of the canonical scores are shown.

5.9 Interpreting the Canonical Variates

Once we have derived the canonical variates and assessed their importance, assuming one or more of the solutions are nontrivial, we can proceed to interpret their meaning. This is done by examining the relationships between the individual variables and the canonical variates.

5.9.1 Standardized Canonical Coefficients

The traditional approach to interpreting the canonical variates is to use the canonical coefficients, since these measure the absolute contribution of each variable to the canonical scores and to the variances of the canonical variates. However, we are usually more interested in knowing the relative importance rather than the absolute importance of each variable in the canonical variates. While the unstandardized canonical coefficients do tell us the absolute contribution of a variable in determining the canonical score, this information can be misleading when a unit change in a variable is not equivalent from one variable to another (i.e., when the standard deviations are not the same because of differences in measurement scale or variances). To assess the relative importance of each variable, we need to look at standardized canonical coefficients (or standardized canonical weights). The matrix of standardized canonical coefficients is called the *canonical pattern* (or *pattern matrix*).

Standardized canonical coefficients are obtained by adjusting the eigenvector coefficients to create standardized weights that, when applied to the original data in standard score form, produce standardized canonical scores. Note that the standardized coefficients are the same regardless of whether the raw covariance matrix or correlation matrix is used in the analysis. The relative contribution of the original variables to each canonical variate may then be gauged by the relative magnitude of the standardized canonical coefficients; the larger the magnitude, the greater that variable's contribution. However, even the standardized canonical coefficients may distort the true relationship among variables in the canonical variates when the correlation structure of the data is complex, and thus it is generally preferable to base the interpretation on the structure coefficients (see Sec. 5.9.2).

Empirical Example. In the example, we conducted CANCOR on the log-transformed bird variables and original vegetation variables and computed the canonical coefficients (Fig. 5.7). The unstandardized coefficients (referred to as raw coefficients) have no interpretive value. The standardized coefficients, on the other hand, could be used to define the relative importance of the original variables in each canonical variate. However, the relative magnitudes of the standardized coefficients do not agree closely with the structure coefficients (see Sec. 5.9.2). It is likely, therefore, that the standardized coefficients are reflecting a good deal of covariance among the variables. Hence, the canonical structure provides a more reliable means of interpreting the canonical variates.

5.9.2 Structure Coefficients

When one canonical coefficient dominates all others in the canonical variate and the correlation structure is simple, it is fairly straightforward to assess the meaning of the canonical variate by the magnitude and sign of the standardized canonical coefficients. However, when the correlation structure is complex and there are several coefficients of significant size, interpretation is not as easy. The difficulty stems from the fact that the original variables in the new canonical variates are correlated; some of them perhaps highly correlated. Individual coefficients, in this case, reflect not only the influence of their corresponding variables, but also the influence of other variables as reflected through the correlation structure of the data (Williams 1980). The effect of a variable on the standardized canonical coefficient is only partially given by the numerical value of the corresponding coefficient. For example, when two or more variables are correlated, the weight (reflected in the standardized canonical coefficients) may be split between them, making the associated coefficients appear relatively small. Alternatively, the coefficient for one variable may be inflated (to consider both variables), while the other variable is assigned a near zero coefficient.

The usual way of dealing with this is to examine groups of highly correlated variables and retain only the variables with the greatest multiple correlation with the opposing set of variables and/or those variables with the greatest ease of ecological interpretation (see Sec. 5.5.2). A better technique, or one to be used in conjunction with other multicollinearity reduction approaches, is to examine the product-moment correlations between the canonical variates and the individual variables included in the analysis. These simple bivariate correlations are not affected by relationships with other variables and therefore reflect the true relationship between each variable and its respective canonical variate. These correlations are referred to as *structure coefficients* (or *canonical loadings*), and the matrix of structure coefficients is called the *canonical structure* (or *structure matrix*). The structure coefficients for variables in the Y-set and X-set, respectively, are given as

$$s_{ij(y)} = \sum_{h=1}^{M} r_{jh(y)} c_{ih(y)}$$

$$s_{ij(x)} = \sum_{h=1}^{P} r_{jh(x)} c_{ih(x)}$$

where $s_{ij(y)}$ is the structure coefficient (correlation) for the ith canonical variate and jth variable in the Y-set, $s_{ij(x)}$ is the structure coefficient (correlation) for the ith canonical variate and jth variable in the X-set, $r_{jh(y)}$ is the total correlation between the jth and hth variables in the Y-set, $r_{jh(x)}$ is the total correlation between the jth and hth variables in the X-set, $c_{ih(y)}$ is the standardized canonical coefficient for the ith canonical variate and hth variable in the Y-set, and $c_{ih(x)}$ is the standardized canonical coefficient for the ith canonical variate and hth variable in the X-set.

A structure coefficient tells us how closely a variable and a canonical variate are related. When the absolute magnitude of the coefficient is large (i.e., approaching a value of one), the canonical variate is carrying nearly the same information as the variable. Conversely, when the coefficient is near zero, the variable and variate have little in common. Hence, we can define a canonical variate on the basis of the structure coefficients by noting the variables that have the largest coefficients. Of course, we still must decide how large a structure coefficient has to be before we consider it significant (see Sec. 5.9.4). Structure coefficients are intuitively appealing because most researchers are familiar with simple correlation coefficients and can visualize differences in the magnitude of the coefficient. Note that squared structure coefficients measure the proportion of variance in a variable accounted for by the respective canonical variate. The canonical structure provides a way to establish an ecological interpretation of each canonical function and is one of the most important pieces of information resulting from canonical correlation analysis.

Empirical Example. In the example, we conducted CANCOR on the log-transformed bird variables and original vegetation variables and computed the canonical structure coefficients (Fig. 5.11). The correlations between the vegetation variables and the first habitat variate indicate that HAB1 represents a gradient in shrub cover, conifer cover, and snag density. Specifically, HAB1 represents a gradient in vegetation structure with sites characterized by an abundance of conifers (BAC = 0.90), low shrubs (LTOTAL = 0.85), snags (BAS = 0.70, SNAGM45 = 0.65), and overstory cover (OTOTAL = 0.64), loading high on one end of the gradient, and sites characterized by an abundance of tall shrubs (TTOTAL = –0.80), midstory cover (MTOTAL = –0.57), and high foliage height diversity (FHD = –0.58) loading high on the opposite end of the gradient. Remember, however, that an analysis of the canonical scores revealed a somewhat discrete distribution of sites along this axis, reflecting the predominance of two habitat types (conifer-dominated upslope sites and hard-

	Correlations Between the VAR Variables and Their Canonical Variables				Correlations Between the WITH Variables and Their Canonical Variables		
	BIRD1	BIRD2	BIRD3		HAB1	HAB2	HAB3
AMRO	-0.2087	0.1067	0.1469	GTOTAL	-0.4386	-0.3385	0.4261
BHGR	-0.2761	0.4366	0.0749	LTOTAL	0.8527	0.0974	-0.2833
BRCR	0.6039	0.2401	0.1160	TTOTAL	-0.8046	0.0827	0.0035
CBCH	0.4149	0.1229	0.0965	MTOTAL	-0.5660	-0.1436	0.2632
DEJU	0.2779	0.1275	0.2851	OTOTAL	0.6419	0.2305	-0.1172
EVGR	0.1621	0.0270	-0.0763	SNAGM1	0.2945	0.1869	-0.1566
GCKI	0.7993	0.1375	0.1471	SNAGM23	0.3133	0.2154	0.0586
GRJA	0.2439	0.6713	0.3897	SNAGM45	0.6499	0.1130	0.0120
HAFL	0.7328	0.0414	-0.2107	SNAGL1	0.1176	-0.0242	0.2953
HAWO	0.0976	0.0286	-0.0233	SNAGL23	0.2926	0.3711	0.6150
HEWA	0.4813	0.0678	0.2002	SNAGL45	0.3940	-0.1443	0.1119
RUHU	-0.1392	0.0853	-0.0118	SNAGS6	0.3913	0.2390	-0.3354
SOSP	0.0515	-0.2055	0.1572	SNAGM6	0.1548	0.0889	-0.2655
STJA	-0.0699	-0.2269	0.2169	SNAGL6	0.3902	-0.1441	-0.0636
SWTH	-0.6289	0.4673	0.0830	BAS	0.6969	0.0412	0.3809
VATH	0.0693	-0.1702	0.3787	BAC	0.9021	-0.0661	-0.2143
WEFL	-0.3195	-0.1715	0.4178	BAH	-0.3286	0.4400	0.4399
WIWA	-0.2159	0.3364	-0.4451	FHD	-0.5755	-0.3550	0.5861
WIWR	-0.4284	-0.1225	0.2755				

FIGURE 5.11 Canonical structure coefficients for the first three pairs of canonical variates.

wood-dominated streamside sites) associated with either end of the gradient, rather than a continuous gradient in vegetation structure. The second and third habitat variates can be interpreted in a similar manner.

Similarly, the correlations between bird abundances and the first bird variate identify a gradient in species abundances that closely parallels the habitat gradient (recall the high canonical correlation between HAB1 and BIRD1, $R_c = 0.94$). Specifically, four bird species show a strong correlation with BIRD1. Three species (GCKI, HAFL, and BRCR) characteristic of upslope, conifer-dominated areas are positively correlated with the first bird variate, while one species (SWTH) is negatively correlated with the first bird variate. Thus, we can conclude that at least four species appear to be closely aligned with the upslope-streamside vegetation structure gradient defined by the first habitat variate. The remaining bird variates can be interpreted in a similar manner. Overall, the interpretations of the canonical variates based on the structure coefficients are intuitively satisfying.

5.9.3 Canonical Cross-Loadings

In addition to canonical loadings, we can also compute canonical *cross-loadings*. A canonical cross-loading gives the correlation between an observed variable from one set with a canonical variate from the other set. In this manner, canonical cross-loadings are similar to redundancy coefficients, except that they measure the correlation between a single variable in one set (rather than the entire variable set) and the corresponding canonical variate from the other set. Cross-loadings are obtained by taking the product of the canonical correlation coefficient and the canonical loading of the observed variable as given by

$$s_{ij(y|x)} = R_{ci} s_{ij(y)}$$

$$s_{ij(x|y)} = R_{ci} s_{ij(x)}$$

where $s_{ij(y|x)}$ is the cross-loading (correlation) for the ith canonical variate in the X-set and jth variable in the Y-set, $s_{ij(x|y)}$ is the cross-loading (correlation) for the ith canonical variate in the Y-set and jth variable in the X-set, R_{ci} is the canonical correlation of the ith pair of canonical variates, $s_{ij(y)}$ is the structure coefficient (correlation) for the ith canonical variate and jth variable in the Y-set, and $s_{ij(x)}$ is the structure coefficient (correlation) for the ith canonical variate and jth variable in the X-set.

Thus, the cross-loading of a particular variable is determined by computing how well it is correlated with its own canonical variate and multiplying this by how well its own canonical variate is correlated with the corresponding variate in the other set (i.e., by multiplying the correlation coefficient). In this manner, the cross-loading is like a conventional loading, but it measures how well a variable loads on the opposing variate instead of its own. Note that squared canonical cross-loading measures the proportion of variance in a variable accounted for by the opposing canonical variate.

When an independent-dependent distinction is made between sets, the cross-loadings provide a direct measure of how well each dependent variable is accounted for by the opposing independent canonical variates. As such, cross-loadings can provide a better measure for interpreting each canonical variate. That is, instead of basing the interpretation of a canonical variate on how well the variables in its own set load, we can base our interpretation on how well the variables in the opposing set load. This is particularly meaningful when an independent-dependent distinction is made. Because cross-loadings are simple bivariate correlations, they have the same intuitive appeal that conventional loadings have, and thus provide another, perhaps more meaningful, way to establish an ecological interpretation of each canonical variate.

Empirical Example. In the example, we conducted CANCOR on the log-transformed bird variables and original vegetation variables and computed the canonical cross-loadings (Fig. 5.12). The cross-loadings merely confirm our interpretation of the variates based on the structure coefficients. Specifically, three bird species (GCKI, HAFL, and BRCR) load positively on the first habitat variate and one species (SWTH) loads negatively. A word of caution concerning cross-loadings is warranted. In this case, the cross-loadings of the bird variables and habitat variates measure how strongly each species is associated with each habitat gradient. It is tempting to interpret these correlations as multiple correlations between each species abundance and the habitat variables. However, it must be remembered that these habitat gradients represent composite variates that were derived to maximally correlate with bird community gradients, not each species individually. If species-specific habitat associations are of primary interest, then other statistical procedures (e.g., multiple regression, logistic regression, discriminant analysis) are more appropriate. Canonical correlation analysis should be reserved for analyzing community-level relationships.

	Correlations Between the VAR Variables and the Canonical Variables of the WITH Variables				Correlations Between the WITH Variables and the Canonical Variables of the VAR Variables		
	HAB1	HAB2	HAB3		BIRD1	BIRD2	BIRD3
AMRO	-0.1957	0.0890	0.1142	GTOTAL	-0.4113	-0.2821	0.3310
BHGR	-0.2589	0.3638	0.0582	LTOTAL	0.7997	0.0811	-0.2201
BRCR	0.5664	0.2001	0.0901	TTOTAL	-0.7546	0.0689	0.0028
CBCH	0.3891	0.1024	0.0750	MTOTAL	-0.5308	-0.1197	0.2045
DEJU	0.2607	0.1063	0.2215	OTOTAL	0.6021	0.1921	-0.0911
EVGR	0.1520	0.0225	-0.0593	SNAGM1	0.2762	0.1557	-0.1217
GCKI	0.7497	0.1146	0.1142	SNAGM23	0.2938	0.1795	0.0455
GRJA	0.2288	0.5595	0.3028	SNAGM45	0.6096	0.0941	0.0094
HAFL	0.6873	0.0345	-0.1637	SNAGL1	0.1103	-0.0202	0.2294
HAWO	0.0915	0.0239	-0.0181	SNAGL23	0.2744	0.3092	0.4777
HEWA	0.4514	0.0565	0.1555	SNAGL45	0.3696	-0.1202	0.0869
RUHU	-0.1305	-0.0711	-0.0092	SNAGS6	0.3670	0.1992	-0.2606
SOSP	0.0483	-0.1713	0.1221	SNAGM6	0.1452	0.0741	-0.2063
STJA	-0.0656	-0.1891	0.1685	SNAGL6	0.3660	-0.1201	-0.0494
SWTH	-0.5899	0.3894	0.0645	BAS	0.6536	0.0343	0.2959
VATH	0.0650	-0.1419	0.2942	BAC	0.8460	-0.0551	-0.1665
WEFL	-0.2996	-0.1429	0.3246	BAH	-0.3082	0.3667	0.3418
WIWA	-0.2025	0.2804	-0.3458	FHD	-0.5398	-0.2958	0.4553
WIWR	-0.4018	-0.1021	0.2141				

FIGURE 5.12 Canonical cross-loadings for the first three pairs of canonical variates.

Overall, based on the canonical correlations, canonical loadings and cross-loadings, and canonical redundancy, we can conclude that there is probably at least one important vegetation structure gradient that influences bird community structure within mature, unmanaged forest stands. Specifically, this gradient is characterized by a contrast between sites with an abundance of low shrubs, conifers, snags, and overstory cover (i.e., conifer- and snag-dominated upslope areas) and sites with an abundance of tall shrubs and midstory cover, and high foliage height diversity (i.e., streamside habitat). The abundance distribution of several bird species, including golden-crowned kinglets, Hammond's flycatchers, brown creepers and Swainson's thrushes, seem to be strongly influenced by this vegetation gradient. However, this gradient is more disjoint than continuous and only explains about 24 percent of the variation in bird abundances; other unmeasured abiotic or biotic factors also significantly influence species distributions.

5.9.4 Significance Tests Based on Resampling Procedures

While the canonical loadings and cross-loadings provide measures for assessing the relative importance of the original variables in the canonical variates, they do not tell us how big a coefficient has to be before it is considered significant. We would like to be able to assess the statistical significance of each variable's contribution to each canonical variate as measured by the loadings or cross-loadings. This would allow us to generate a more objective ecological interpretation of each canonical variate based on the significant variables in the respective variate. To do this, we can employ a jackknife, bootstrap, or randomization resampling procedure (see Chapter 2, Sec. 2.8.5 for a general discussion of these procedures). These same procedures can be applied here by substituting the appropriate parameter (e.g., structure coefficient or cross-loading coefficient). In each case, the null hypothesis is that the particular structure coefficient or cross-loading coefficient is approximately equal to zero. Thus, rejection of the null hypothesis implies that a particular coefficient probably represents a real association between the corresponding variable and canonical variate. Recall from our discussion in Chapter 2 (Sec. 2.9.2) that some form of experimentwise error control, such as the Bonferroni procedure, is necessary because multiple tests are performed. Also, remember that statistical significance does not guarantee ecological meaningfulness. Thus, these significance tests should be used only in conjunction with other performance criteria.

5.10 Validating the Canonical Variates

Once we have derived the canonical variates, assessed their significance, and interpreted their meaning, we can test the validity of our findings to insure that they are reliable. Validating the results of CANCOR is an important and often overlooked step. Results from CANCOR are reliable only if the parameters have been estimated accurately and precisely, particularly when the objective is to generalize the conclusions to other areas. Of course, the best assurance of reliable results is an intelligent sampling plan and a large sample. We know, for example, that sampling variability increases rapidly with increases in numbers of variables and with decreases in sample size. When the number of parameters to be estimated approaches the number of samples, there is a strong likelihood that any patterns exhibited by individual canonical coefficients are fortuitous and therefore of no ecological consequence. Thus, validation becomes increasingly important as the sample size decreases relative to dimensionality (number of variables). We briefly discuss two common validation approaches.

5.10.1 Split-Sample Validation

The most common validation procedure when the sample size is large enough is to randomly divide the total sample of entities into two groups. One subset, referred to as the *analysis*, *training*, or *calibration* sample, is used to derive the canonical variates and the other, referred to as the *holdout*, *test*, or *validation sample*, is used to test the variates. This method of validating the variates is referred to as *split-sample validation* or *cross-validation*. Alternatively, we could collect a fresh new sample of entities to serve as a validation data set.

Briefly, the split-sample validation procedure involves randomly dividing the total sample into two subsamples and then conducting separate canonical analyses. To measure stability in the canonical solution, the simple product-moment correlation between the respective sets of canonical scores is computed; a high correlation is indicative of good stability (Dillon and Goldstein 1984). Alternatively, the sample can be split, but instead of conducting separate canonical analyses, the canonical weights (canonical coefficients) derived from one subsample can be applied to the other sample to calculate canonical scores (Dillon and Goldstein 1984). The correlations between the P (or M) pairs of canonical variates for the second subsample (i.e., the canonical correlations R_{ci}) are then determined and compared with the canonical correlations for the first subsample. Similar canonical correlations between subsamples would be evidence of coefficient stability.

No guidelines have been established for determining the appropriate sizes for the analysis and holdout samples. Some recommend that they be equal, while others prefer more entities in the analysis sample. Of course, the total sample size must be large enough to divide into two sets, both of which meet the minimum sample size requirements (see Sec. 5.6).

Another strategy is to follow the split-sample procedure several times. Instead of randomly dividing the total sample into analysis and holdout samples once, the total sample is randomly divided into analysis and holdout samples several times, each time testing the stability of the coefficients. The coefficient estimates would be averaged and a measure of variance (e.g., coefficient of variation) would be used to indicate the precision with which the coefficients have been estimated. Actually, this procedure would be very similar to the resampling procedure described in the next section.

5.10.2 Validation Using Resampling Procedures

As should be apparent from previous discussions, resampling procedures such as the jackknife, bootstrap, and randomization tests can be used in a variety of ways to assess the validity of CANCOR results (see Chapter 2, Sec. 2.8.5 for a general discussion of these procedures). In the present context, we can use the jackknife or bootstrap resampling procedure to judge the reliability of the canonical results.

Although generally inferior to the bootstrap procedure, the most common validation approach based on resampling the original data involves the jackknife procedure, particularly when small sample sizes prohibit the use of the split-sample approach. In the present context, jackknife validation proceeds as follows: (1) omit a single sample from the data set; (2) derive the canonical coefficients; (3) repeat the process sequentially N times, each time leaving out a different sampling entity; (4) calculate the jackknife estimate of each canonical coefficient and its variance (see Sec. 2.8.5); and (5) judge the stability of the canonical coefficients by the magnitude of the coefficient of variation in the jackknife estimate of the coefficient. Alternatively, test the null hypothesis that each observed coefficient from the full data set is equal to the jackknife estimate of the coefficient using a t-statistic, distributed with $N-1$ degrees of freedom, as given by

$$\frac{a_{ij}^* - a_{ij}}{\frac{SE(a_{ij}^*)}{\sqrt{N}}}$$

where a_{ij}^* is the jackknife estimate of the canonical coefficient for the ith canonical variate and jth variable in the X-set, a_{ij} is the canonical coefficient for the ith canonical variate and jth variable in the X-set, derived from the full sample, $SE(a_{ij}^*)$ is the standard error of the jackknife estimate, and N is the sample size.

If we reject the null hypothesis, then we conclude that the coefficients are unstable. We can test the stability of other canonical parameters (e.g., structure coefficients, cross-loadings, redundancy coefficients) in a similar manner.

5.11 Limitations of Canonical Correlation Analysis

Most of the limitations of CANCOR have been discussed already. Nevertheless, we briefly summarize some of these limitations here, other than those associated with the assumptions, to emphasize the importance of their consideration when interpreting the results of canonical correlation analysis.

- One objective of CANCOR is to reduce the dimensionality of the multivariate data set into one or a few important joint gradients. Thus, rarely do we retain and interpret more than a few pairs of canonical variates. However, by only interpreting the first few canonical variates, one may overlook a later axis that accounts for most of the predictive power in some variable. Consequently, even though this variable has a significant univariate relationship (i.e., significant simple linear regression with one or more variables from the opposing set), this relationship is

lost in the canonical transformation. Examination of all the canonical variates is necessary to assess this problem.
- Like other eigenanalysis techniques discussed in this book (PCA and DA), CANCOR is only capable of relating gradients that are intrinsic to the data sets. There may exist other, more important gradients not measured using the selected variables, and these dominant, undetected gradients may distort or confuse any relationships that are intrinsic to the data.
- As with DA, the stability and consistency of the relationships identified by CANCOR, as reflected in the canonical coefficients, should be validated. Ideally, a fresh sample of observations should be taken and analyzed in a similar manner and the resulting coefficients compared. Although, if the original data set is large, then cross-validation and jackknife validation may be employed to assess the stability of the canonical solution. An important point to keep in mind is that the larger the sample-to-variable ratio, the more stable canonical coefficients become.
- Perhaps the greatest limitation of CANCOR is the difficultly of interpreting the results. The canonical correlations may be very high, but the redundancy may be quite low. Several parameters must be assessed in order to get a complete and accurate picture of how well the canonical transformation worked and, more important, how to interpret the findings ecologically. Nevertheless, with proper attention and understanding, CANCOR can provide an effective tool for making sense out of what otherwise might be an unwieldy number of bivariate correlations between sets of variables.

Bibliography

Procedures

Alpert, M.I., Peterson, R.A. and Martin, W.S. 1975. Testing the significance of canonical correlations. *Proceedings of the American Marketing Association* 37:117–119.
Barcikowski, R.S., and Stevens, J.P. 1975. A Monte Carlo study of the stability of canonical correlations, canonical weights and canonical variate-variable correlations. *Multivariate Behavioral Research* 10:353–364.
Bartlett, M.S. 1941. The statistical significance of canonical correlations. *Biometrika* 32:29–37.
Dillon, W.R., and Goldstein, M. 1984. *Multivariate Methods: Methods and Applications*. New York: Wiley and Sons.
DiPillo, P.J. 1976. The application of bias to discriminant analysis. *Communications in Statistics. A Theory and Methods* 5:834–844.
Gauch, H.G., Jr., and Chase, G.B. 1974. Fitting the Gaussian curve to ecological data. *Ecology* 55:1377–1381.
Gauch, H.G., Jr., and Wentworth, T.R. 1976. Canonical correlation analysis as an ordination technique. *Vegetatio* 33:17–22.
Johnson, D.H. 1981. The use and misuse of statistics in wildlife habitat studies. In *The Use of Multivariate Statistics in Studies on Wildlife Habitat*. ed. D.E. Capen, pp 11–19. U.S. Forest Service General Technical Report RM–87.

Kettenring, J.R. 1971. Canonical analysis of several sets of variables. *Biometrika* 58:433–451.
Lawley, D.N. 1959. Tests of significance in canonical analysis. *Biometrika* 46:59–66.
Levine, M.S. 1977. *Canonical Analaysis and Factor Comparison.* Sage University paper series *Quantitative Applications in the Social Sciences,* Series No. 07–006. Beverly Hills and London: Sage Publications.
Marcus, M., and Minc, H. 1968. *Elementary Linear Algebra.* New York: Macmillan.
McDonald, L.L. 1981. A discussion of robust procedures in multivariate analysis. In *The Use of Multivariate Statistics in Studies on Wildlife Habitat.* ed. D.E. Capen, pp 242–242. U.S. Forest Service General Technical Report RM–87.
McKeon, J.J. 1965. Canonical analysis: some relations between canonical correlation, factor analysis, discriminant function analysis, and scaling theory. *Psychometric Monograph 13.*
Meents, J.K., Rice, J., Anderson, B.W., and Ohmart, R.D. 1983. Nonlinear relationships between birds and vegetation. *Ecology* 64:1022–1027.
Noy-Meir, I., Walker, D., and Williams, W.T. 1975. Data transformations in ecological ordination. II. On the meaning of data standardization. *Journal of Ecology* 63:779–800.
Smidt, R.K., and McDonald, L.L. 1976. Ridge discriminant analysis. *Technical Report No. 108,* Department of Statistics, University of Wyoming, Laramie.
Smith, K.G. 1981. Canonical correlation analysis and its use in wildlife habitat studies. In *The Use of Multivariate Statistics in Studies on Wildlife Habitat.* ed. D.E. Capen, pp 80–92. U.S. Forest Service General Technical Report RM–87.
Stewart, D.K., and Love, W.A. 1968. A general canonical correlation index. *Psychological Bulletin* 70:160–163.
Thorndike, R.M. 1978. *Correlation Procedures for Research.* New York: Gardner Press.
Thorndike, R.M., and Weiss, D.J. 1973. A study of the stability of canonical correlations and canonical components. *Educational and Psychological Measurement* 33:123–134.
Wollenberg, A.L. van den. 1977. Redundancy analysis: an alternative for canonical correlation analysis. *Psychometrika* 41:207–219.

Applications

Andries, A.M., Gulinck, H., and Herremans, M. 1994. Spatial modelling of the barn owl (Tyto alba) habitat using landscape characteristics derived from SPOT data. *Ecography* 17(3):278–287.
Barkham, J.P., and Norris, J.M. 1970. Multivariate procedures in an investigation of vegetation and soil relations of two beech woodlands, Cotswold Hills, England. *Ecology* 51:630–639.
Beyer, D.E., Jr., Costa, R., Hooper, R.G., and Hess, C.A. 1996. Habitat quality and reproduction of red-cockaded woodpecker groups in Florida. *Journal of Wildlife Management* 60(4):826–835.
Boyce, M.S. 1981. Robust canonical correlation of sage grouse habitat. In *The Use of Multivariate Statistics in Studies on Wildlife Habitat.* ed. D.E. Capen, pp 152–159. U.S. Forest Service General Technical Report RM–87.
Calhoon, R.E., and Jameson, D.L. 1970. Canonical correlation between variation in weather and variation in size in the Pacific tree frog, Hyla regilla, in southern California. *Copeia* 1970:124–134.
Carnes, B.A., and Slade, N.A. 1982. Some comments on niche analysis in canonical space. *Ecology* 63(4):888–893.

Folse, L.J., Jr. 1981. Ecological relationships of grassland birds to habitat and food supply in East Africa. In *The Use of Multivariate Statistics in Studies on Wildlife Habitat*. ed. D.E. Capen, pp 160–166. U.S. Forest Service General Technical Report RM–87.

Herrera, C.M. 1978. Ecological correlates of residence and non-residence in a Mediterranean passerine bird community. *Journal of Animal Ecology* 47:871–890.

Human, K.G., Weiss, S., Weiss, A., Sandler, B., and Gordon, D.M. 1998. Effects of abiotic factors on the distribution and activity of the invasive Argentine ant (Hymenoptera, Formicidae). *Environmental Entomology* 27(4):822–833.

James, F.C., and Porter, W.P. 1979. Behavior-microclimatic relationships in the African rainbow lizard, Agama agama. *Copeia* 1979:585–593.

Jameson, D.L., Mackey, J.P., and Anderson, M. 1973. Weather, climate, and the external morphology of Pacific tree toads. *Evolution* 27:285–302.

McIntire, C.D. 1978. The distributions of estuarine diatoms along environmental gradients: a canonical correlation. *Estuarine and Coastal Marine Science* 6:447–457.

Poore, G.C.B., and Mobley, M.C. 1980. Canonical correlation analysis of marine macrobenthos survey data. *Journal of Experimental Marine Biology and Ecology* 45:37–50.

Vogt, T., and Jameson, D.L. 1970. Chronological correlation between change in weather and change in morphology of the Pacific tree frog in southern California. *Copeia* 1970:135–144.

Weixelman, D.A., Zamudio, D.C., Zamudio, K.A., and Tausch, R.J. 1997. Classifying ecological types and evaluating site degradation. *Journal of Range Management* 50(3):315–321.

Willms, W., Hudson, R.J., and McLean, A. 1979. Assessment of variability among diets of individual deer with the aid of canonical analysis. *Canadian Journal of Zoology* 57(10):1856–1862.

Appendix 5.1

SAS program statements used to conduct canonical correlation analysis on the empirical data set presented in Figure 5.2. Everything given in lower case varies among applications with regard to characteristics of the specific data set (e.g., variable names) and the personal preferences of the user (e.g., naming output files).

The following header information creates a library named "km" for SAS files in the specified directory and provides page formatting information:

```
LIBNAME km 'd:\stats\cancor';
OPTIONS PS=60 LS=80 REPLACE OBS=MAX;
```

The following procedure is used to drop two vegetation variables involved in known singularities and log-transform the bird abundance variables in order to improve the distribution of each variable.

```
DATA A; SET km.cancor;
  DROP snagt bat;
  ARRAY XX amro--wiwr;
    DO OVER XX;
      XX = LOG10(XX+1);
    END; RUN;
```

Appendix 5.1   231

The following procedure is used to compute descriptive univariate statistics and plots for each of the original vegetation variables and log-transformed bird abundance variables in order to assess univariate normality:

```
PROC UNIVARIATE DATA=A NORMAL PLOT;
   TITLE 'Assessing Univariate Normality of the Original Variables';
   VAR gtotal--wiwr; RUN;
```

The following procedure is used to inspect the standardized data for possible outliers, including only the subset of variables identified from the stepwise procedure:

```
PROC STANDARD DATA=A OUT=standard MEAN=0 STD=1;
    VAR gtotal--wiwr; RUN;
DATA B; SET standard;
   ARRAY XX gtotal--wiwr;
      DO OVER XX;
         IF XX GT 2.5 THEN OUTLIER=1;
         ELSE XX='.';
      END; RUN;
DATA C; SET B;
   IF OUTLIER=1;
   TITLE 'List of Potential Outlier Observations';
   PROC PRINT;
      VAR id gtotal--wiwr; RUN;
```

The following procedure is used to conduct canonical correlation analysis on the vegetation structure variables and log-transformed bird abundance variables, including computing the pairwise correlations between variables and the squared multiple correlation coefficients for predicting the habitat variables from the bird variables, eigenvalues, canonical correlations, significance tests associated with each canonical correlation, canonical coefficients, canonical redundancy analysis, canonical loadings and cross-loadings:

```
PROC CANCORR DATA=A VPREFIX=bird WPREFIX=hab OUT=scores SMC ALL;
   TITLE 'Canonical Correlation Analysis of Vegetation Structure and
Bird Variables';
      WITH gtotal--fhd; RUN;
```

The following procedure is used to compute descriptive univariate statistics and plots for each canonical variate in order to assess the multivariate normality assumption:

```
PROC UNIVARIATE DATA=scores NORMAL PLOT;
      TITLE 'Assessing Multivariate Normality of the Canonical
Variates';
      VAR bird1--bird3 hab1--hab3; RUN;
```

The following procedure is used to generate a scatter plot for each of the first three pairs of canonical variates:

```
PROC PLOT DATA=scores;
   TITLE 'Scatter Plots of the Canonical Variates';
   PLOT bird1*hab1=id;
   PLOT bird2*hab2=id;
   PLOT bird3*hab3=id; RUN;
```

The following procedure is used to list the canonical scores for the first three pairs of canonical variates:

```
PROC PRINT DATA=scores;
   TITLE 'List of Canonical Scores';
   VAR id bird1--bird3 hab1--hab3; RUN;
```

6
Summary and Comparison

6.1 Objectives

By the end of this chapter, you should be able to do the following:

- Differentiate among four multivariate techniques (ordination, cluster analysis, discriminant analysis, and canonical correlation analysis) on the basis of:
  — Primary objective of the technique
  — Statistical procedure employed
  — Source of variation emphasized
  — Type of technique (dependence or interdependence)
  — Characteristics of the variable set
  — Sampling design assumed
- Explain the relationship between various univariate and multivariate statistical techniques based on the type of variables involved (continuous versus categorical) and the relationship of the variables to each other (independent versus dependent).
- Recognize the appropriate form (layout) of a two-way data matrix for each of the four multivariate techniques.
- Recognize the types of research questions best handled with each of the four multivariate techniques.
- Describe three research scenarios that involve the complementary use of more than one multivariate technique.

6.2 Relationship Among Techniques

Understanding the relationships among available multivariate techniques is an absolute necessity for multivariate statistics to be fully appreciated and understood. Unfortunately, there is no one way of effectively displaying all the relationships among the various techniques. Therefore, it is necessary to approach the task from several different perspectives.

In Table 6.1 (tables appear on pages 243 to 248), we summarize and compare the techniques covered in this book with respect to major objective, statistical procedure employed, source of variation emphasized, type of technique (dependence or interdependence), characteristics of the variables involved, and sampling design assumed by the technique. From Table 6.1, it should be clear that while these multivariate techniques have many features in common, they also differ in certain important aspects. Table 6.2 provides a key for selecting the appropriate multivariate technique based on the research objectives or questions of interest. A few statistical techniques not covered in detail in this book are also included for comparative purposes. Table 6.3 depicts the relationships among various univariate and multivariate statistical techniques based on the types of variables (continuous or categorical) and the variables' relationship to each other (independent or dependent). Table 6.4 depicts a hypothetical two-way data matrix and the relationship among the various multivariate statistical techniques in the structure of the data matrix.

6.2.1 *Purpose and Source of Variation Emphasized*

Perhaps the most useful way to compare multivariate techniques is by looking at the types of research questions the techniques were designed to answer (Tables 6.1 and 6.2). Understanding the primary objective of each technique is without a doubt the single most important step in fully understanding the relationship among techniques. Obviously, the guidelines for choosing a technique of analysis in Table 6.2 are most useful in *designing* a study. The tables are also useful for depicting general conceptual differences among multivariate techniques.

Ordination (e.g., principal components analysis, or PCA) is used when the objective is to identify and describe the major ecological *gradients* of variation among sampling entities, presumably corresponding to the major underlying ecological or environmental factors that govern the distribution of these entities and portray the distribution of entities along "continuous" gradients variation. For example, ordination can be used to describe the major environmental variation among sites (sampling entities) based on a set of environmental characteristics (variables). This contrasts with cluster analysis (CA), which is used for finding groups of ecologically similar sampling entities and portraying the entities in "discrete" groups. For example, CA can be used to establish groups of sites based on similarity in environmental characteristics in an effort to create plant associations or habitat types. Hence, ordination and CA differ in that the former emphasizes continuous variation among sampling entities, whereas the latter emphasizes discrete variation among sampling entities. Thus, these techniques are complementary and are often used to help assess whether the pattern of variation among sampling entities is more continuous or discrete. Both techniques emphasize the variation among individual sampling entities (i.e., entity-to-entity variance structure). Ordination seeks to *define* the entity-to-entity variance structure of the data set, whereas CA capitalizes on the entity-to-entity variation to *form* groups with less entity-to-entity variation within groups than among groups. Thus, ordination can be used to describe how entities differ most and CA can be used to establish groups of entities that differ least within groups and most among groups.

Canonical analysis of discriminance (CAD) is used when the objective is to describe the major ecological differences among two or more prespecified and well-defined groups of sampling entities. The key feature here is that *groups* of sampling entities are clearly identified, and it is the differences between or among these groups in terms of the variables that is of primary interest. For example, CAD can be used to describe the major differences in habitat preferences between two species (groups) based on a suite of habitat characteristics (discriminating variables). Like ordination, CAD also seeks to define gradients of continuous variation. However, CAD seeks gradients of maximum among-group variation rather than entity-to-entity variation as in ordination. Hence, CAD emphasizes the variation among groups of sampling entities (i.e., among-group variance structure). In this manner, CAD and CA are conceptually related because both are appropriate when the group structure is of primary interest. They differ in the following manner: in CA, the group structure is artificially superimposed on the data; that is, no obvious and well-defined groups exist a priori. In CAD, however, the group structure exists a priori and is usually obvious and well-defined. Cluster analysis is used to *find* groups, while CAD is used to *describe* group differences. Hence, these two techniques can be used in a complementary manner when well-defined groups do not exist. Cluster analysis is used to create groups and CAD is used subsequently to describe how those groups differ in ecological terms. For example, CA can be used to *establish* groups of sites (habitat types) based on major discontinuities in environmental characteristics, and CAD can be used to describe how these established groups differ most (i.e., which environmental characteristics most strongly drive the habitat classification system).

Classification is the predictive side of discriminant analysis; it is used when the objective is to develop a model to predict the likelihood that an entity of unknown origin will belong to a particular group based on a suite of discriminating characteristics. For example, classification can be used to develop a model for predicting the likelihood that a beaver will build a dam at a particular site and flood an area of interest based on characteristics of the site. Like CAD, classification is appropriate only when two or more prespecified and well-defined groups of sampling entities exist. Once the classification rule has been derived, it then can be used to predict group membership for a new set of sampling entities in which group membership is unknown. Although classification can be used when the predictive model is the sole desired outcome, more often in wildlife research it is used to evaluate how much canonical discrimination is achieved by CAD. In this role, following CAD, a classification procedure is employed and a measure of classification accuracy (a correct classification rate) is used to help judge the canonical performance. Logistic regression analysis is an alternative to classification when the objective is to predict the likelihood of an entity belonging to one of two groups (two-group case only) based on a suite of independent characteristics. It is perhaps the more robust technique when using mixed data sets (i.e., both continuous and categorical independent variables) and when the multivariate distribution of the data is nonnormal.

Canonical correlation analysis (CANCOR) is used when the objective is to describe the major ecological relationships between two or more sets of related variables. For example, CANCOR can be used to examine the relationship between animal abundance patterns at the community (multispecies) level and environmental patterns. Conceptually, CANCOR is related to ordination because both techniques attempt to define continuous gradients of maximum variation among sampling entities. Both CANCOR and constrained ordination techniques, such as canonical correspondence analysis (CCA), deal with two sets of related variables. The latter technique defines gradients of maximum total sample variance under the constraint that the gradients extracted be independent, linear combinations of the second set of variables. Canonical correspondence analysis is essentially a reciprocal averaging ordination which is constrained by multiple least squares regression to yield axes that define the largest gradients in the dependent variables explainable by linear combinations of the independent variables. In contrast, CANCOR defines gradients of maximum sample variance in one set of variables that are maximally correlated with gradients of maximum sample variance in another set of variables, and it is not necessary to make the distinction between a dependent variable set and an independent variable set. In this manner, performing CANCOR is like doing a separate ordination on two (or more) sets of related variables with the added constraint that the composite variables (canonical variates) from one set be derived so that they are maximally correlated with the composite variables from the other set. In wildlife research, CANCOR has been most frequently used to generate environmental gradients that are maximally correlated with animal abundance gradients. In this case, given the logical dependent-independent relationship among variable sets, CCA is a reasonable alternative and is often the best choice. Canonical correspondence analysis is not hampered by CANCOR's most limiting assumptions of linearity, multivariate normality, and absence of multicollinearity, and, consequently, is more robust than CANCOR. Moreover, CCA produces results that are much more readily interpretable in terms of species optima in environmental space, and relationships between environmental and compositional gradients. In general, however, CCA and CANCOR have similar objectives and therefore can be used as complementary analyses.

6.2.2 *Statistical Procedure*

The statistical procedures employed by most of the multivariate techniques are closely related. In particular, PCA, CAD, and CANCOR are all eigenanalysis problems. They differ merely in the form of the characteristic equation solved. In PCA, the eigensolution is derived from the total sample covariance or correlation matrix (recall that the solution differs depending on which matrix is used). In CAD, the eigensolution is derived from the within-groups and among-groups covariance matrices. The CANCOR eigensolution is derived from within-variable set and between-variable set covariance or correlation matrices. In each case, the eigensolution is derived from the matrix or matrices that define the variance structure sought by the technique. Moreover, in each case, the eigensolution includes one or more eigenvectors that represent weighted linear combinations of the original variables. The linear combinations are derived to maximize different criteria. In PCA, the eigenvectors (principal components) represent linear combinations of variables that define gradients of maximum sample variance. In contrast, CCA eigenvectors describe the linear combinations of the independent variables (usually environmental factors) that explain the largest gradients of independent variation among the dependent variables (usually species abundance). In CAD, the eigenvectors (canonical functions) represent linear combinations of variables that define gradients of maximum among-group variance. And in CANCOR, the eigenvectors (canonical variates) represent linear combinations of variables that define gradients of maximum within-set sample variance given that the canonical variates from one set are maximally correlated with the variates from the other set.

6.2.3 Type of Statistical Technique and Variable Set Characteristics

When considering the application of multivariate statistical techniques, the first question to be asked is, Can the data variables be divided into independent and dependent classifications? The answer to this question indicates whether a dependence or interdependence technique should be utilized. A *dependence* technique may be defined as one in which a variable or set of variables is identified as the dependent variable(s) to be predicted or explained by other independent variables. Discriminant analysis, CANCOR, and CCA are examples of dependence techniques (Table 6.1). In DA, group membership is used either to predict the values of several discriminating variables (MANOVA analogy) or, vice versa, several discriminating variables are used to predict group membership (MRA analogy; see Chapter 4, Sec. 4.2.3). Whether the grouping variable is treated as the dependent or independent variable depends on the sampling design (see Sec. 6.2.5) and research situation. In either case, a clear distinction between an independent and dependent variable set is made, and it is this dependence relationship that is being analyzed.

Similarly, in CANCOR, one variable set is typically treated as an independent set and the other as a dependent set, and it is the dependence or relationship between sets that is of interest. Even if no distinction between independent and dependent sets is made, it is still the relationship between variable sets that is of interest. In this case, however, we are interested in how each variable set influences the other, not just in how the independent set influences the dependent set. In CCA, there is always a distinction between dependent and independent variables. Typically the dependent set comprises abundance data for a number of species at a number of sites and the independent set comprises the values of a number of ecological variables at the same sites. The technique aims to quantify how much of the variability in the dependent set is due to variability in the independent set, describe the nature of the largest gradients in the dependent set accounted for by the independent set, and locate the maxima of each dependent variable in an ecological space defined by combinations of the independent variables.

In contrast, an *interdependence* technique is one in which no single variable or group of variables is defined as being independent or dependent. Rather, the procedure involves the analysis of all variables in the set simultaneously in an effort to give meaning to the entire set of variables or sampling entities. Unconstrained ordination (e.g., PCA) and CA are examples of interdependence techniques (Table 6.1). In both cases, the complete set of variables is used to evaluate relationships among sampling entities, and it is the relationships among the entities as defined by the variables that is of primary interest.

Understanding the relationships among multivariate techniques in terms of the type of statistical technique (dependence or interdependence) and the types of variables involved (continuous or categorical) and their relationship to each other (independent or dependent) is an important step toward fully understanding the conceptual relationships among techniques (Tables 6.1 and 6.3). Perhaps equally important, however, is understanding the relationships between common univariate and multivariate statistical techniques. Table 6.3 displays the relationships among multivariate techniques and corresponding univariate techniques based on the types of variables involved and their relationships. Several things are evident in Table 6.3. First, with the exception of DA (MRA analogy), univariate techniques (in the classical sense) are generally distinguished from the classical multivariate techniques on the basis of the number of dependent variables; univariate techniques involve one dependent variable, whereas multivariate techniques involve more than one dependent variable (e.g., DA–MANOV analogy, CANCOR, CCA) or no dependent variable at all (e.g., unconstrained ordination, CA). Second, each univariate technique has a multivariate counterpart obtained by adding additional dependent variables. Third, multivariate techniques are largely distinguished from each other by the type (continuous or categorical) and number of independent variables involved. Fourth, unconstrained ordination and CA are unique in that they involve a single set of variables, and thus do not fit a definition of multivariate techniques based on multiple dependent variables.

6.2.4 Data Structure

The conceptual relationship among multivariate techniques can be further understood by examining the structure of the data matrix employed by each procedure. Table 6.4 depicts a hypothetical data matrix partitioned into four submatrices (A, B, C, and D). Submatrices A and C can be considered together to form submatrix AC, and submatrices A and B can be considered together to form submatrix AB, and so forth. Submatrices A and C also contain a categorical grouping variable that defines populations (or groups) of sampling entities.

Unconstrained ordination and CA techniques operate on data matrices in the form of (for example) submatrix B or D; that is, they operate on a single, undifferentiated set of variables relating to the phenomenon under investigation measured on a single, undifferentiated sample of entities. Unconstrained ordination seeks linear combinations of the B (or D) variables that describe maximum variation among the sample entities. Cluster analysis combines the entities into classes or groups based on their similarity in values on the B (or D) variables.

Discriminant analysis operates on data matrices in the form of submatrix AC; that is, a single set of *X* variables (A and C combined) measured on two (or more) distinct groups of entities as defined by a single, categorical grouping variable. Canonical analysis of discriminance seeks linear combinations of the X variables that maximally differentiate between the A entities and the C entities.

Canonical correlation analysis (CANCOR) and CCA operate on data matrices in the form of submatrix AB (or CD); that is, they operate on two sets of related variables measured on a single, undifferentiated sample of entities. Canonical correlation analysis seeks linear combinations of the A (or C) variables and linear combinations of the B (or D) variables that are maximally correlated with each other. In contrast, CCA finds the linear combinations of the B (or D) variables that explain the *largest* linear combinations of the A or (C) variables.

From the previous discussion is should be evident that unconstrained ordination and CA explain within-set data structure, DA explains the relationship between sets of entities, and CANCOR and CCA explain the relationship between sets of variables.

6.2.5 *Sampling Design*

Each technique generally assumes a particular sampling design, that is, the manner in which the sample of entities has been drawn from the underlying population. Most of the techniques assume that a single random sample of entities has been drawn, but differ in whether that sample is assumed to represent one or more known or an unknown number of populations (Table 6.1). Unconstrained ordination, for example, assumes that a single random sample (N) has been drawn from either a known mixture of populations or an unknown mixture of one or more populations. In the former case, a sample is drawn from a known mixture of populations, but without any attempt to statistically assess the among-population structure of the sample. In the latter case, a sample is drawn without any prior knowledge of whether there is more than one distinct underlying population being sampled.

For example, in an investigation of bird-habitat relationships in managed forests, ordination could be used to describe the dominant habitat gradients among sampling points (e.g., sites). However, sampling points may be distributed across several statistically distinct habitat types (e.g., grassland, shrubland, forest). In this case, the sample represents a known mixture of statistically distinct populations. Nevertheless, ordination can still be used to evaluate gradients of variation among sampling points. Some of the dominant gradients may (and will likely) be directly related to the discrete habitat classes, while others may be related to other underlying environmental gradients not related to the habitat type scheme (e.g., moisture gradient).

Regardless of whether the sample represents a single population or a known or unknown mixture of populations, it is the entity-to-entity variance structure that is of primary interest, not differences among separate known or hypothesized populations. Of course, if more than one distinct population is sampled, then the gradients (e.g., the principal components) can be severely distorted if the relative sample sizes are very different among the respective populations. In such cases, some form of stratified random sampling scheme is generally more appropriate.

Cluster analysis also assumes that a single sample (N) has been drawn, but in this case, from an *unknown* number of distinct populations. Moreover, in CA, it is usually assumed or hypothesized that more than one underlying population exists and that the sample characteristics can be used to help determine how many real and distinct populations may exist. For example, in clustering bird species on the basis of similarities in habitat associations, we usually assume that species can be aggregated into groups, where each group more or less represents a distinct underlying population (or habitat association). In CA we attempt to determine the most likely number of distinct underlying populations.

Discriminant analysis also typically assumes that a single random sample (N) has been drawn, but in this case, from a *known* mixture of two or more populations (MRA analogy). In this case, the populations are well-defined and are specified prior to sampling, but the actual group membership of each sampling entity is not determined until during the sampling process or after the sample has been collected. The grouping variable is treated as dependent upon the independent discriminating characteristics. For example, we could randomly sample a number of sites and, as a result of the sampling process, determine whether species A is present or absent. We might hypothesize that the habitat characteristics at any particular site determine whether species A will occupy the site. In other words, the presence or absence of species A is dependent upon the habitat characteristics.

Alternatively, DA can be approached in an ANOVA-like design where several independent, random samples ($N_1, N_2, ..., N_G$) of two or more (say G, where G = number of groups) distinct populations are taken (MANOVA analogy). In this case, a random sample is drawn from each population separately. The grouping variable is considered the independent variable, and it defines the populations to be compared with respect to several dependent discriminating characteristics. For example, we could randomly sample a fixed number of nest sites of species A and a fixed number of nest sites of species B. Each species is treated as a separate statistical population and is sampled independently. In both sampling approaches, one of the primary purposes of DA is to evaluate how distinct the two or more prespecified populations actually are.

Canonical correlation analysis and CCA also assume that a single random sample (N) has been drawn, but in this case, from two (or more) jointly distributed populations. Each sampling entity is assumed to represent two (or more) related statistical populations, where each population is defined by a suite of related variables. The primary purpose of CANCOR is to evaluate the strength of association between the two (or more) population distributions. For example, we could randomly sample a number of sites and use CANCOR to relate gradients of habitat variation to community-level gradients of species abundances. A single random sample has been drawn, but each sampling entity represents two underlying statistical populations, one pertaining to the habitat and the other to the animal community.

6.3 Complementary Use of Techniques

Although Tables 6.1 to 6.3 display the multivariate techniques as separate and distinct techniques, it should be emphasized that most multivariate problems are best investigated using more than one technique. Many of the techniques are tightly related or complementary and, when used in combination, offer an effective analytic strategy. Many of these complementary relationships have been discussed in previous chapters. Nevertheless, we briefly review a few of the most common complementary uses.

Hotelling's T^2, MANOVA, CAD are essentially one procedure. The former two procedures represent the inferential, hypothesis-testing side of the analysis, while the later procedure represents the descriptive side of the analysis. Hotelling's T^2 and MANOVA represent the inferential counterparts of two-group CAD and multiple CAD, respectively. Both procedures calculate canonical variates; Hotelling's T^2 and MANOVA merely conduct t- or F-tests on the canonical variates, while CAD describes the canonical variates in terms of the original variables. Canonical analysis of discriminance has the added distinction of computing additional canonical variates when more than two groups exist. Ideally, these procedures should be used together to test and then describe group differences. In practice, however, field studies rarely meet the rigorous statistical assumptions required for hypothesis testing.

Classification and CAD procedures are also quite complementary, since both procedures attempt to *discriminate* among groups. The former represents the predictive side of discrimination, while the latter represents the descriptive side of discrimination. In practice, when the objective is primarily descriptive, classification is used to evaluate the effectiveness of the canonical transformation; that is, to provide a measure of reliability for descriptive results. Conversely, when the objective is primarily predictive, CAD is used to describe the group differences on which the classification is based.

Ordination and CA are also quite complementary when the research objective is to evaluate whether sampling entities are distributed *continuously* along gradients of variation or whether they are better described as belonging to *discrete groups* along discontinuous or disjoint gradients of variation. In other words, is the ecological variation among entities more continuous or more discrete? Because ordination procedures emphasize continuous variation and CA procedures emphasize discrete variation, insight into this question can be obtained by using both techniques on the same data set.

Last, unconstrained and constrained ordination are complementary techniques and a comparison of results obtained from these techniques can be very useful. With samples-by-species data unconstrained ordination techniques, like PCA and DCA, extract the largest gradients in the species data without regard to secondary ecological variables, while constrained ordination, like CCA, extracts the largest gradients in the species data explainable by the measured explanatory variables. When the analyses produce similar results, we assume that the measured ecological variables account for the main variation in the dependent variables. When the constrained ordination accounts for substantially less variance than the unconstrained ordination, we infer that other unaccounted factors are also important in driving the patterns in the dependent variables. We believe that comparing unconstrained and constrained ordinations is valuable whenever a constrained ordination is conducted, and recommend DCA also be used whenever CCA is performed.

As demonstrated by the empirical example used throughout this book, most multivariate data sets can be adapted to meet the requirements of several different techniques. Indeed, legitimate research questions can be asked of the same data set whose answers require using several different multivariate techniques. Furthermore, techniques such as ordination and CA often serve as precursors to other statistical procedures such as MRA, MANOVA, and DA. For example, CA might be used to define groups of ecologically similar entities, followed by CAD to describe the major ecological differences among the newly created groups. Alternatively, PCA might be used to generate new, uncorrelated variables, followed by MRA to explain the variation in some response variable based on a few important new principal components.

TABLE 6.1 Summary and comparison of multivariate techniques

	Ordination	Cluster Analysis (CA)	Discriminant Analysis (DA)	Canonical Correlation Analysis (CANCOR)
Objective	To describe the major ecological gradients of variation among sampling entities	To find groups of ecologically similar entities, and/or to determine whether entities are better described as belonging to disjoint groups or as being distributed continuously along gradients	*CAD*: To describe the major ecological differences among two or more groups of entities. *Classification*: To predict the likelihood that an entity of unknown origin will belong to a particular group based on a suite of discriminating characteristics	To describe the major ecological relationships between two (or more) sets of variables
Variation emphasis	Emphasizes variation among individual sampling entities by defining gradients of maximum total sample variance; describes the inter-entity variance structure	Emphasizes both differences and similarities among individual sampling entities by clustering entities based on inter-entity resemblance	Emphasizes variation among groups of sampling entities; describes the inter-group variance structure	Emphasizes variation among individual sampling entities within each set of variables such that the variance structures of the sets are maximally related to each other; describes the joint inter-entity variance structure of two (or more) sets of variables

(continued)

TABLE 6.1 Summary and comparison of multivariate techniques (*continued*)

	Ordination	Cluster Analysis (CA)	Discriminant Analysis (DA)	Canonical Correlation Analysis (CANCOR)
Statistical procedure	Eigenanalysis procedure that creates weighted linear combinations of the original variables (i.e., principal components) that maximize the variation among sampling entities	Variety of procedures that create groups using one of many different clustering strategies; these strategies, in general, maximize within-group similarity and minimize between-group similarity	*CAD*: Eigenanalysis procedure that creates weighted linear combinations of the original variables (i.e., canonical functions) that maximize differences among prespecified groups; that is, the procedure maximizes the ratio of among- to within-group variance (F-ratio) on the canonical function *Classification*: Variety of procedures using one of several alternative classification criteria; in general, these criteria maximize the classification accuracy of the classification criterion	Eigenanalysis procedure that creates weighted linear combinations of the original variables (i.e., canonical variates) in each set of variables that are maximally correlated with each other; that is, the procedure maximizes the correlation of the canonical variates from one set with the canonical variates from the other set(s)

TABLE 6.1 Summary and comparison of multivariate techniques (*continued*)

	Ordination	Cluster Analysis (CA)	Discriminant Analysis (DA)	Canonical Correlation Analysis (CANCOR)
Relationship among variables	Interdependence	Interdependence	Dependence	Dependence
Variable set	One set of variables; no independent-dependent distinction	One set of variables; no independent-dependent distinction	One categorical grouping variable and a set of discriminating variables; grouping variable can be considered independent or dependent, depending on sampling design and research situation	Two sets of variables; independent-dependent distinction is typically made, although it is not required
Sampling design	Single random sample (N) of a known or unknown number of populations	Single random sample (N) of an unknown number of populations	*MRA Analogy:* Single random sample (N) of a mixture of two or more distinct populations. *MANOVA Analogy:* Several independent, random samples ($N_1, N_2, ..., N_g$) of two or more distinct populations	Single random sample (N) of two jointly distributed populations

TABLE 6.2 Key for selecting the appropriate multivariate statistical technique based on the objective of the research. Note: A few techniques not discussed in detail in this book are included here for comparison.

If the research objective is to:	Then use:
— Describe the major ecological gradients of variation among individual sampling entities, and/or to portray sampling entities along continuous gradients of maximum sample variation	Ordination (e.g., principal components analysis)
— Establish artificial classes or groups of similar entities where prespecified, well-defined groups do not already exist, and/or to portray sampling entities in disjoint groups	Cluster analysis
— Differentiate among prespecified, well-defined classes or groups of sampling entities, and determine if classes or groups are significantly different; and if	
— two classes or groups exist	Hotelling's T^2
— more than two classes or groups exist	Multiple analysis of variance
— Describe the major ecological differences among classes or groups	Canonical analysis of discriminance or logistic regression analysis
— Develop a classification rule to predict group membership of future observations	Classification or logistic regression analysis
— Explain the variation in a continuous, dependent variable using two or more continuous, independent variables, and/or to develop a model for predicting the value of the dependent variable from the values of the independent variables	Multiple regression analysis
— Explain the variation in a dichotomous, dependent (grouping) variable using two or more continuous and/or categorical independent variables, and/or to develop a model for predicting the group membership of a sampling entity from the values of the independent variables	Logistic regression analysis
— Describe the major ecological relationships between two or more sets of variables	Canonical correlation analysis or constrained ordination (e.g., canonical correspondence analysis)

TABLE 6.3 Relationships among various statistical techniques based on the types of variables (continuous or categorical) and variable relationship (independent or dependent).

	Independent variables				
	Categorical			Continuous	
Dependent variables	1 dichotomous (d)	1 polytomous (p)	>1 d and/or p	1	>1
Categorical					
1 dichotomous (d)	Contingeny tables	Contingency tables	Contingency tables Two-group DA	—	Two-group DA Logistic RA
1 polytomous (p)	Contingency tables	Contingency tables	Contingency tables MDA	—	MDA
>1 d and/or p	Contingency tables Two-group DA	Contingency tables MDA	Contingency tables CCA CANCOR	—	CCA CANCOR
Continuous					
1	T-test	ANOVA	Higher order ANOVA	Correlation SRA	MRA
>1	Hotelling's T^2 Two-group DA	MANOVA MDA	Higher order MANOVA CCA CANCOR	—	CCA CANCOR

ANOVA   Analysis of variance   MRA   Multiple regression analysis
MANOVA   Multivariate analysis of variance   RA   Regression analysis
DA   Discriminant analysis   CCA   Canonical correspondence analysis
MDA   Multiple discriminant analysis   CANCOR   Canonical correlation analysis
SRA   Simple regression analysis

TABLE 6.4 Hypothetical two-way data matrix depicting the relationship between principal components analysis, cluster analysis, discriminant analysis, and canonical correlation analysis in the layout of the data matrix.

		Variables	
Sample	Population	X-set	Y-set
1	A	$a_{11}\, a_{12}\, a_{13} \ldots a_{1P}$	$b_{11}\, b_{12}\, b_{13} \ldots b_{1M}$
2	A	$a_{21}\, a_{22}\, a_{23} \ldots a_{2P}$	$b_{21}\, b_{22}\, b_{23} \ldots b_{2M}$
3	A	$a_{31}\, a_{32}\, a_{33} \ldots a_{3P}$	$b_{31}\, b_{32}\, b_{33} \ldots b_{3M}$
⋮	⋮	⋮	⋮
n	A	$a_{N1}\, a_{N2}\, a_{N3} \ldots a_{NP}$	$b_{N1}\, b_{N2}\, b_{N3} \ldots b_{NM}$
$n+1$	C	$c_{11}\, c_{12}\, c_{13} \ldots c_{1P}$	$d_{11}\, d_{12}\, d_{13} \ldots d_{1M}$
$n+2$	C	$c_{21}\, c_{22}\, c_{23} \ldots c_{2P}$	$d_{21}\, d_{22}\, d_{23} \ldots d_{2M}$
$n+3$	C	$c_{31}\, c_{32}\, c_{33} \ldots c_{3P}$	$d_{31}\, d_{32}\, d_{33} \ldots d_{3M}$
⋮	⋮	⋮	⋮
N	C	$c_{N1}\, c_{N2}\, c_{N3} \ldots c_{NP}$	$d_{N1}\, d_{N2}\, d_{N3} \ldots d_{NM}$

Appendix
Acronyms Used in This Book

ANOVA	Analysis of variance
CA	Cluster analysis
CAD	Canonical analysis of discriminance
CANCOR	Canonical correlation analysis
CC	Composite clustering
CCA	Canonical correspondence analysis
CCC	Complemented coefficient of community
CCJ	Complemented coefficient of Jaccard
CSMC	Complemented simple matching coefficient
CV	Coefficient of variation
D^2	Mahalanobis distance
DA	Discriminant analysis
DCA	Detrended correspondence analysis
ED	Euclidean distance
ED	Euclidean distance
FA	Factor analysis
HC	Hierarchical clustering
MANOVA	Multivariate analysis of variance
MDA	Multiple discriminant analysis
MRA	Multiple regression analysis
MVP	Minimum-variance partitioning
NHC	Nonhierarchical clustering
NMMDS	Nonmetric multidimensional scaling
OSP	Ordination space partitioning
PAHC	Polythetic agglomerative hierarchical clustering
PANHC	Polythetic agglomerative nonhierarchical clustering
PCA	Principal components analysis

PD	Percentage dissimilarity
PDHC	Polythetic divisive hierarchical clustering
PDNHC	Polythetic divisive nonhierarchical clustering
PO	Polar ordination
PSF	Pseudo F-statistic
PST2	Pseudo t^2-statistic
R^2	Coefficient of determination
RA	Reciprocal averaging
RSQ	Coefficient of determination; R^2 in SAS
SAHN	Sequential, agglomerative, hierarchical, and nonoverlapping
SAS	Statistical analysis system
SSW	Sums-of-squares within
TWINSPAN	Two-way indicator species analysis
UPGMA	Unweighted pair-group method using arithmetic averages

Glossary

Agglomerative clustering
　　Clustering process in which each entity begins in a class of its own and then entities are fused together (agglomerated) into larger and larger classes, until all entities belong to a single cluster.

Agglomeration schedule (or table)
　　Table showing the agglomeration sequence and the corresponding dissimilarity values at which entities and clusters combine (fuse) to form new clusters in an agglomerative clustering procedure.

Arch effect
　　An "arched" or curved distribution of sample points on the second and later axes of an ordination space resulting from pronounced nonlinearities in the data set (e.g., nonlinear responses to underlying environmental gradients).

Association coefficient
　　Measure of the agreement (or resemblance) between two data rows (representing two entities) in a two-way data matrix when the data are categorical.

Average linkage (or unweighted pair-group average)
　　Clustering strategy in which distance values between entities and clusters are computed as the average dissimilarity between clusters. An entity's dissimilarity to a cluster is defined to be equal to the average of the distances between the entity and each point in the cluster. When two clusters agglomerate, their dissimilarity is equal to the average of the distances between each entity in one cluster with each entity in the other cluster.

Bootstrap procedure
　　A resampling procedure in which sample entities (i.e., rows of the data matrix)

are randomly drawn, with replacement after each drawing, from the original two-way data matrix to generate "bootstrap" samples, from which confidence intervals may be constructed based on the repeated recalculation of the statistic under investigation.

Box plot

A graphical display that provides information about a sample frequency distribution, with emphasis on the tails of the distribution. The box plot provides a quick method of comparing the sample distribution to a normal distribution. Specifically, the center of the distribution is indicated by a median line and a box containing the central 50% of the observations, 25% on each side of the median line. The tails of the distribution are shown by lines (whiskers) extending from the box as far as the data extend, to a distance of at most 1.5 interquartile ranges (an interquartile range is the distance between the 25th and 75th sample percentiles). Any values more extreme than this are marked as points and are interpreted as outliers.

Broken-stick criterion

Ordination criterion for determining how many principal components to retain for interpretation based on a comparison of the observed eigenvalues with those expected under the "broken-stick" distribution (i.e., eigenvalues generated from random data with the same total variation). Observed eigenvalues that exceed the eigenvalues expected under the broken-stick distribution are considered meaningful and are retained for interpretation.

Canonical analysis of discriminance (CAD)

Descriptive part of discriminant analysis; an analytical procedure that describes the relationships among two or more groups of entities based on a set of two or more discriminating variables. The procedure creates linear combinations (i.e., canonical functions) of the original variables that maximize differences among the prespecified groups, maximizing the F-ratio (or t-statistic for two groups) on the canonical function(s). When two groups are involved, it is referred to as two-group CAD; when more than two groups are involved, it is referred to as multiple CAD.

Canonical coefficient (or weight)

The coefficients of the original variables in the linear equations that define the canonical functions (or variates), as given by the elements of each eigenvector. In raw form, these are weights to be applied to the variables in raw-score scales. As such, they are totally uninterpretable as coefficients, and the scores they produce for entities have no intrinsic meaning because they are affected by the particular unit (i.e., measurement scale) used for each variable.

Canonical correlation

In discriminant analysis, it is a measure of the multiple correlation between the set of discriminating variables and the corresponding canonical function (similar to multiple correlations in regression). It ranges between 0 and 1; a value of 0 denotes no relationship between the groups and the canonical function, while large values represent increasing degrees of association. In canonical correlation analysis, it is a measure of the correlation between each pair of canonical variates (linear combinations of the original variables) where one variate is from each set of variables.

Canonical correlation analysis (CANCOR)
Analytical technique used to describe the relationship between two (or more) sets of variables, where the underlying principle is to develop linear combinations (i.e., canonical variates) of variables in each set (both dependent and independent if such a distinction is made) such that the correlation between the composite variates is maximized.

Canonical correspondence analysis (CCA)
Constrained ordination technique in which the dominant gradients of variation in one set of variables (i.e., dependent variables, usually species abundances) are computed as linear combinations of explanatory variables (usually environmental characteristics) in a second set. Hence, CCA extracts the major gradients in the data that can be accounted for by the measured explanatory variables. This is very different from unconstrained ordination techniques like principal components analysis, where, for example, the axes are major gradients within the species data themselves, irrespective of any ecological explanatory variables.

Canonical cross-loading
In canonical correlation analysis, a cross-loading is the product-moment correlation between an original variable from one set of variables with a canonical variate from the other set. It is obtained by multiplying the canonical correlation coefficient with the canonical loading of the observed variable, and measures how well a variable loads on the canonical variate from the opposing variable set instead of its own.

Canonical function (or variate)
Weighted, linear combinations of the original variables that define new axes that maximize some objective function. In discriminant analysis, the canonical functions define linear combinations of the two or more discriminating variables that will discriminate "best" among the a priori defined groups. In canonical correlation analysis, the canonical functions define linear combinations of variables in two (or more) sets of variables such that the correlation between the functions is maximized.

Canonical loading (or structure coefficient)
Product-moment correlations between canonical functions (or variates) and the individual original variables included in the analysis. These simple bivariate correlations are not affected by relationships with other variables, and therefore reflect the true relationship between each variable and the canonical function. The full set of canonical loadings is referred to as the canonical structure or structure matrix.

Canonical pattern (or pattern matrix)
Pattern in the relationship between the original variables and the canonical functions (or variates), as given by the standardized canonical coefficients. The matrix containing these coefficients is called the pattern matrix. The relative contribution of the original variables to each canonical function may be gauged by the relative magnitude of the standardized canonical coefficients; the larger the magnitude, the greater that variable's contribution. However, the standardized canonical coefficients may distort the true relationship among variables in the canonical functions when the correlation structure of the data is complex, and thus, interpretation of the canonical functions is usually based on the structure coefficients.

Canonical plot
Scatter plots (for two or more canonical axes) and histograms (for one canonical axis) of the canonical scores used to graphically illustrate the relationships among entities, since entities in close proximity in canonical space are ecologically similar with respect to the environmental gradients defined by the canonical axes.

Canonical redundancy
Measure of the amount of variance in one set of original variables that is explained by (or "redundant" with) a canonical variate from the other data set. Note that the relationship is between "original variables" in one set and a "canonical variate" from the other set; it provides a summary measure of the ability of one set of variables (taken as a set) to explain the variation in the other variables (taken one at a time).

Canonical score
Each sampling entity has a single composite canonical score associated with each canonical function (or variate), derived by multiplying the sample values for each variable by a corresponding weight and adding these products together. These scores represent an entity's location along each canonical axis and are standardized such that they represent the number of standard deviations an entity is from the mean along each axis.

Canonical structure (or structure matrix)
Pattern of product-moment correlations between canonical functions (or variates) and the individual original variables included in the analysis. The matrix containing these correlation coefficients (i.e., canonical loadings) is called the structure matrix. The canonical structure is used to interpret the meaning of the canonical functions (or variates).

Categorical variable
Values are assigned for convenience only and are not useful for quantitative expressions.

Centroid
Location (or score) of the average sample entity in a P-dimensional space. Equivalent to the mean when there is more than one dimension.

Centroid linkage (or unweighted pair-group centroid)

Clustering strategy in which an entity's dissimilarity to a cluster is defined to be equal to its dissimilarity to the cluster centroid; when two clusters agglomerate, their dissimilarity is equal to the dissimilarity between cluster centroids.

Chaining

Characteristic of "space-contracting" fusion strategies in cluster analysis in which groups appear, on formation, to move nearer to some or all of the remaining entities; the chance that an individual entity will add to a preexisting group rather than act as the nucleus of a new group is increased, and the system is said to "chain." This effect is usually apparent in the dendrogram as a stair-step sequence of fusions.

Characteristic equation

Equation used in an eigenanalysis from which the eigenvalues (characteristic roots) and eigenvectors are derived. The form of the characteristic equation varies among techniques depending on the variance structure the technique seeks to define.

Classification

Classification represents the predictive side of discriminant analysis; it is the process by which a decision is made that a specific entity "belongs to" or "most closely resembles" one particular group.

Classification coefficient

The coefficients of the original variables in the equations that define the classification functions used to classify entities into groups.

Classification function

A function used to classify sample entities into groups based on a combination of the discriminating variables that maximizes group differences while minimizing variation within the groups.

Classification matrix (or confusion matrix)

Summary table (or matrix) of the classification results, which provides the number and percent of sample entities classified correctly or incorrectly into each group. The numbers on the diagonal represent correct classifications; off-diagonal numbers represent incorrect classifications.

Classification rate

Number and/or percent of sample entities classified correctly or incorrectly into each group in discriminant analysis. The correct classification rate (i.e., the percentage of samples classified correctly) is the most direct measure of discrimination when the purpose is classification of sample entities into prespecified, well-defined groups.

Cluster

An aggregation or grouping of sample entities based on their similarity or likeness on one or more variables, such that within-cluster homogeneity is maximized according to some objective criterion.

Cluster analysis (CA)
Family of analytical procedures whose main purpose is to develop meaningful aggregations, or groups, of sample entities based on a large number of interdependent variables. Specifically, the objective is to classify a sample of entities into a smaller number of usually mutually exclusive groups based on the multivariate similarities among entities.

Cluster membership table
Table identifying which cluster each entity belongs to for any specified number of clusters.

Coefficient of community
An association coefficient developed for binary data (i.e., dichotomous variables) that omits consideration of negative matches, yet gives more weight to matches than to mismatches.

Coefficient of Jaccard
An association coefficient developed for binary data (i.e., dichotomous variables) that omits consideration of negative matches, and gives equal weight to matches and mismatches.

Coefficient of variation
Measure of relative variability, defined as the ratio of the standard deviation to the mean expressed as a percentage. It is only meaningful if the variable is measured on a ratio scale (i.e., true origin) and is unaffected by the magnitude of the sample values. Hence, it is useful for comparing the variability of variables with different units of measurement.

Communality
The amount of variance an original variable shares with all other variables included in the analysis. In principal components analysis, final communality estimates represent the proportion of a variable's variance that is accounted for by the retained principal components; that is, the squared multiple correlation for predicting the variable from the principal components. Final communality estimates are useful in evaluating how well the retained components account for the total variation in the original variables.

Comparisonwise error rate
Type I error rate associated with each individual hypothesis test in an experiment involving many tests (or multiple comparisons).

Complemented coefficient of community (CCC)
The complement of the coefficient of community (i.e., one minus the coefficient of community).

Complemented coefficient of Jaccard (CCJ)
The complement of the coefficient of Jaccard (i.e., one minus the coefficient of Jaccard).

Complemented simple matching coefficient (CSMC)
The complement of the simple matching coefficient (i.e., one minus the simple matching coefficient).

Complete linkage (or furthest-neighbor)
Clustering strategy in which distance values between entities and clusters are computed as the dissimilarity between the two most widely separated entities. An entity's dissimilarity to a cluster is defined to be equal to its dissimilarity to the furthest entity in that cluster; similarly, when two clusters agglomerate, their dissimilarity is equal to the greatest dissimilarity for any pair of entities with one in each cluster.

Composite clustering
Polythetic, agglomerative, nonhierarchical clustering approach, including several different algorithms (e.g., COMPCLUS, FASTCLUS), designed for the rapid initial clustering of large data sets. The single criterion achieved is within-cluster homogeneity, and for each resulting cluster, a composite sample is produced by averaging the entities it contains.

Composite variable (or variate)
A weighted combination of the original variables derived as a product of an eigenanalysis.

Compression effect
The distortion (i.e., compression) in the relative positioning of sample points on the ordination axes associated with the second and later axes as a consequence of pronounced nonlinearities in the data set (e.g., nonlinear responses to underlying environmental gradients).

Confidence interval
An interval (around a sample estimate of a parameter) that includes the true value of a population parameter with a specified probability and for which the interval length provides a quantitative measure of the uncertainty. For example, a 95% confidence interval specifies the range of values containing the true population value with a 95% probability (i.e., it will miss the true value 5% of the time, equivalent to a 5% Type I error rate).

Constrained ordination
Ordination technique (e.g., canonical correspondence analysis) in which the ordination of one set of variables (e.g., dependent variable set, often species abundances) is constrained by the relationship to a second set (e.g., explanatory variables, often environmental characteristics). That is, the ordination axes in the first variable set are described directly in terms of the second (explanatory) set.

Continuous variable
Values are useful for quantitative expressions and can assume values at any point in continuum of possible values.

Cophenetic correlation
Correlation between input dissimilarities (in the entities-by-entities dissimilarity matrix) and the output dissimilarities implied by the resulting dendrogram (using the lowest level required to join any given entity pair in the dendrogram). Cophenetic correlation is essentially a measure of how faithfully the dendrogram (i.e., results of the cluster analysis) represents the information in the original data, and thus provides a way of objectively comparing different clustering strategies.

Correlation (or correlation coefficient)
Measure of the strength of association between two variables. It represents the similarity in the average *profile shapes* of two vectors when the data are continuous. Generally given as a product-moment correlation coefficient that ranges from 0 to 1; the correlation is 1 whenever two profiles are parallel, irrespective of how far apart they are in absolute terms. Unfortunately, the converse is not true, because two profiles may have a correlation of 1 even if they are not parallel. All that is required for perfect correlation is that one set of scores be linearly related to a second set.

Correlation matrix
Square, symmetrical matrix containing correlation coefficients for all pairwise combinations of variables, where the diagonal elements equal 1 (i.e., because a variable's correlation with itself is equal to 1) and the off-diagonals represent intercorrelations among all variables.

Count variable (or meristic data)
Values occur in discrete, indivisible units of the same size (e.g., integer-valued).

Covariance
Measure of the association between two vectors when the data are continuous. Specifically, the covariance between two variables (vectors) measures the absolute magnitude of change in one variable for a given unit change in the other. In contrast to correlation coefficients, covariance is an absolute measure of association that depends on the scale and unit of measurement of the variables involved.

Covariance-controlled partial F-ratio (or partial F-ratio)
An F-ratio (i.e., ratio of among-group variation to within-group variation) that is computed for a variable on the residual variation remaining after accounting for the variation explained by the other variables in the model. Partial F-ratios help to reveal whether any one variable is a significant contributor to the model after all the other variables have been accounted for. In the context of discriminant analysis, the partial F-ratio is an aggregative measure in that it measures the amount of total discrimination across functions associated with each variable after accounting for the discrimination achieved by the other variables.

Covariance matrix (or variance-covariance matrix)
Square, symmetrical matrix containing a measure of association for all pair-

wise combinations of variables, where the diagonal elements equal the variances of the variables and the off-diagonals represent covariances. Note that the covariance matrix computed from standardized data (variables with zero mean and unit variance) is equivalent to the correlation matrix. Otherwise, if computed from the raw data, it is usually referred to as the "raw" covariance matrix.

Data cloud

The geometric distribution of sample entities in multidimensional space, where each dimension, or axis, is defined by one of the variables. Each sample point has a position on each axis and therefore occupies a singular location in this space. Collectively, all the sample points form a cloud of points in this multidimensional space.

Data space

A geometric space occupied by sample points, where each dimension, or axis, is defined by one of the P original variables. Each sample point has a position on each axis and therefore occupies a singular location in this P-dimensional space.

Dendrogram

A treelike plot (vertical or horizontal) depicting the agglomeration sequence in cluster analysis, in which entities are enumerated (identified) along one axis and the dissimilarity level at which each fusion of clusters occurs on the other axis. Dendrograms are the most frequently presented result of hierarchical clustering, since they display relationships effectively.

Dependence technique

Technique in which a variable or set of variables is identified as the dependent variable(s) to be predicted or explained by other independent variables.

Dependent variable (Y's)

Variable presumed to be responding to a change in an independent variable; free to vary in response to controlled conditions.

Detrended correspondence analysis (DCA)

Unconstrained ordination technique similar to reciprocal averaging (RA) but with two additional steps, detrending and rescaling, added to remove RA's two major faults, the arch effect and axis compression. The orthogonality criterion for second and higher axes in RA (which is where arch and compression effects cause distortions) is replaced with the stronger criterion that the second and higher axes have no systematic relationship of any kind to lower axes.

Detrending

Technique used in detrended correspondence analysis to eliminate the arch effect; accomplished by dividing the first axis into a number of equal segments and within each segment adjusting the ordination scores to a mean of zero.

Dichotomous variable

Categorical variable with only two possible values (or states).

Discriminant analysis (DA)
Refers to a couple of closely related analytical procedures whose main purpose is to *describe* the differences among two or more well-defined groups (canonical analysis of discriminance) and *predict* the likelihood that an entity of unknown origin will belong to a particular group based on a suite of discriminating characteristics (classification). When two groups are involved, it is referred to as two-group DA; when more than two groups are involved, it is referred to as multiple DA.

Discriminant function
Linear equation representing a weighted combination of the original variables that define the canonical and classification functions in discriminant analysis.

Discriminating variable
A variable used to discriminate among groups in discriminant analysis.

Dissimilarity matrix
A resemblance matrix containing dissimilarity coefficients for every pair of entities.

Dissimilarity (or distance) coefficient
Measure of the dissimilarity (or distance) between two entities based on one or more variables. Dissimilarity coefficients represent the complement of similarity coefficients.

Dissimilarity space
Resemblance space where each dimension, or axis, is defined by the dissimilarity to one of the other N sample entities based on a dissimilarity coefficient.

Divisive clustering
Clustering process in which all entities begin in a single class and then are divided into progressively smaller classes, stopping when each class contains a single member or when the predetermined limit of some "stopping rule" is reached.

Eigenanalysis
Analytical process of decomposing a matrix (usually a secondary matrix summarizing the variance-covariance structure of the original data set) into its eigenvalues (or latent roots) and eigenvectors (or latent vectors).

Eigenvalue (or characteristic root or latent root)
The scalar product of an eigenanalysis; each eigenvalue (or latent root) defines the variance associated with a corresponding eigenvector.

Eigenvector
The vector product of an eigenanalysis; each eigenvector (or latent vector) defines the coefficients of the original variables in the linear equations that define the new axes (or variates) that maximize some objective criterion (e.g., maximum variance).

Environmental gradient
An environmental characteristic that varies among sample entities, usually in a continuous manner and often spatially across the study area, and that presumably

exerts an importance influence on the ecological phenomenon (e.g., species' abundances) under consideration.

Euclidean distance (*ED*)
Measure of the dissimilarity (or distance) between two entities based on the average score on one or more variables. Essentially, *ED* is a measure of the length of a straight line drawn between two entities in multidimensional space.

Exclusive clustering (or nonoverlapping clustering)
Clustering process in which each entity is placed in one group only.

Experimentwise error rate
The overall experiment-wide probability of committing a type I error in an experiment involving many tests (or multiple comparisons). The probability of getting a significant result in an individual test or comparison due to chance variation alone (Type I error) increases rapidly as the number of tests increases. The experimentwise error rate measures the likelihood of such an occurrence.

Factor analysis (FA)
Family of unconstrained ordination techniques that provide explicitly for a separation of the shared and unique variance associated with the P original variables. All FA models separate the "unique" variance from "common" variance and make the assumption that the intercorrelations among the P original variables are generated by some smaller number of hypothetical (latent) variables (common factors). Under this assumption, the "unique" part of a variable is factored out and does not contribute to the relationships among the variables, for which description only the M (usually $\ll P$) common factors are employed.

Factor structure
Pattern of product-moment correlations between the common factors and the individual original variables included in the factor analysis. The matrix containing these correlation coefficients is called the structure matrix. The factor structure is used to interpret the meaning of the common factors.

Fisher's linear discriminant function analysis
Classification technique first proposed by Fisher (1936) whereby classification of sample entities is based on a linear combination of the discriminating variables that maximizes group differences while minimizing variation within the groups; separate "classification functions" are derived for each group using the pooled within-groups covariance matrix, and each sample entity is classified into the group with the highest score.

***F*-to-enter**
F-ratio criterion (i.e., ratio of among-group to within-group variation) used to determine whether a variable has enough explanatory power to be considered for entry into the model during a stepwise variable selection procedure.

***F*-to-remove**
F-ratio criterion (i.e., ratio of among-group to within-group variation) used to

determine whether a variable has enough explanatory power to remain in the model during a stepwise variable selection procedure.

Gaussian
Bell-shaped or normal distribution of a variable, as in a Gaussian species' response to an underlying environmental gradient.

Hierarchical clustering (HC)
Clustering process in which similar entities are grouped together such that the groups can be arranged into a hierarchy that expresses the relationships among groups.

Icicle plot
Treelike plot (vertical or horizontal) depicting cluster membership in relation to the number of clusters, in which entities are enumerated (identified) along one axis and the number of clusters (cluster level) along the other axis. Icicle plots are helpful in understanding the consequences of retaining varying numbers of clusters.

Identity matrix (I)
Diagonal matrix (i.e., symmetric matrix with all off-diagonal elements equal to zero) with all of its diagonal entries being one; it has the property that the product of any matrix X by I leaves X unchanged, and is thus analogous to unity in scalar algebra.

Independent variable (X's)
Variable presumed to be a cause of any change in a dependent variable; often regarded as fixed, either as in experimentation or because the context of the data suggests they play a causal role in the situation under study.

Inferential technique
Technique involving hypothesis testing in which inferences about the characteristics of a population are made from the characteristics of a sample.

Interdependence technique
Technique in which no single variable or group of variables is defined as being independent or dependent. Rather, the procedure involves the analysis of all variables in the set simultaneously in an effort to give meaning to the entire set of variables or sampling entities.

Interval scale
Continuous variable containing an arbitrary zero point (i.e., no true origin).

Jackknife procedure
A resampling procedure to determine the effect of each sampling entity on a statistic by iteratively removing successive sampling entities (i.e., rows of the data matrix) from the original two-way data matrix and recalculating the desired statistic. The resulting "pseudoestimates" can be used to construct confidence intervals about the statistic under investigation.

Joint plot
An ordination plot generated from reciprocal averaging (or correspondence analysis) and detrended correspondence analysis in which both sample scores and species scores (for a samples-by-species matrix) are plotted simultaneously.

Kappa statistic (or Cohen's Kappa statistic)
A measure of chance-corrected classification (i.e., a standardized measure of improvement over random assignment) when prior probabilities are estimated by group sample sizes.

Kurtosis
A measure of the heaviness of the tails in a distribution of a variable; that is, for the tails of the distribution about the mean to contain more observations than would be expected under a normal distribution. Population kurtosis must lie between −2 and positive infinity, inclusive. Kurtosis equals zero for a normal distribution; larger positive values indicate heavier tails (i.e., longer) than would be expected under a normal distribution.

Latent root criterion (or Kaiser-Guttman criterion)
Criterion for determining the number of latent roots (e.g., principal components) to retain for interpretation. When the correlation matrix is used in the eigenanalysis, components with eigenvalues <1 are dropped from further scrutiny. These components represent less variance than is accounted for by a single original variable and therefore should not be interpreted.

Linear dependency
Refers to the linear association between two variables. A perfect linear dependency between two variables means that one variable can be derived exactly from the other variable; that is, they have a perfect association and therefore a correlation coefficient of 1. Near linear dependencies lead to multicollinearity and the inaccurate estimation of coefficients in dependence techniques such as discriminant analysis, regression, and canonical correlation analysis.

Linearity
Refers to the underlying assumption of many multivariate techniques (e.g., principal components analysis) that variables change linearly along underlying gradients and that linear relationships exist among variables such that the variables can be combined in a linear fashion (e.g., to create principal components).

Mahalanobis distance
Measure of the dissimilarity (or distance) between two entities based on a standardized form of Euclidean distance. Data are standardized by scaling responses in terms of standard deviations, and adjustments are made for intercorrelations between the variables. Mahalanobis distance is equal to Euclidean distance when the variables are standardized and the intercorrelations are 0.

Maximum chance criterion
A measure of chance-corrected classification (i.e., a standardized measure of

improvement over random assignment) based on the premise that classification based on the discriminating variables should exceed that obtained by simply assigning all samples to the group with the largest size. It is determined by computing the percentage of the total sample represented by the largest of the two or more groups.

Median linkage (or weighted pair-group centroid)
Clustering strategy in which an entity's dissimilarity to a cluster is defined to be equal to its dissimilarity to the cluster centroid; when two clusters agglomerate, their dissimilarity is equal to the dissimilarity between cluster centroids. Median linkage is similar to centroid linkage except that the centroids of newly fused groups are positioned at the median between old group centroids.

Meristic data
Pertains to count data, typically species abundances across multiple samples.

Minimum variance partitioning
A polythetic divisive nonhierchical clustering technique in which entities are split into a specified number of groups in order to minimize within-cluster variation.

Monothetic clustering
Clustering process in which only the information from a single variable is considered in the process of deriving cluster assignments. "Monothetic" techniques can only be divisive and are often used in community ecology studies involving samples-by-species data, where sets of samples are divided according to the presence or absence of a single species.

Monotonic (or monotonicity)
Refers to a unidirectional relationship between two variables (e.g., the change in a variable along an environmental gradient). Although the relationship may be nonlinear, a monotonic relationship never changes from positive to negative or vice versa over the full range of values.

Multicollinearity
Near multiple linear dependencies (i.e., high correlations) among variables in the data set, which can cause problems in the interpretation of the derived coefficients in dependence techniques such as discriminant analysis, regression, and canonical correlation analysis. When the original variables are highly correlated, individual coefficients in the canonical functions (or variates), for example, measure not only the influence of their corresponding original variables, but also the influence of other variables as reflected through the data correlation structure. Under these conditions, the canonical coefficients can be incorrectly estimated and even have the wrong signs, and interpretation of the coefficients becomes difficult.

Multiple regression analysis (MRA)
Statistical technique used to determine the relationship between a single dependent variable and multiple independent variables.

Glossary 265

Multivariate analysis of variance (MANOVA)
Statistical technique used to determine if samples came from populations with equal centroids. MANOVA employs multiple dependent variables to compare populations, rather than a single dependent variable as in ANOVA. Hotelling's T^2 test is the special case of MANOVA when there exists only two populations.

Multivariate normal (or multivariate normality)
A generalization of the univariate normal distribution to the case of P variables. A multivariate normal distribution exists when the data cloud (in P-dimensional space) is hyperellipsoidal with normally varying density around the centroid. Such a distribution exists when each variable has a normal distribution about fixed values on all others. A multivariate normal distribution is a basic assumption required for the validity of the significance tests in certain procedures, such as multivariate analysis of variance.

Noise
Refers to the inherent variation in the data set due to unaccounted-for sources of variation, such as other unmeasured factors, measurement error, and sampling error.

Nominal scale
Categorical variable in which the order of categories is meaningless; numbers utilized in a nominal scale are arbitrary.

Nonexclusive clustering (or overlapping clustering)
Clustering process in which each entity can be placed in more than one group.

Nonhierarchical clustering (NHC)
Clustering process in which similar entities are merely assigned to groups to achieve within-cluster homogeneity. There is not necessarily any interesting structure within clusters or definition of relationships among clusters.

Nonmetric multidimensional scaling (NMMDS)
Family of related unconstrained ordination techniques that use only rank order information in the dissimilarity matrix to locate sample entities in a low-dimensional ordination space such that the intersample distances in the ordination have the same rank order as do the intersample dissimilarities in the dissimilarity matrix.

Nonsingular matrix
See singular matrix.

Normality
Refers to the shape of the underlying distribution of a variable. A "normal" or Gaussian distribution is a smooth symmetric function often referred to as "bell-shaped." Its skewness and kurtosis are both zero.

Normal probability plot
Normal probability plots are plots of sample points against corresponding per-

centage points of a standard normal variable. If the data are from a normal distribution, the plotted values will lie on a straight line.

Oblique rotation
Rotation of orthogonal ordination axes about the centroid such that the orthogonal (i.e., right-angle) relationships among axes are lost (i.e., angles become oblique).

Ordinal scale
Categorical variable in which the order of categories is meaningful and indicates relative differences; numbers utilized in the ordinal scale, however, are nonquantitative.

Ordination
Family of techniques whose main purpose is to organize sampling entities along a meaningful continuum or gradient based on the interrelationships among a large number of interdependent variables. Specifically, the objective is to condense the information contained in the original variables into a smaller set of dimensions (e.g., principal components), defined as linear combinations of the original variables, that describe maximum variation among individual sampling entities.

Ordination space
Geometric space occupied by sample points, where each dimension, or axis, is defined by a linear combination of the original variables (e.g., principal component). Each sample point has a position on each axis based on its score on that ordination axis and therefore occupies a singular location in this P-dimensional space (where P equals the number of variables).

Ordination space partitioning (OSP)
A polythetic, divisive, hierarchical clustering technique in which entities are first positioned in a low-dimensional ordination space (usually using detrended correspondence analysis) and then successive partitions are drawn in the ordination space to generate a divisive, hierarchical clustering.

Orthogonal (or orthogonality)
A mathematical constraint specifying that variables are independent of each other. Geometrically, axes are said to be orthogonal if they are at right angles to each other. Orthogonal axes are statistically independent; that is, they are oriented in directions such that they explain independent variation in the data.

Orthogonal rotation
Rotation of orthogonal ordination axes about the centroid such that the orthogonal (i.e., right-angle) relationships among axes are maintained.

Outlier
Sample point that lies distinctly apart from all others with respect to the variables in the data set, usually resulting either from unusual environmental or historical conditions, from aberrations of the organism responses, or from measurement error. Outliers can exert undue influence on the analysis and are therefore often

eliminated prior to the final analysis, although it is important to distinguish between extreme observations and true "ecological" outliers.

Parsimony (or parsimonious, or parsimoniously)
Refers to the notion that "simpler" is better. Parsimony is an unstated goal of most statistical analyses. For example, a parsimonious model contains the fewest number of variables sufficient to explain the patterns in the data set.

Partial F-test (or partial F-ratio)
Simply an F-test (i.e., ratio of among-group to within-group variance) for the additional contribution (in terms of explanatory power) of a variable above the contributions of those variables already in the model.

Percentage dissimilarity (PD)
Measure of the dissimilarity (or distance) between two entities based on the percent similarity in scores on one or more variables. Percentage dissimilarity is the most commonly employed distance measure for species abundance data (i.e., meristic data).

Polar ordination (PO), or Bray and Curtis ordination
Relatively simple and intuitively straightforward unconstrained ordination technique in which two entities representing opposite ends of an environmental gradient are selected to serve as "poles" and then each entity is projected onto the axis formed by connecting the two endpoints based on a measure of dissimilarity.

Polythetic agglomerative hierarchical clustering (PAHC)
Family of clustering techniques in which each entity is assigned as an individual cluster and then, based on a resemblance measure computed from all the variables, entities are agglomerated (fused) in a hierarchy of larger and larger clusters until finally a single cluster contains all entities.

Polythetic agglomerative nonhierarchical clustering (PANHC)
Family of clustering techniques in which similar entities are merely assigned to groups to achieve within-cluster homogeneity based on a resemblance measure computed from all the variables. There is not necessarily any interesting structure within clusters or definition of relationships among clusters.

Polythetic clustering
Clustering process in which all the information for each entity (i.e., all the variables) is considered in the process of deriving cluster assignments.

Polythetic divisive hierarchical clustering (PDHC)
Family of clustering techniques that begin with all sample entities together in a single cluster and then successively divide the entities into a hierarchy of smaller and smaller clusters until finally each cluster contains only one entity or some specified number of entities.

Polythetic divisive nonhierarchical clustering (PDNHC) (or K-means clustering)
Clustering technique in which entities are split into a specified number of groups

using some criterion, for example, maximizing the ratio of among- to within-cluster variation, or simply minimizing within-cluster variation.

Polytomous variable
Categorical variable with more than two possible values (or states).

Principal component
Weighted, linear combinations of the original variables that define new axes that maximize the variation among sample entities; that is, they define the principal (major) gradients of variation among sample entities.

Principal component loading (or structure coefficient)
Product-moment correlations between the principal components and the individual original variables included in the analysis. These simple bivariate correlations are not affected by relationships with other variables, and therefore reflect the true relationship between each variable and the principal component. The full set of loadings is referred to as the principal component structure or structure matrix.

Principal components analysis (PCA)
Unconstrained ordination technique whose main purpose is to organize sampling entities along meaningful gradients (i.e., principal components) based on the interrelationships among a large number of interdependent variables. Specifically, the objective is to condense the information contained in the original variables into a smaller set of dimensions (i.e., principal components), defined as linear combinations of the original variables, that describe maximum variation among individual sampling entities.

Principal component score
Each sampling entity has a single composite score associated with each principal component, derived by multiplying the sample values for each variable by a corresponding weight and adding these products together. These scores represent an entities' location along each principal component axis and are standardized such that they represent the number of standard deviations an entity is from the mean along each axis.

Principal component structure (or structure matrix)
Pattern of product-moment correlations between the principal components and the individual original variables included in the analysis. The matrix containing these correlation coefficients (i.e., loadings) is called the structure matrix. The principal component structure is used to interpret the meaning of the principal components.

Principal component weight (or scoring coefficient)
The coefficients of the original variables in the linear equations that define the principal components, as given by the elements of each eigenvector derived from the characteristic equation. In raw form, these are weights to be applied to the variables in raw-score scales. As such, they are totally uninterpretable as coeffi-

cients, and the scores they produce for entities have no intrinsic meaning because they are affected by the particular unit (i.e., measurement scale) used for each variable.

Prior communality

Refers to the proportion of a variable's variance that is accounted for by the full set of principal components (principal components analysis; PCA) or common factors (factor analysis; FA). In PCA, prior communalities are equal to 1 because 100% of the variation in each variable is accounted for by the P principal components; whereas, in FA, prior communalities are estimated from the data or from ancillary information.

Prior probabilities (or priors)

Probability of a sample entity belonging to each group in discriminant analysis. It is not required that prior probabilities be equal among groups, but that the prior probabilities of group membership be known. Prior probabilities influence the forms of both classification and canonical functions, and an incorrect or arbitrary specification of prior probabilities can distort or otherwise obscure any underlying structure of the data.

Product-moment correlation (or Pearson's product-moment correlation)

See correlation coefficient.

Proportional chance criterion

A measure of chance-corrected classification (i.e., a standardized measure of improvement over random assignment) based on the premise that classification based on the discriminating variables should exceed that obtained by randomly assigning samples to groups in proportion to group sizes.

Pseudoreplication

The testing for treatment effects with an error term inappropriate to the hypothesis being considered. Pseudoreplication refers not to a problem in experimental design (or sampling) per se, but rather to a particular combination of experimental design (or sampling) and statistical analysis that is inappropriate for testing the hypothesis of interest. For example, in a manipulative experiment, pseudoreplication results when the treatments are not replicated (though the samples may be) or the replicates are not statistically independent.

Q-factor analysis

Refers to an ordination procedure based on the correlation or covariance structure of the sampling entities (rows) rather than the variables (columns). Q-factor analysis can be used if the objective is to combine or condense large numbers of sample entities (rows) into distinctly different groups within a larger population, similar to cluster analysis, but is generally inferior to the latter and therefore rarely employed.

Randomization procedure

A resampling procedure in which the data matrix is randomly reshuffled to remove any "real" structure in order to generate random permutations, from which the expected distribution of the statistic under the null hypothesis of no "real" data structure can be generated. By comparing the observed value of the statistic to the permutation distribution, we can determine directly the probability of observing that value if the original sample was actually drawn from a population defined by the sample data characteristics, but without any "real" structure (i.e., Type I error rate).

Random sample (or random sampling)
A subset of entities drawn from a population in which each entity has an equal chance of being drawn.

Ratio scale
Continuous variable containing an absolute zero point (i.e., true origin).

Reciprocal averaging (RA) (or correspondence analysis)
Unconstrained dual ordination procedure in which samples (rows) and variables (columns) are ordinated simultaneously on separate but complementary axes. Reciprocal averaging is an eigenanalysis problem where the eigenvalues reflect the degree of correspondence between variables and sample scores; a high eigenvalue indicates a long and important gradient.

Redundancy
Refers to identical information shared by two or more variables. Two variables that are highly correlated are redundant, because the values of either variable can be estimated reasonably well by knowing the values of the other variable.

Redundancy analysis
Technique developed for canonical correlation analysis whereby one can calculate the amount of variance in one set of variables that is explained by (or redundant with) a linear combination of variables (i.e., canonical variate) of another set. By calculating redundancy for all variates of a data set and summing the results, the proportion of variance of one set that is accounted for by the other set can be calculated.

Redundancy coefficient
The amount of variance in one set of variables explained by a linear combination of variables (i.e., canonical variate) from another set in canonical correlation analysis. The redundancy coefficient is the equivalent of computing the squared correlation coefficient between one set of variables (represented by a canonical variate) and each individual variable in the other set, and then averaging these squared coefficients to arrive at an average squared correlation coefficient. It provides a summary measure of the ability of one set of variables (taken as a set) to explain the variation in the other variables (taken one at a time).

Relative percent variance criterion (or percent trace)
Criterion for determining the number of eigenvector solutions (e.g., principal

components) to retain for interpretation based on the percentage of total variation accounted for by each eigenvalue.

Resampling procedure (or resampling technique)
Refers to a variety of procedures that involve drawing new samples from the existing data set from which confidence intervals may be constructed based on the repeated recalculation of the statistic under investigation or from which the expected distribution of the statistic under the null hypothesis of no "real" data structure can be generated.

Resemblance coefficient (or measure)
Measure of the resemblance between any two entities based on their similarity or dissimilarity in scores on the measured variables.

Resemblance matrix
Matrix of resemblance coefficients for every pair of entities in which coefficients are arranged in a square, symmetric matrix, with diagonal elements for self-comparisons. The resemblance matrix can be given as a similarity matrix, where the coefficients represent the similarity between pairs of entities, although the resemblance matrix usually is given as a dissimilarity matrix by taking the complement of each similarity coefficient.

Resemblance space
Geometric space occupied by sample points, where each dimension, or axis, is defined by the resemblance to one of the other N sample entities. Each sample point has a position on each axis based on some resemblance measure and therefore occupies a singular location in this N-dimensional space.

Residual discrimination
The ability of the variables in a discriminant analysis to discriminate among groups after accounting for the discrimination achieved by the previously computed canonical functions.

Reversals
The condition in which a fusion (i.e., joining of two entities or clusters into a single cluster) takes place at a lower dissimilarity than a prior fusion during an agglomerative hierarchical clustering procedure.

R-factor analysis
Refers to an ordination procedure based on the correlation or covariance structure of the variables (i.e., columns of a two-way data matrix). This is the most common ordination approach. In principal components analysis (PCA), for example, we create components out of the environmental variables (columns) and position sample entities (rows) in the reduced ordination space defined by the principal components. In this case, PCA is computed on the correlation or covariance matrix of the environmental variables (columns).

Rotation
Geometrically, a rotation of the ordination axes about the centroid to maximize

some objective criterion, usually for the purpose of obtaining a simpler, more interpretable solution.

Scatter plot
A two-dimensional plot of the distribution of sample entities, where each sample entity is displayed as a point based on its score (or value) on each axis.

Scree plot
An ordination scree plot is a plot of the eigenvalues (latent roots) against the ordination axis (e.g., principal component) number in their order of extraction. The shape of the resulting curve is used to evaluate the appropriate number of components to retain. A cluster analysis scree plot is a plot of the fusion level (dissimilarity values) against the number of clusters. In this case, the shape of the resulting curve is used to evaluate the appropriate number of clusters to retain.

Scree test criterion
Criterion for determining the number of latent roots (e.g., principal components) to retain for interpretation based on the scree plot. Typically, the scree plot slopes steeply down initially and then asymptotically approaches zero. The point at which the curve first begins to straighten out is considered to indicate the maximum number of components to extract.

Secondary data matrix
Any matrix derived from the original data matrix. A correlation matrix, covariance matrix, and dissimilarity matrix are examples of secondary data matrices.

Sequential clustering
Clustering process involving a recursive sequence of operations to the set of entities to form clusters.

Similarity coefficient
Measure of the similarity between two entities based on one or more variables.

Similarity matrix
Resemblance matrix containing similarity coefficients for every pair of entities.

Similarity space
Resemblance space where each dimension, or axis, is defined by the similarity to one of the other N sample entities based on a similarity coefficient.

Simple matching coefficient
An association coefficient developed for binary data (i.e., dichotomous variables) that gives equal weight to positive and negative matches.

Simultaneous clustering
Clustering process involving a single nonrecursive operation to the entities to form clusters.

Single linkage (or nearest-neighbor)
Clustering strategy in which distance values between entities and clusters are

computed as the dissimilarity between the two closest entities. An entity's dissimilarity to a cluster is defined to be equal to its dissimilarity to the closest entity in that cluster; similarly, when two clusters agglomerate, their dissimilarity is equal to the smallest dissimilarity for any pair of entities with one in each cluster.

Singularity
Linear dependency between two or more variables resulting in a singular matrix that cannot be inverted and thus decomposed into its eigenvalues and eigenvectors.

Singular matrix
Matrix that does not have an inverse. Conversely, an invertible matrix is nonsingular. The invertibility of the secondary matrix (e.g., correlation or covariance matrix) is necessary for its decomposition in the eigenanalysis (i.e., the calculation of its eigenvalues and eigenvectors).

Skewness
A measure of the tendency for asymmetry in the distribution of a variable; that is, for the tail of the distribution about the mean to be longer in one direction than in the other. Skewness can be positive or negative and is unbounded. Skewness is zero for a normal distribution; large positive or negative values indicate a longer tail than would be expected under a normal (i.e., perfectly symmetrical) distribution.

Space-conserving
Clustering strategy that preserves the multidimensional structure defined by the relationships among sampling entities in the original data matrix. Thus, the distances between entities in dissimilarity space are roughly preserved during the fusion process so that there is a high correlation between the input dissimilarities (in the original dissimilarity matrix) and the output dissimilarities defined by the lowest dissimilarity required to join any given entity pair during the fusion process.

Space-contracting
Clustering strategy in which groups appear, on formation, to move nearer to some or all of the remaining entities; the chance that an individual entity will add to a preexisting group rather than act as the nucleus of a new group is increased, and the system is said to "chain." This effect is usually apparent in the dendrogram as a stair-step sequence of fusions.

Space-dilating
Clustering strategy in which groups appear to recede on formation and growth; individual entities not yet in groups are more likely to form nuclei of new groups. The effect is usually apparent in the dendrogram as tight clusters connected by relatively long branches.

Space-distorting
Clustering strategy in which the model behaves as though the space in the immediate vicinity of a group has been contracted or dilated. As a result, space-

distorting strategies may not faithfully represent the spatial relationships in the input data.

Split-sample validation (or cross-validation, or data splitting)
A common validation procedure used in several multivariate techniques to assess the stability (i.e., reliability and robustness) of the solution. In discriminant analysis, for example, the procedure involves dividing the total sample of entities randomly into two (or more) groups. One subset, referred to as the "analysis," "training," or "calibration" sample, is used to derive the classification functions, and the other, referred to as the "holdout," "test," or "validation" sample, is used to test the stability of the functions. The functions derived from the analysis set are used to classify samples from the test set, and the resulting correct classification rate is used to judge the reliability and robustness of the classification criterion.

Squared canonical correlation
In discriminant analysis, the squared canonical correlation is equal to the ratio of among-group to pooled within-group variation for the corresponding canonical function. In practical terms, it represents the proportion of total variation in the corresponding canonical function (measured by the eigenvalue) explained by differences in group means, in other words, how much of the canonical variation is due to group differences. In canonical correlation analysis, it equals the proportion of variance in the canonical variate of one set that is accounted for by the variate of the other set. Note that the relationship being described here is between the variates, not between the original variables.

Square matrix
A matrix with the same number of rows and columns.

Standard deviation
A measure of the absolute variability about the mean. Standard deviation equals the square root of the variance, or root-mean-square deviation from the mean, in either population or sample, and is intuitively interpreted as the average deviation of samples from the mean. Approximately 68% of the values in a normal distribution are within one standard deviation of the mean; approximately 95% are within two standard deviations of the mean; and about 99.7% are within three standard deviations of the mean.

Standard error
A measure of the variability associated with a sample estimate of a population parameter. Standard error provides a quantitative measure of the likelihood that the true population value is within a certain distance of the observed sample value; hence, it provides the basis for a statistic about which hypotheses can be tested. The standard error of the mean is equal to the standard deviation divided by the square root of the sample size.

Standardization
The process whereby raw data are transformed into new variables with a mean of zero and standard deviation of one. When variables are measured on different

units and, as a result, have vastly different variances in absolute terms, standardization allows for the equal weighting of the variables in the analysis.

Standardized (or standardization)
A standardized variable is one with a zero mean and unit variance. Standardization is achieved by subtracting the mean from the raw-score values and dividing by the standard deviation.

Standardized canonical coefficient (or weight)
These adjusted canonical coefficients represent the weights that would be applied to the original variables in standardized form (i.e., zero mean and unit variance) to derive the standardized canonical functions (or variates). These coefficients produce canonical scores that are measured in standard deviation units. As such, the relative contribution of the original variables to each canonical variate may be gauged by the relative magnitude of the standardized canonical coefficients; the larger the magnitude, the greater that variable's contribution, but this interpretation can be distorted when the correlation structure of the data is complex.

Standardized principal component weight (or scoring coefficient)
These adjusted principal component weights (or scoring coefficients) represent the weights that would be applied to the original variables in standardized form (i.e., zero mean and unit variance) to derive the standardized principal components. These coefficients produce principal component scores that are measured in standard deviation units. As such, the relative contribution of the original variables to each principal component may be gauged by the relative magnitude of the standardized principal component weights; the larger the magnitude, the greater that variable's contribution, but this interpretation can be distorted when the correlation structure of the data is complex.

Stem-and-leaf plot
A graphical display of a sample frequency distribution. Stem-and-leaf plots provide a quick pictorial representation of the distribution. Specifically, each "stem" defines a class of the grouped frequency distribution, represented usually by the first one or two digits of the observed values, and the "leaves" represent the individual values of the observations, represented usually by the last digit.

Stepwise selection of variables
A process used to select a subset of variables from a data set in order to achieve a more parsimonious final model (i.e., the fewest variables possible that explain the data structure). Stepwise methods enter and remove variables one at a time, selecting them on the basis of some objective criterion. Although stepwise procedures produce an "optimal" set of variables, they do not guarantee the best (maximal) combination.

Stopping rule
A heuristic or statistical rule for determining how many ordination axes (e.g., principal components) or clusters to retain for interpretation.

Stratified random sample (or stratified random sampling)
A random sample of a population in which a random subset of entities is drawn separately from each well-defined stratum in the population.

Symmetric matrix
A square matrix (i.e., same number of rows as columns) such that $x_{ij} = x_{ji}$ for all i and j. In other words, one off-diagonal side of the matrix is the mirror image of the other side.

Tau statistic
A measure of chance-corrected classification (i.e., a standardized measure of improvement over random assignment) when prior probabilities are known or are not assumed to be equal to sample sizes.

Tolerance
The proportion of variation in the variables that is not explained by the variables already in the model. The tolerance for a variable not yet in the model is 1 minus the squared multiple correlation between that variable and all variables already in the model. Tolerance represents the percentage of variance in a variable *not* accounted for by the variables already entered. Thus, a small tolerance indicates that most of the variation in that variable is already accounted for; that is, the variable is a near-linear combination of the variables already in the model. Conversely, a tolerance of 1 means that a variable is totally independent of other variables already in the model.

Trace
The sum of the diagonal elements of a square matrix. The trace of a correlation matrix equals the number of variables because the diagonal elements equal 1. Similarly, the trace of a covariance matrix equals the total variance in the data set, because the diagonal elements equal the variances associated with the original variables.

Triplot
An ordination diagram produced in canonical correspondence analysis depicting the joint distribution of species and sample points along the dominant ecological gradients (i.e., canonical functions) and depicting the explanatory variables from the second matrix as arrows emanating from the origin (i.e., the grand mean of all explanatory variables).

Two-way data matrix
Rectangular array of real numbers arranged in rows and columns, where the rows typically represent observation vectors associated with each sampling entity and columns represent variables of some sort (e.g., species abundances, environmental characteristics).

Two-way indicator species analysis (TWINSPAN)
A polythetic, divisive, hierarchical clustering technique involving the joint clustering of samples and species by successive partitions of ordination axes generated at each step by reciprocal averaging.

Type I error
The probability of rejecting the null hypothesis when it should be accepted (i.e., when the null hypothesis is true). A small Type I error rate, denoted by alpha (usually referred to as the P-value), leads to the rejection of the null hypothesis in favor of the alternative hypothesis.

Unconstrained ordination
Ordination techniques (e.g., principal components analysis) in which the ordination is conducted on a single set of interdependent variables with the purpose of extracting the major patterns among those variables, irrespective of any relationship to variables outside that set. In other words, the ordination of sampling entities is not constrained by any relationship to variables outside the set; any relationship to another set of (explanatory) variables must be determined by a separate, secondary analysis.

Unstandardized canonical coefficient (or weight)
These adjusted canonical coefficients represent the weights that would be applied to the original variables in raw-scale form to derive the standardized canonical functions (or variates). Although these coefficients produce canonical scores that are measured in standard deviation units, they are uninterpretable as coefficients. This is because they tell us the "absolute" contribution of a variable in determining the canonical score, yet this information can be misleading when a unit change in a variable is not equivalent from one variable to another (i.e., when the standard deviations are not the same because of differences in measurement scale or variances).

Variable splitting
A validation procedure used in cluster analysis to assess the stability (i.e., reliability and robustness) of the solution. Variable splitting involves subdividing the variables into two or more sets, conducting separate analyses on each set, and comparing the solutions.

Vector
A set of real numbers or, alternatively, a matrix with only one row or only one column. Typically, we refer to a vector of scores associated with each sampling entity (a row of the data matrix) where each element represents the value or score on a variable (e.g., species abundance, environmental characteristic).

Ward's minimum-variance linkage (or minimization of within-group dispersion)
Clustering strategy in which the fusion of entities and clusters is based on minimizing the increase in within-group dispersion (variance). Thus, minimum-variance linkage is similar to average-linkage fusion, except that instead of minimizing an average distance it minimizes a squared distance weighted by cluster size.

Wilk's Lambda statistic
Likelihood ratio statistic for testing the null hypothesis that group centroids are equal in the population; Lambda approaches zero if any two groups are well separated.

Index

A
Agglomeration schedule 101
Agglomeration table 103
Agglomerative clustering 90, 91
Arch effect 33, 64, 66, 68, 72
Arranged data matrix 120
Association coefficient 98, 99, 100
Average linkage 108

B
Bootstrap procedure 45, 46, 47, 48, 52, 113, 168, 175, 178, 217, 225, 227
Botryology 11
Box plot 28, 145, 152, 199, 202
Box's M-test 143
Bray and Curtis ordination 11, 63
Bray–Curtis dissimilarity 98
Broken-stick criterion 43, 49
Broken-stick model 43

C
Canberra distance 98
Canonical analysis of discriminance 11, 15, 131, 132, 133, 134, 235, 236, 237, 239, 241
Canonical coefficient 159, 208
Canonical correlation 162, 194, 210, 211
Canonical correlation analysis 11, 14, 16, 191, 192, 193, 236, 237, 239, 240
Canonical correlation coefficient 71, 162
Canonical correlation criterion 209
Canonical correspondence analysis 11, 14, 22, 25, 61, 63, 69, 70, 71, 72, 190, 194, 236, 237, 239, 240, 242
Canonical cross-loading 223
Canonical discriminant function 132
Canonical function 15, 131, 132, 133, 134, 135, 148, 155, 161, 169, 176, 237
Canonical loading 172, 221
Canonical pattern 160, 220
Canonical plot 142
Canonical redundancy 212
Canonical score 133, 145, 152, 160, 169, 193, 199, 202, 218
Canonical structure 172, 221
Canonical variate 16, 132, 192, 193, 205, 209, 220, 237
Canonical variates analysis 11
Canonical weight 159, 208
Categorical data 98, 99
Categorical variable 12, 13, 86, 138, 196
Centroid 24, 28, 31, 40, 58, 133, 136, 137
Centroid linkage 107
Chain 106, 107
Chance-corrected classification rate 164
Characteristic equation 38, 158, 206
Characteristic root 38, 158, 206

280   Index

Classification 10, 11, 15, 84, 131, 132, 134, 163, 235, 241
Classification coefficient 135
Classification criterion 163
Classification function 132, 135
Classification matrix 163
Classification rate 163, 164
Clumping 11
Cluster 85, 86, 94, 124
Cluster analysis 10, 11, 14, 15, 62, 82, 83, 84, 85, 86, 87, 88, 91, 121, 122, 123, 131, 191, 234, 235, 238, 239, 240, 241, 242
Cluster membership table 112
Coefficient of community 63
Coefficient of dissimilarity 96
Coefficient of similarity 96
Coefficient of variation 119, 226, 227
Cohen's *Kappa* statistic 165
Communality 56
Comparisonwise error rate 52, 114
COMPCLUS 92, 93
Complemented coefficient of community 92, 101
Complemented coefficient of Jaccard 100
Complemented simple matching coefficient 100
Complete linkage 107
Composite clustering 92, 93
Composite variable 9, 21
Composite variate 16, 23
Compression effect 33, 64, 66, 68
Confidence ellipse 142
Confidence interval 46, 47
Confusion matrix 163
Constrained ordination 23, 25, 63, 69, 72, 73, 190, 236, 241, 242
Continuous data 98
Continuous variable 12, 13, 86, 138, 196
Cophenetic correlation 108, 116, 117
Correlation 16, 36, 37, 38, 39, 40, 50, 53, 54, 56, 62, 67, 68, 70, 71, 72, 99, 101, 105, 108, 117, 142, 148, 149, 150, 151, 156, 157, 171, 172, 192, 193, 194, 195, 200, 201, 202, 206, 207, 217, 221, 222, 223, 224, 226
Correlation coefficient 98, 101
Correlation matrix 35, 36, 37, 38, 39, 41, 43, 51, 205

Correspondence analysis 11, 62, 67
Count data 98
Count variable 12, 13, 139, 196
Covariance 28
Covariance-controlled partial F-ratio 173, 174
Covariance matrix 35, 36, 38, 39, 135, 136, 142, 205
Cross-validation 177
Cubic clustering criterion 114, 115

D

Data cloud 3, 4, 8, 25, 28, 36, 59, 86
Data space 3, 8, 9, 25
Data splitting 121
Dendrogram 101, 103, 104, 112, 117, 120
Dependence 234
Dependence technique 237
Dependent variable (Y's) 13, 14
Descriptive discriminant analysis 10, 11, 132
Detrended correspondence analysis 11, 22, 33, 61, 63, 68, 69, 120, 121, 190
Detrended principal components analysis 34
Detrending 68, 72
Dichotomous variable 12, 13, 99, 100
Discriminant analysis 10, 11, 14, 15, 84, 85, 118, 131, 132, 191, 192, 193, 237, 239, 240, 242
Discriminant function 132
Discriminant function analysis 11, 132
Discriminating variable 137, 139
Dissimilarity matrix 35, 38, 39, 63, 66, 93, 96, 101, 102
Dissimilarity space 63, 92, 105
Distance coefficient 98
Distance measure 98
Divisive clustering 90, 91

E

Eigenanalysis 37, 39, 41, 50, 67, 153, 158, 159, 204, 205, 206, 208, 228
Eigenvalue 35, 37, 38, 39, 41, 43, 60, 67, 158, 161, 206, 207
Eigenvector 35, 39, 50, 159, 208
Environmental gradient 3, 8, 57, 61
Equal frequency ellipse 142, 169
Equimax rotation 59
Euclidean distance 63, 92, 95, 96, 98, 99, 101, 102

Exclusive clustering 89, 91
Experimentwise error rate 2, 9, 52, 114, 175

F
Factor analysis 11, 22, 56, 63, 64, 65
Factor structure 64, 65
FASTCLUS 92
Final communality 56, 61
Fisher's linear discriminant function analysis 11, 135
F-to-enter 156
F-to-remove 156
Furthest neighbor 107

G
Gaussian 33, 68, 70
Grouping 11

H
Hierarchical clustering 90, 91, 92, 94, 95
Homogeneity of variance 143
Hotelling's T^2 241

I
Icicle plot 112, 116
Identity matrix 38, 39
Independent variable (X's) 13, 14
Inferential technique 2
Interdependence 234
Interdependence technique 238
Interset correlation 72
Interval scale 12, 13
Intraset correlation 72

J
Jackknife procedure 45, 46, 48, 52, 122, 168, 175, 178, 217, 225, 227
Jackknife validation 178
Joint plot 67

K
Kaiser–Guttman criterion 41
Kappa statistic 165, 166
K-means clustering 93, 108
Kurtosis 28, 145, 199

L
Latent 65
Latent root 37, 49

Latent root criterion 41, 42
Levene's test of homogeneity of variance 143
Linear classification criterion 163
Linear dependency 23
Linearity 28, 33, 57, 61, 66, 142, 153, 203
Logistic regression analysis 236

M
Mahalanobis distance 95, 98, 99, 135
Manhattan or city-block distance 98
Maximum chance criterion 164
Median linkage 107
Meristic data 98, 99
Minimization of within-group dispersion 108
Minimum variance partitioning 93
Monothetic clustering 91
Monotonicity 33, 66, 107, 108
Morphometrics 11
Multicollinearity 72, 146, 148, 149, 150, 155, 172, 192, 199, 200, 221
Multiple analysis of variance 240, 241, 242
Multiple canonical analysis of discriminance 134, 136, 143, 152, 153, 161, 241
Multiple discriminant analysis 11, 132
Multiple regression analysis 15, 16, 24, 131, 136, 137, 192, 193, 237, 240, 242
Multivariate analysis of variance 15, 24, 131, 136, 137, 141, 192, 193, 237
Multivariate normality 28, 59, 144, 198

N
Natural cluster 86
Nearest neighbor 106
Noise 66, 72, 83, 92
Nominal scale 12, 13
Nonexclusive clustering 89
Nonhierarchical clustering 90, 91, 92, 94
Nonlinearity 153
Nonmetric multidimensional scaling 11, 22, 63, 66
Nonsingular 146
Normality 28, 145, 199
Normal probability plot 28, 145, 152, 199, 202
Nosography 11
Nosology 11
Numerical taxonomy 11

O

Oblique rotation 59
Optimization-partitioning technique 93
Ordinal scale 12, 13
Ordination 11, 14, 20, 22, 26, 63, 82, 92, 121, 122, 123, 131, 192, 193, 234, 235, 241, 242
Ordination space 23, 50, 57, 61, 66, 67, 68, 120, 122
Ordination space partitioning 120, 121, 123
Orthogonality 25, 28, 38, 59, 68, 138
Orthogonal rotation 59
Outlier 31, 32, 66, 69, 83, 92, 123, 151, 152, 202

P

Parsimonious 37
Partial F-ratio 174, 175, 176
Partial F-test 156
Partitioning 11
Pattern matrix 160, 220
Pearson product-moment correlation coefficient 36, 101
Percent dissimilarity 63, 92, 98, 99
Percent of remaining variance 44
Percent trace 43
Polar ordination 11, 22, 63, 64
Polythetic agglomerative hierarchical clustering 95, 96, 121
Polythetic agglomerative nonhierarchical clustering 92, 93, 101
Polythetic clustering 91
Polythetic divisive hierarchical clustering 120, 121
Polythetic divisive nonhierarchical clustering 93, 94, 123
Polytomous variable 12, 13, 99
Potency index 175, 176
Predictive discriminant analysis 10, 11, 132
Principal component 21, 23, 24, 25, 28, 32, 33, 34, 35, 36, 38, 39, 41, 43, 49, 50, 56, 57, 58, 61, 116, 122, 149, 237
Principal components analysis 11, 14, 22, 23, 24, 25, 26, 27, 28, 31, 33, 34, 35, 36, 37, 41, 49, 51, 56, 61, 62, 63, 64, 65, 118, 149, 190, 192, 234, 236, 237, 238, 242
Principal component loading 39, 40, 50, 51, 52, 59
Principal component scores 29, 32, 34, 57

Principal component structure 50, 51, 53, 59
Principal component weight 39, 65
Prior communality 64
Prior probability (or priors) 135, 136, 152, 153
Product-moment correlation 50, 172, 210
Product-moment correlation coefficient 37
Proportional chance criterion 164, 165
Pseudo F-statistic 114
Pseudoreplication 8
Pseudo t^2-statistic 114

Q

Q-analysis 11
Q-factor ordination 61, 62
Quadratic classification criterion 163
Quartimax rotation 59

R

Randomization procedure 48, 49
Randomization resampling procedure 175, 225
Randomization test 45, 47, 168, 178, 217, 227
Random sample 31
Ratio scale 12, 13
Raw canonical coefficient 209
Raw covariance matrix 36, 38, 39, 43, 50, 51, 205
Reciprocal averaging 11, 22, 61, 62, 63, 67, 68, 118, 120, 121
Reciprocal averaging ordination 236
Redundancy 2, 20, 21, 23, 25, 28, 37, 83, 85, 92, 131, 132, 138, 192, 194, 212, 214
Redundancy analysis 214
Redundancy coefficient 212, 213, 214, 223
Relative percent variance 43, 44, 45, 49, 162
Relative percent variance criterion (or percent trace) 43, 44, 49, 161
Resampling procedure (or resampling technique) 34, 45, 48, 52, 154, 168, 175, 178, 217, 225, 227
Resemblance coefficient 96
Resemblance matrix 95, 96, 97
Resemblance measure 96, 98, 99
Resemblance space 97, 98
Residual discrimination 168
Reversal 107, 108

R-factor ordination 61, 62
Ridge regression 149
Rotation 58, 60

S
Scatter plot 32, 34, 57, 152, 153, 169, 202, 203, 219
Scoring coefficient 39
Scree plot 41, 42, 43, 115
Scree test criterion 41
Secondary data matrix 35, 36, 96
Segmentation analysis 11
Sequential clustering 89, 91
Similarity coefficient 98
Similarity space 97
Simultaneous clustering 89
Single linkage 105, 106
Singularity 146, 148, 199, 200
Skewness 28, 145, 199
Space-conserving 105, 106, 107, 108
Space-contracting 106, 107
Space-dilating 106, 107, 108
Space-distorting 105, 106
Split-sample validation 177, 226
Squared canonical correlation 163, 206, 211, 212
Standard deviation 32, 46, 47, 202
Standard error 46, 47, 62
Standardization 32, 36, 93, 96, 152
Standardized canonical coefficient 148, 150, 160, 171, 201, 209, 220
Standardized canonical score 160, 208
Standardized canonical weight 160, 220
Standardized principal component score 40
Standardized principal component weight 39, 57
Standardized scoring coefficient 40, 57
Stem-and-leaf plot 28, 145, 152, 199, 202
Stepwise selection of variables 155
Stopping rule 41, 90, 113
Stratified random sampling 31
Structure coefficient 150, 172, 201, 212, 221, 222
Structure matrix 50, 53, 59, 172, 221

Symmetric matrix 35, 96
Systematics 11

T
Tau statistic 165
Taximetrics 11
Taxonorics 11
Tolerance 156
Total structure coefficient 171
Trace 38
Triplot 70
Two-group canonical analysis of discriminance 134, 142, 241
Two-group discriminant analysis 132
Two-way data matrix 8, 26, 62, 87, 139, 197
Two-way indicator species analysis 120, 121, 123
Type I error 9, 45
Typology 11

U
Unconstrained ordination 22, 23, 25, 63, 68, 69, 70, 72, 190, 191, 238, 239, 241
Unconstrained ordination technique 242
Unstandardized canonical coefficient 160, 171, 209, 220
Unstandardized canonical weight 160
Unsupervised pattern recognition 11
Unweighted pair-group average 108
Unweighted pair-group centroid 107

V
Variable splitting 122
Variance-covariance matrix 35, 141, 143, 205
Varimax rotation 59
Vector 8, 38

W
Ward's minimum-variance linkage 108
Weighted pair-group centroid 107
Wilks' Lambda statistic 156